24 Springer Series in Chemical Physics
Edited by Robert Gomer

Springer Series in Chemical Physics
Editors: V. I. Goldanskii   R. Gomer   F. P. Schäfer   J. P. Toennies

Desorption Induced by Electronic Transitions DIET I

Proceedings of the First International Workshop,
Williamsburg, Virginia, USA, May 12–14, 1982

Editors: N. H. Tolk   M. M. Traum   J. C. Tully
T. E. Madey

With 112 Figures

Springer-Verlag Berlin Heidelberg GmbH

Drs. Norman H. Tolk, Morton M. Traum †, and John C. Tully

Bell Laboratories, 600 Mountain Avenue, Murray Hill, NJ 07974, USA

Dr. Theodore E. Madey

National Bureau of Standards, Chemistry Building,
Washington, DC 20234, USA

ISBN 978-3-642-45552-0
DOI 10.1007/978-3-642-45550-6

ISBN 978-3-642-45550-6 (eBook)

Library of Congress Cataloging in Publication Data. Main entry under title: Desorption induced by electronic transitions. (Springer series in chemical physics ; v. 24). 1. Electron-stimulated desorption–Congresses. I. Tolk, N. H. (Norman H.) II. Series. QD547.D47 1982   541.3'453   82-19657

We dedicate this volume to the memory of our friend and co-editor

MORTON M. TRAUM

who died on December 1, 1982

Mort Traum was a pioneer in the field of desorption induced by electronic transitions. His intelligence, enthusiasm, compassion and sense of humor made working with him an intensely stimulating experience. He will be remembered with esteem and affection.

Preface

The Workshop on Desorption Induced by Electronic Transitions (DIET) took place May 12-14, 1982, in Williamsburg, Virginia. The meeting brought together, for the first time, most of the leading workers in the fields of electron and photon stimulated desorption from surfaces, as well as many workers in related fields, including sputtering, gas-phase photodissociation and solid-state theory. The emphasis of the workshop was on the microscopic mechanism of stimulated desorption. Many possible mechanisms have been proposed, and a few new ones emerged at the meeting. Though no consensus was reached, many views were espoused and criticized, frequently with considerable enthusiasm. The result was an appraisal of our current understanding of DIET, and a focus on the experimental and theoretical efforts most likely to lead to new insights.

This volume is an attempt to record the information exchanged in this very successful workshop and, perhaps, convey some of the excitement of the field of DIET. The book is a collection of papers written by participants in the DIET workshop, including in addition a contribution from Dietrich Menzel, who was unable to attend. Thus, this book represents a complete statement of the state of the art of experimental and theoretical studies of DIET and related phenomena. More importantly, it addresses the interesting unsolved problems, and suggests strategies for unraveling them.

We acknowledge the assistance given by the other members of the organizing committee, A.E. de Vries, R. Gomer, M.L. Knotek, D. Menzel and D.P. Woodruff, in developing the workshop program, and the generous support provided by the U.S. National Bureau of Standards and by the U.S. Office of Naval Research through Dr. L.R. Cooper.

Murray Hill, N.J. · Washington, D.C.   *N.H. Tolk · M.M. Traum · J.C. Tully*
December 1982   *T.E. Madey*

Contents

Part 4 Molecular Dissociation

Part 5 Ion-Stimulated Desorption

Part 6 Electronic Erosion

Part 7 Condensed Gas Desorption

Introduction

N.H. Tolk, M.M. Traum, J.C. Tully
Bell Laboratories, Murray Hill, NJ 07974, USA

T.E. Madey
Chemistry Building, Surface Science Division,
National Bureau of Standards, Washington, DC 20234, USA

The study of Desorption Induced by Electronic Transitions (DIET) has enjoyed rapid growth over the past two decades. DIET processes have been experimentally and theoretically investigated by many different researchers representing diverse disciplines. It is the purpose of this book to examine and synthesize information and concepts from a number of research directions that in the past have been widely separated but are suddenly converging on identical theoretical and in some cases experimental challenges. Included are electron-stimulated desorption (ESD), photon-stimulated desorption (PSD), bond distance measurements by the surface X-ray absorption fine structure method (SEXAFS), excited neutral desorption, and bombardment-induced erosion of condensed gases and alkali halides.

On a fundamental level, DIET studies are intimately concerned with the most basic questions of surface physics and chemistry, relating to geometrical structure, electronic structure, and the dynamics of bond making and breaking on surfaces. In addition, these studies are important in understanding the static and dynamic electronic behavior of atoms and molecules near surfaces. The contributions in this volume deal not only with the central issues, but also with related areas, including molecular dissociation and ion-induced desorption. The major focus of this book is on identifying and elucidating the underlying electronic mechanisms associated with the desorption of atomic and molecular particles from surfaces due to electronic processes stimulated by electron, photon, and even ion bombardment. Emphasis is placed on (a) elementary excitations, (b) transformation of electronic to kinetic energy, and (c) particle-surface electronic interactions. Though progress has been made in these areas, many fundamental questions remain unanswered.

This book is very timely in view of the recent increased emphasis on surface characterization. An understanding of how and why atoms are displaced through electronic interactions on surfaces and in the surface region of solids is critical to the development of a number of vital areas, including interface growth (where defects can play a dominant role in determining the behavior of electrical or photoelectrical devices), catalysis (where small amounts of impurities can govern overall reactions), and erosion (where many material problems arise because of non-momentum-exchange processes). Electronically induced desorption can also be a major barrier to quantitative surface analysis using electron, photon, and ion beam spectroscopies. This volume provides a stimulating opportunity for a critical examination of these aspects of surface characterization.

As even a cursory reading of this book makes clear, the DIET field has progressed in an exciting manner to a state of dynamic adolescence. It remains a challenge to both experimentalists and theoreticians to boldly shape this subject to an even more exciting and dynamic maturity.

Part 1

Fundamental Excitations

1.1 Fundamental Excitations in Solids Pertinent to Desorption Induced by Electronic Transitions

J.W. Gadzuk

National Bureau of Standards, Washington, DC 20234, USA

Abstract

Various aspects of the dynamics of time-dependent localized potentials and interactions in solids and at surfaces, as they might relate to the fundamental processes involved in desorption induced by electronic transitions (DIET) are explored.

1. Introduction

The first stage of a DIET necessarily involves the time-dependent behavior of the fundamental excitations in solids. In order to gain insights into this behavior, special emphasis in this chapter is placed on physical pictures and phenomenological modeling and as such, draws heavily on previous papers in this vein [1].

The general scheme of a DIET is as follows. Initially the target solid is subjected to a source of excitation, usually an electron or photon beam. The beam can create both electronic and ion core excitations of the solid, either localized or delocalized. Low energy ($\sim 10^{-2}$ eV) delocalized nuclear excitations or phonons are generally of relevance in thermal desorption [2] but not in a DIET. Likewise coherent direct excitation up a vibrational ladder provided by the localized chemical bond between "desorbate" and substrate [3], in analogy with laser-induced molecular photodissociation [4], while leading to desorption, is not DIET.

On the other hand, direct coupling to electronic excitations is the central focus of this workshop. Creation of delocalized electronic excitations, either single particle electron-hole pairs or collective plasmons, is not of primary importance in desorption since the energy associated with these excitations is spread throughout the solid and thus is not concentrated in the localized region where chemical bond breaking needed for desorption must occur. Nonetheless, delocalized electronic excitations can be quite relevant in a secondary way, as we will soon see. First consider the possibilities which follow localized electronic excitation. An electron can be excited from a quasi-localized virtual resonance state such as a bonding orbital between an adsorbed atom or molecule and a surface. The resulting "final state" of the total system is less bonding, perhaps anti-bonding, in which case the ionized particle is repelled and then ejected from the surface, in the spirit of the MENZEL-GOMER-REDHEAD (MGR) picture [5], if the hole state is sufficiently long lived for the particle to exit. Another possibility exists when the initial excitation involves a core state. Again the particle may be ejected à la MGR. However a new richness of possibilities arise due to the extra-atomic coupling of the resulting ion with the surrounding solid.

4

First, in analogy with core hole creation in x-ray photoelectron spectroscopy, it is well known that switching on a localized potential within an excitable medium necessarily results in relaxation or polarization shifts and shakeup satellites due to the response of the surroundings to the transient potential [1.e., 6]. The relative intensities of the spectrum depend crucially on the temporal nature of the potential. For a metallic substrate, the shakeup satellites can include delocalized pair, plasmon, phonon excitation ("p^3 satellites") as well as localized resonance excitation and surface plasmons. Low-energy pair and phonon excitation results in characteristic lineshapes [7], whereas plasmon excitation yields discrete satellite peaks, all of which we will return to [8]. Resonance excitation is directly related to the so-called "unfilled orbital screening mechanism" [1d,9]. Each one of these possible final states may (or may not) result in a repulsive potential felt by the ion, in which case many different (or none) desorption channels are opened up, basically still in the spirit of MGR. The core hole has its own independent time dependence in which radiative or Auger decay may occur, opening up still more desorption channels. One channel of very special interest involves inter-atomic Auger decay in which a highly repulsive state is produced when two electrons from a given atom Auger discharge their core-hole-containing neighbor. The resulting desorption, which displays an initial excitation energy threshold equal to the binding energy of the core level, is said to occur via the KNOTEK-FEIBELMAN (KF) mechanism [10]. As can be seen through-out the discussions in this volume, considerable effort is being invested in sorting out the relative roles of MGR vs. KF pictures of DIETS. This question will not be addressed in this chapter but instead we will focus on common physical points which must be dealt with in either case.

From a dynamics perspective, the essential issue is the determination of the time evolution of the nuclear coordinates of candidate "desorbates" follow-ing an initial excitation in which the total system, desorbate plus desorber, is placed in some non-stationary excited state. Within a Born-Oppenheimer separation of nuclear and electronic motion, the electronic state of the system defines a potential energy surface for nuclear motion. Transitions within the manifold of electronic states, for instance diabatic (or non-adiabatic) transitions associated with curve crossings [11] or radiative and Auger decay leading to finite lifetimes for core holes, require that the nuclear motion is over an effectively time-dependent potential surface where the time dependence is dictated by the evolution of the electronic system. In general the temporal coherence of the various electronic transitions must be maintained. The formidable task to be confronted is the construction of a reasonably useful picture of the dynamics of such a coupled system, given the potential surfaces corresponding to the allowed electronic states. This task is identical for MGR and KF systems, the distinction between the two being only in the specifics of the potential surfaces, both geometrical and temporal. Those specifics must be regarded as independent input data characterizing the system. Obtaining such "input data" is indeed a formidable task in itself, but one which is distinctly different from that of constructing a manageable theory of the dynamics of desorption, including the role played by the fundamental excitations of solids.

In order to present this case, the remainder of the chapter is structured as follows. Some underlying philosophical dogma is put forth in Section 2. Reduction of the desorption yield to a convolution of "slow system" Franck-Condon factors multiplicatively scaled by a "fast system" initial-excitation transition probability is dealt with in Section 3. Section 4 is concerned with general time-dependent quantum mechanics and its simplification by phenomenological modeling. Aspects of the time-dependent displaced harmonic oscillator, particularly as they clarify the observable manifestations of the

5

p^3 excitations are brought out in Section 5. The electron gas/metal conduction band excited states and their screening response are the focus in Section 6. Section 7 contains some thoughts on the relations between transient potentials and adiabatic to sudden limits. Finally, these ideas are put together to form some realizable dynamic scenarios in Section 8, first in the context of currently-observed nuclear-motion-induced features in photoelectron spectroscopy and secondly, in the very related theoretically observable desorbate energy distributions.

2. Fundamental Dogma

Most of what follows in this chapter relies on the two propositions given below:

A) Since the time dependence of non-interacting "p^3" excitations is given as $\rho_p(t) \sim e^{-i\omega t}$ ($\ddot{\rho} + \omega^2 \rho = 0$), mastery of the time-dependent forced harmonic oscillator is a necessary step towards mastery of most of the dynamics problems needed here.

B) Initial excitation of a core hole creates a localized charge distribution (only $|\phi_c(r)|^2$, not $\phi_c(r)$, enters our considerations). Subsequently, the localized charge distribution, which may have its own independent time evolution, acts Coulombically on the elementary excitations providing a time-dependent perturbation

$$H_{int}(r,t) = e^2 \int \frac{d^3r' \; \rho_{LCD}(r',t)}{|r - r'|} \quad . \tag{1}$$

3. Initial Excitation

Almost all tractable theories involving the time development of an initially excited or prepared system have built into them the equivalent of the following reduction scheme.

A) Born-Oppenheimer separability: It is assumed that the quantum system of interest can be segmented into two components, a "fast" system described by a manifold of spatial coordinates $\{r\}$ and a "slow" system similarly described by a manifold $\{R\}$, such that any transitions within the $\{r\}$ subspace occur over such a short time duration that $\{R\}$ can be regarded as stationary. Furthermore, the slow systems are non-interacting amongst themselves. Thus the total system wavefunction

$$\Psi = \Psi (r, R_i, \ldots R_j)$$

can be separated into the form

$$\Psi \approx \psi_\alpha (r; R_i, \ldots R_j) \; \chi_{\beta\alpha} (R_i) \cdots \chi_{\delta\alpha} (R_j) ,$$

where in ψ_α, r is a dynamic variable whereas $\{R\}$ are parameters and the state of the slow systems $\chi_{\beta\alpha} \cdots \chi_{\delta\alpha}$ depend upon the current state α of the fast system.

6

B) Franck-Condon or sudden approximation: The initial excitation processes represented by some operator $O(r,R)$ act only on the fast
system and thus the operator can be replaced by $O(r;R)$ where again
{R} is a set of <u>parameters</u> determined by the initial state.

C) Slowly varying with {R} approximation: It is assumed that the
variation of the initial excitation matrix element

$$<\psi_{\alpha'fin}(r;R)|O(r;R)|\psi_{\alpha\ in}(r;R)>,$$

is small over the range of {R} spanned by the relevant χ's. Consequently the golden-rule transition rate out of the initial state,
due to the initial excitation O, is given by

$$R = \sum_{fin} \delta(E_{fin}- E_{in})|< \psi_{fin}|O|\psi_{in}>|^2$$

$$\approx \sum_{\alpha'} |<\psi_{\alpha'}(R_c)|O(R_c)|\psi_{\alpha}(R_c)>|^2 \sum_{\beta'...\delta'} \delta(\varepsilon_{\alpha'}+\varepsilon_{\beta'}...+ \varepsilon_{\delta'}- \varepsilon_{\alpha}- \varepsilon_{\beta}-...-\varepsilon_{\delta})$$

$$\times|<\beta';\alpha'|\beta;\alpha>|^2 \ .... \ \times|<\delta';\alpha'|\delta;\alpha>|^2 \ , \tag{2}$$

where {R_c} is some "most likely" set of {R}, as (not always unambiguously) determined by the specifics of the problem. For instance with regards to x-ray produced core level spectroscopies
[12], the question of a common and single valued set of {R_c} is
equivalent to the choice between wavefunctions with either a
screened or unscreened core hole potential or no hole potential
at all. Similarly in surface-induced diabatic transitions on an
incident particle [13], setting a common {R_c} usually implies that
the fast system diabatic transition rate is taken to be given by
its value at the curve crossing point where $\varepsilon_{\alpha'}= \varepsilon_{\alpha}$.

The moral of this exercise is that from the point of view of the slow
systems, an initial excitation of the fast system appears as a change in the
Hamiltonians of the slow system (given by (1)) which allows for a redistribution of energy amongst the various subsystems, with the probability for
various partitionings of energy weighted by the slow system Franck-Condon
factors. The <u>initial excitation matrix element</u> sets the absolute-value scale
for the process but does not determine the relative partitioning amongst the
slow systems. On the other hand, both the initial and final fast system
<u>states</u> (but not how they were connected) do enter significantly in this deter-
mination through the indices α and α' in the Franck-Condon factors, as the
states α and α' determine, the potential energy surfaces entering the slow
system Hamiltonians.

4. Time-Dependent Quantum Mechanics

At this point we wish to obtain a description of the time development of the
slow system, given that it is subjected to a time-dependent perturbation
resulting from initial excitation (and possible decay) of the fast system.
Going back to basics the time-dependent Schrödinger equation for the slow
system is written as

$$i\hbar \ \frac{\partial \psi_{ss}(t)}{\partial t} \ = H(t) \ \psi_{ss}(t) \ , \qquad\qquad (3)$$

with $H(t) = H_0 + H_{int}(t)$ where $H_{int}(t)$ is the dynamic perturbation due to the independently-evolving fast system. Integration of (3) yields:

$$\psi_{ss}(t) = \exp \ (-\frac{i}{\hbar} \int_{-\infty}^{t} H(t) \ dt) \ \psi_0 \ (t=-\infty). \qquad\qquad (4)$$

The probability that the slow system ultimately ends up in some final eigenstate $|f,n>$ is simply

$$P_n = \lim_{t\to\infty} \ |<f,n|\psi_{ss}(t)>|^2 \qquad\qquad (5)$$

and the energy distribution of final states is

$$P(\varepsilon) = \lim_{t\to\infty} \ \sum_n \ \delta(\varepsilon-\varepsilon_n)|<f,n|\psi_{ss}(t)>|^2. \qquad\qquad (6)$$

Eqs. (4) - (6) are quite manageable for some simple forms of $H_{int}(t)$ which can be chosen to simulate particular excitation and decay processes. Some examples are now given.

A) Sudden switch on: Suppose that the time dependence of $H(t)$ is given by a single piecewise-continuous step function jump; that is $H(t) = H_0$, $t < 0$; $= H_1$, $t>0$. Eigenstates for each Hamiltonian are given by the eigenvalue equations

$$H_0|0,m > = \varepsilon_0^m \ |0,m> \qquad\qquad (7a,b)$$
$$\text{and} \quad H_1|1,n> = \varepsilon_1^n|1,n> \ ,$$

where the index 0 or 1 denotes the state of the fast system, or equivalently, the Hamiltonian acting on the slow system. If at $t = -\infty$ the slow system is in its ground state, then (4) becomes (for $t>0$)

$$\psi_{ss}(t) = e^{-iH_1 t}|0,0>. \qquad\qquad (8)$$

The resulting amplitude for the nth final state is

$$A_n = <1,n|e^{-iH_1 t}|0,0> = e^{-i\varepsilon_1^n t} \ <1,n|0,0> \qquad\qquad (9)$$

and thus, from (6), the energy distribution of final states is

$$P(\varepsilon) = \sum_n \delta(\varepsilon-\varepsilon_n) \ |<1,n|0,0>|^2 \qquad\qquad (10)$$

which is time-independent provided $t>0$. Eq. (10) gives a series of delta functions, placed at the eigenenergies of the final state

Hamiltonian, whose strength is determined by overlap integrals between the initial and particular final slow system wavefunctions, or in other words, the Franck-Condon factors of (2). An important physical realization of this model is the permanent switching on of a localized (core) hole potential as in photoemission. If the ionized object is a molecule, intra-molecular vibrational modes could be the slow system, whereas for solids, all of the "p^3 excitations" are so regarded.

B) Coherent double switching: Now suppose that we have the same problem as in part A, but in addition at $t=\tau$ allow for yet another sudden change; i.e., $H(t) = H_2$, $t>\tau$. As before, the amplitude for the n'th eigenstate of H_2 is

$$<2,n|e^{-iH_2(t-\tau)} e^{-iH_1\tau}|0,0>$$

$$= \sum_* e^{-i\varepsilon_2(t-\tau)} <2,n|1,*> e^{-i\varepsilon_1^*\tau} <1,*|0,0> ,$$

where the complete set of intermediate H_1 eigenstates $|1,*>$, have been inserted. The final state energy distribution is

$$P(\varepsilon) = \sum_n \delta(\varepsilon-\varepsilon_n)| \sum_* <2,n|1,* >e^{-i\varepsilon_1^*\tau} <1,*|0,0>|^2$$

$$\equiv \sum_n \delta(\varepsilon-\varepsilon_n)P_n(\tau) \tag{11}$$

which demonstrates interferences between various intermediate state paths connecting initial and final states, depending on the "lifetime" τ of the intermediate state. From (11), we should expect that if $\tau>>1/\varepsilon_1^*$, then the sum on a relatively dense set of ε_1^* should wipe out the interference terms (except in perfectly periodic systems) due to rapid oscillations in the phase factor. Dephasing and thus loss of coherence between the $t=0$ and $t=\tau$ switches the result. Such double switching has many physical realizations such as coherent XPS-Auger decay [14], resonance Raman spectroscopy [15], and atom-metal scattering [16]. The resonance Raman literature is particularly rich in this area as much detailed analysis on the relative Raman intensities, as a function of intermediate state dynamics, has been put forth.

As a final aside, we note that $H_2>H_1$ could correspond to a one to two core-hole transition as in CCC Auger transitions, whereas $H_2= H_0$ might describe a CVV transition, in which case the probability for return to the slow system ground state (no-loss line), from (5) and (11) with $|2,n> = |0,0>$, is

$$P_0 = |\sum_* |<0,0| 1,*>|^2 e^{-i\varepsilon_1^*\tau}|^2 \tag{12}$$

which is independent of the phase of the overlap integrals.

C) Fourier Decomposition: From the most cursory understanding of Fourier analysis one knows that the likelihood that a time-dependent

9

potential $V(t)$ will excite a mode of frequency ω depends upon the strength of the ω'th Fourier component of $V(t)$, that is

$$V(\omega) = \int_{-\infty}^{\infty} dt \; e^{-i\omega t} \; V(t).$$

In fact, for a harmonic oscillator of frequency ω, the energy transfer to the oscillator is given by $\Delta E_{osc}(\omega) = |V(\omega)|^2/2m$, hence the importance of $V(\omega)$ in modeling "p^3" excitation probabilities.

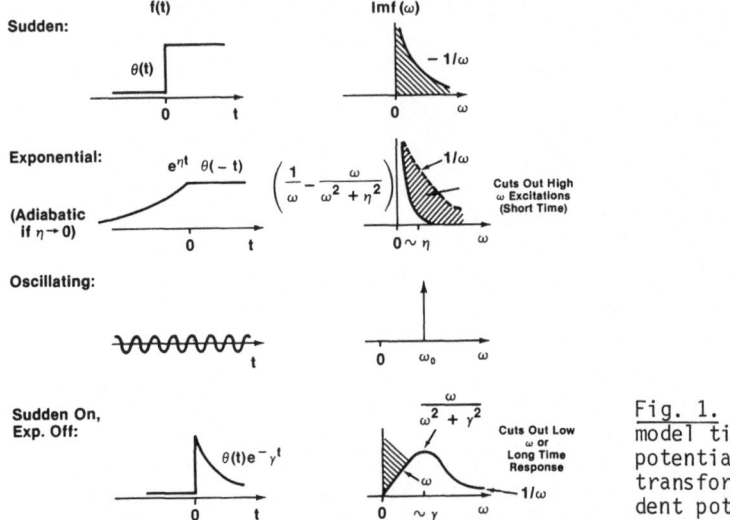

Fig. 1. Left: Various model time-dependent potentials. Right: Fourier transforms of time-dependent potentials

A number of model time dependent potentials are shown in Fig. 1, (with $V(t)=V_0 f(t)$), together with their Fourier transforms and algebraic statements. Several points should be noted. For the sudden switching already discussed, the Fourier components are abundant at all frequencies. On the other hand, for the exponentially switched on potential, high frequency or short-time response is not possible since the system cannot be excited before the potential is switched on. Finally, the sudden on-exponential off potential, which could be a model for core hole creation followed by exponential decay, shows drastic reduction in the low-frequency excitations which correspond to frequencies much less than γ, the inverse lifetime, saying that the long-time response is small to a potential that does not survive for the requisite long time.

5. Forced Harmonic Oscillator

One of only a handful of exactly soluble time-dependent quantum mechanical models is that of a harmonic oscillator subjected to a separable interaction of the form

$$H_{int}(z,t) = z \; g(t), \tag{13}$$

10

varying linearly in oscillator displacement. The time dependence is arbitrary [17]. Often, particularly in the case of localized hole creation, an additional potential acting on the effective oscillator (such as (1)) is switched on and this potential is expressed as a series expansion about the equilibrium oscillator configuration in the absence of the potential, that is

$$V_{int}(z) = V_{int}(z_0) + z \left. \frac{dV_{int}(z)}{dz} \right|_{z=z_0} + \ldots \tag{14}$$

If we rewrite g(t) in Eq. 13 as

$$g(t) = \left. \frac{dV_{int}(z)}{dz} \right|_{z=z_0} f(t) , \tag{15}$$

where $f(t \to -\infty) = 0$, $f(t \to +\infty) = 1$, then the asymptotic Hamiltonian of the forced oscillator is $H_0 + V'z$ which is that of an oscillator with the same frequency ω, but whose minimum has been displaced along the z axis by an amount $\delta z = V'/m\omega^2$ and on the V axis by $\Delta V = V'^2/(2m\omega^2)$, easily demonstrated by completing the square of $H(t \to \infty)$.

With regards the switching function $f(t)$, this could be given by the special forms shown in Fig. 1, chosen to simulate some particular core hole dynamics. For the special case of sudden switching on, the overlap integrals required in (10) are those of displaced Gaussians and Hermite polynomials all of which are systematically analytic with [18]

$$<0,r|1,r+n> = (-1)^n <0,r+n|1,r>$$

$$= (-1)^n \exp(-\beta/2)\beta^{n/2} \left[\frac{r!}{(n+r)!}\right]^{1/2} L_r^{(n)} (\beta) , \tag{16}$$

where $L_r^{(n)}$ is a generalized Laguerre polynomial (=1 for r=0) and

$$\beta = \frac{\hbar}{2m\omega} \left(\frac{V'}{\hbar\omega}\right)^2 \equiv \frac{\Delta\varepsilon_{rel}}{\hbar\omega} \tag{17}$$

with $\Delta\varepsilon_{rel} \equiv \left(\left.\frac{dV_{int}}{dz}\right|_{z=z_0}\right)^2/2m\omega^2$, often called the relaxation energy. If the initial state is the ground state, i.e., $|0,r=0>$, then (10) and (16) reduce to a Poisson distribution

$$P(\varepsilon) = e^{-\beta} \sum_{n=0}^{\infty} (\beta^n/n!) \; \delta(\varepsilon - n\hbar\omega + \beta\hbar\omega) , \tag{18}$$

where $\varepsilon_n = n\hbar\omega$ for a quantum oscillator. In the large β limit, the envelope of the Poisson distribution becomes a Gaussian with FWHM = 2.35 $\beta^{1/2}$ $\hbar\omega$. As a final technical point, we note that the displaced oscillator problem is often considered within a second quantized, Green's function framework [1c-f, 8] in which one deals with Hamiltonians of the form $H = \hbar\omega b^+ b + \lambda(b+b^+)$, where b^+ and b are Boson creation and annihilation operators and λ is a coupling "constant" (with dimensions of energy). When written in this form, $\beta = \lambda^2/(\hbar\omega)^2$ and obviously $\Delta\varepsilon_{rel} = \lambda^2/\hbar\omega$. Comparison with Eq. (17) provides the connection between λ and a physical quantity. Furthermore, comparison with Fig. 1 shows that $\beta = |\lambda(\omega)|^2/\hbar^2$ for the sudden limit. This relationship holds beyond the sudden limit.

11

The form of the Poisson distribution, for various oscillator displacements and hence β values, is shown in Fig. 2, where it is apparent that the larger the displacement, the greater the overlap of the ground-state oscillator wavefunction with highly excited states of the final displaced oscillator.

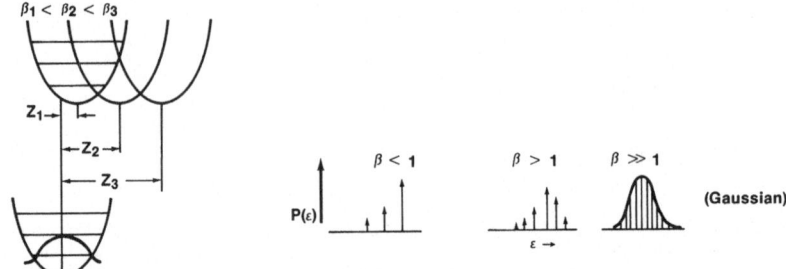

Fig. 2. Potential diagrams for the displaced harmonic oscillator. As the displacement increases, so does the Poisson parameter. On the right are shown three characteristic Poisson distributions, from (18), for various extremes of β

All of these characteristic forms are seen in photoelectron spectroscopy. As a rough estimate, the relaxation energy inferred from (1) and (17) arise from the Coulomb interaction between an induced charge, localized within a characteristic atomic radius of the core hole, and the hole itself. The magnitude is of order 1-4 eV, whereas the plasmons or density fluctuations providing this charge typically have energies $\hbar\omega \sim 5{-}15$ eV. Consequently $\beta_{plasmon} \sim 0.1 - 0.5$, and the theoretical XPS spectrum would look like the far left one [1e,8]. Intramolecular vibrational modes fall in the range $\hbar\omega \sim 0.1$ eV and $\Delta\varepsilon_{rel}$ is somewhat reduced, yielding $\beta_{mv} \sim 1{-}4$ and photoelectron distributions similar to the middle one result when photoionization is from an orbital participating in the molecular bonding [19]. Finally, phonons in solids show $\hbar\omega \sim 10^{-2}$ eV, in which case $\beta \sim 20{-}40$, manifesting itself as a Gaussian broadened line shape [7], as shown on the right.

Another interesting class of time-dependent perturbations involve two sudden changes, as might occur for core hole creation followed, after a time interval τ, by Auger decay to a zero or two hole state. The energy distribution of the final oscillator state is given by (11), using the overlap integrals of (16). Depending upon specifics, different values for $\beta_{0\to1}$ and $\beta_{1\to2}$ might be required. In fact if 0,1,2 are the number of core holes, then $\beta_{0\to1} = \beta_{1\to2}$ whereas if the 1$\to$2 transition represents a return to the no-hole state, then $\beta_{1\to2} = \beta_{1\to0} = -\beta_{0\to1}$. The role of the intermediate state in determining the degree of excitation of the final state is nicely illustrated with the classical picture shown in Fig. 3. Initially the oscillator is in its ground state, as represented by the ball at the bottom of the parabolic well labeled |0>. At time t=0, the undisplaced ball ("vertical transition") finds itself in an excited state of the intermediate parabolic well |1> and thus starts rolling. After a time interval τ, another switch is made to the final parabolic well |2>. Depending upon the role of the intermediate state lifetime τ relative to the frequency of |1>, the 1$\to$2 transition will occur when the ball is in different positions and hence the degree of final excitation in well 2 will be affected. As can be seen, if $\tau = 2\pi n/\omega$ (n=0,1,...), the ball will be

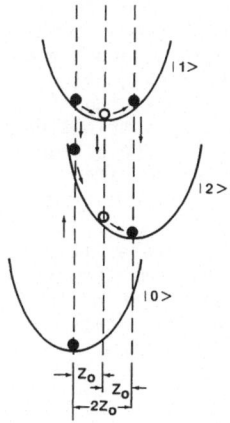

Fig. 3. Classical picture showing the role of time spent as an intermediate state harmonic oscillator $|1>$, in determining the degree of excitation of the final state oscillator $|2>$, given an initially unexcited oscillator $|0>$

in the far left and $|2>$ will be maximally excited, as if $|1>$ never existed. In contrast, if $\tau = 2\pi(1/2 + n)/\omega$, the ball will be at the far right and $|2>$ will be in its ground state. Varying degrees of excitement follow as τ falls between these extrema.

The quantum mechanical generalization is depicted in Fig. 4. The original Gaussian wavefunction of the ground state $|0,0>$, at time $\tau = 0$ is coherently projected onto the complete set of stationary states of H_1, each one of which acquires a different phase before the sudden switch to state, $|2>$ after the

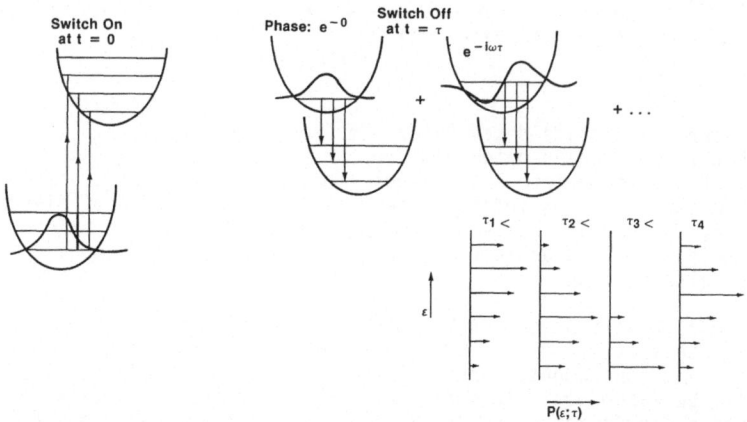

Fig. 4. Quantum mechanical picture showing the role of the intermediate state in determining the energy distribution of the final state oscillator. At t = 0, the intermediate state is switched on and the original ground-state oscillator wavefunction projects onto each of those of the intermediate state. At time t = τ, the final state potential is switched on and each of the intermediate state wavefunctions projects onto those of the final state. However each n'th intermediate state acquires a different phase $e^{-in\omega\tau}$ depending upon the lifetime τ, which is reflected in the series of final state energy distributions $P(\epsilon;\tau)$ for the range $\tau_1 \leq \tau < 2\pi/\omega + \tau_1$

13

time τ. Each of these components projects onto the set of oscillator states of |2> but with different phase factors, as reflected in the coherent summation in (11). The energy distribution of the final state again depends upon the value of τ relative to ω_1, as shown in Fig. 4. While the final state Hamiltonian determines the energy positions of the delta functions, the intermediate state dynamics plays a key role in determining the distribution of energy amongst these eigenstates. Considerations of this sort have been at the heart of much theoretical work in the analysis of resonance Raman intensities [15].

In real situations involving decay of an initially prepared state, it should be recognized that not all decay processes occur at the same time τ, but instead are exponentially spread out. Thus we should average $P_n(\tau)$ over all possible τ's weighted by the probability that the intermediate state still exists at time τ, in the manner suggested by ALMBLADH and MINNHAGEN with regards to x-ray absorption and emission edges [20]. Defining

$$f(*,n) = <2,n|1,*> <1,*|0,0>,$$

(11) can be written as a sum of incoherent plus coherent parts,

$$P_n(\tau) = \sum_* |f(*,n)|^2 + \sum_{*\neq*'} 2\, f(*,n)f^\dagger(*',n)\cos(\varepsilon_1^* - \varepsilon_1^{*'})\tau.$$

The averaged probability is then

$$<P_n> = \Gamma \int_0^\infty d\tau\, e^{-\Gamma\tau}\, P_n(\tau)$$

$$= \sum_* |f(*,n)|^2 + \sum_{*\neq*'} \frac{2\Gamma^2\, f(*,n)f^\dagger(*',n)}{\Gamma^2 + (\varepsilon_1^* - \varepsilon_1^{*'})^2}\,, \tag{19}$$

where Γ is the intermediate state decay rate. In the form shown in (19), it is easy to see that interferences between different paths connecting 0 to 2 are important only for intermediate states whose energy lies within $\sim \Gamma$ of each other, which certainly makes intuitive sense. Thus in the case of plasmon oscillators where $\varepsilon_1^* \sim 10$ eV and core-hole widths are ~ 1 eV, such interferences are negligible, whereas for intramolecular vibrations, pairs, and phonons $\varepsilon_1^* \leq \Gamma$ and the opposite is true.

Lastly we should note limitations on the picture presented here in terms of purely harmonic oscillators [21]. In reality, the oscillator potentials would be anharmonic and consequently, the recurrences suggested in Figs. 3 and 4 could drastically be wiped out. In fact there is a significant possibility that not even quasi-periodic motion in the intermediate state would occur [21]. The resulting chaotic behavior would then have to be dealt with, which while fascinating, is beyond the scope of this exercise.

6. The Electron Gas

The fundamental excited states of the electron gas/metal conduction band are most easily discussed by referring to the textbook picture of the excitation spectrum shown in Fig. 5.

14

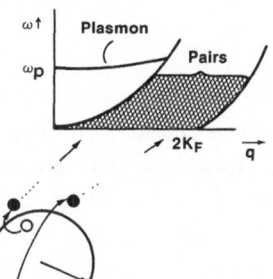

Excitation:
Spectrum:

Fermi:
Sphere:

Fig. 5. Excitation spectrum for a free electron gas. Typically the discrete plasmon mode falls in the 5-15 eV range. The continuum of electron-hole pairs consists of excitations in which electrons are removed from beneath the Fermi sphere and placed above. Examples of low energy long (left) and short (right) wavelength pairs are shown

The discrete plasmon mode, with $\hbar\omega_p = \hbar\sqrt{4\pi ne^2/m} = 47/r_s^{3/2}$ eV, is a coherent collective charge density fluctuation and the continuum of electron-hole pair excitations are single particle excitations formed by taking an electron from within the Fermi sphere and placing it in some unoccupied state above. Possible long and short wavelength low-energy pairs are also shown in Fig. 5. For an infinitely extended metal, there is no gap in the excitation spectrum; thus it is possible to create states containing large numbers of low-energy pairs at very little cost in energy. A reasonable approximation for the pair density of states is given by [22]

$$\rho_{pair}(\omega) \simeq \int\limits_{\varepsilon_1 > \varepsilon_F} d\varepsilon_1 \int\limits_{\varepsilon_2 < \varepsilon_F} d\varepsilon_2 \ \rho(\varepsilon_1) \ \rho(\varepsilon_2) \ \delta(\varepsilon_1 - \varepsilon_2 - \omega)$$

$$\simeq \hbar\omega \ \rho^2_{elec}(\omega_f) \ ,$$

where $\rho(\varepsilon_1)$ and $\rho(\varepsilon_2)$ are the electron and hole density of states above and below the Fermi level. The final result in which the pair density of states varies linearly with ω and with $\rho^2(\varepsilon_f)$ the proportionality constant, implies that any process dependent upon pair excitation should be extremely sensitive to the Fermi-level density of states.

We note that not only plasmons (or surface plasmons) but also the pair excitations can be described in terms of harmonically oscillating charge density fluctuations [23], suggesting that both plasmons and pairs are most easily excited by an external transient <u>charge fluctuation</u> such as a switched on core hole potential.

Independent of the details of the time dependence, the Anderson Orthogonality Theorem [24] states that the ground state of an N-Fermion system is orthogonal to the ground state of the same system containing a finite range scattering potential (such as the core hole), as N→∞. Put another way, it is impossible to switch on or off a core hole without <u>necessarily</u> creating some pair excitations. Within the sudden switching limit, the energy distribution of pair excitations is given by (10). Exact evaluation of the Franck-Condon factors right at the threshold, has been carried out by NOZIERES and de DOMINICIS [25] and the validity extended beyond threshold by others [26], with the result

$$P(\varepsilon) = \eta \frac{1}{\varepsilon^{1-\alpha}} \quad (0<\varepsilon\leq\varepsilon_c) , \qquad (20)$$

where $\eta = \alpha/\varepsilon_c^\alpha$, ε_c is a cutoff energy of order the conduction bandwidth, and $\alpha = \Sigma_\ell (\delta_\ell/\pi)^2 \sim \rho_{elec}^2(\varepsilon_f)|\Delta V|^2$ with δ_ℓ the ℓ'th partial wave Fermi-level electron phase shift associated with the localized hole potential ΔV. Typically $\alpha \simeq 0.1$ and $\varepsilon_c \simeq 10$ eV so with (20), the mean excitation energy is

$$<\varepsilon> = \int_0^{\varepsilon_c} \varepsilon \, \rho(\varepsilon) \, d\varepsilon = (\frac{\alpha}{1+\alpha})\varepsilon_c \simeq 1 eV.$$

Taking into account the finite lifetime of the core hole by convolving the "x-ray edge function", (20) with a "lifetime Lorentzian", the resulting energy distribution is the by-now-venerable "Doniach-Sunjić asymmetric lineshape [1a,1d,7,27]

$$P_{DS}(\varepsilon) = \frac{\eta}{\pi} \int_0^\infty d\varepsilon' \, (\frac{1}{\varepsilon'^{1-\alpha}}) \, \frac{\Delta}{(\varepsilon-\varepsilon')^2 + \Delta^2}$$

$$\frac{\cos[(1-\alpha) \tan^{-1} (\varepsilon/\Delta) -\pi\alpha/2]}{(\varepsilon^2 + \Delta^2)^{(1+\alpha)/2}} , \qquad (21)$$

where Δ is the natural linewidth resulting from x-ray or Auger decay. Furthermore, if the core hole (or other localized charge fluctuation) occurs outside the surface, i.e., on an adsorbate, the resulting expression for the lineshape is [1a,1d]

$$P_{ads}(\varepsilon) \sim \frac{\exp(-\varepsilon/\omega_m) \cos[(1-\alpha(s)) \tan^{-1}(\varepsilon/\Delta) - \pi\alpha(s)/2 - \gamma/\omega_m]}{(\varepsilon^2+\Delta^2)^{(1-\alpha(s))/2}} \qquad (22)$$

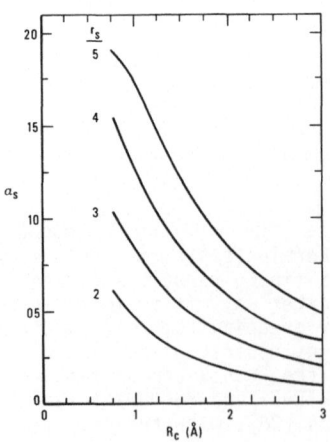

Fig. 6. Asymmetry index for a suddenly-switched-on point charge, located a distance R_c outside an electron gas, characterized by the density parameter r_s. For a point hole within an electron gas, with s-wave scattering only, $\alpha = 0.25$, which indicates that the results shown here are not inconsistent with bulk requirements [13]

for $|\varepsilon| \lesssim 3\Delta$, where now α is a function of s, the hole location with respect to an effective image plane, as shown in Fig. 6, and $\omega_m \simeq v_F/2s$ with v_F the substrate Fermi velocity [13,28]. In the limit that $\omega_m \to \infty$, $P_{ads}(\varepsilon) \to P_{DS}(\varepsilon)$.

Concerning the creation of a localized charge in a metal, we know that this charge must be totally screened. Up to this point it has been implicitly assumed that this screening charge could be described in terms of some wave-packet state formed by a coherent superposition of plasmons. Another possibility exists in terms of conduction band hopping into an unfilled atomic or molecular orbital of the ionized particle, if this is energetically more favorable than the dielectric plasmon response. The mechanism is illustrated in Fig. 7.

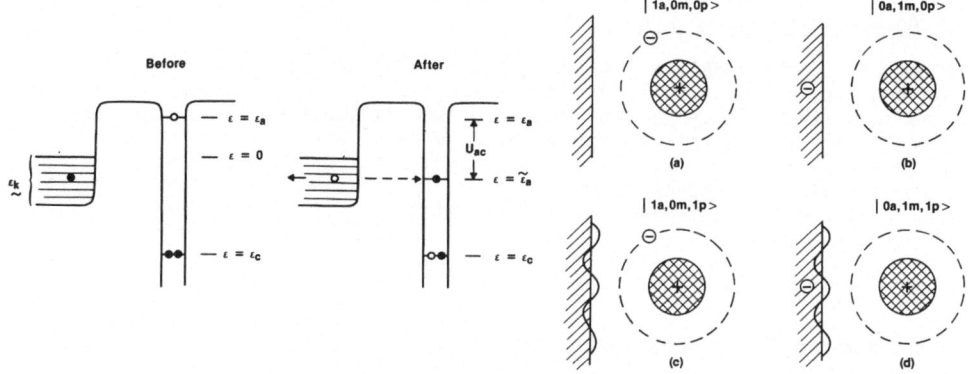

Fig. 7. Left: Energy diagram illustrating the relevant energy levels of the adsorbate and substrate, both before and after photoionization. Full (open) circles represent occupied (vacant) electron states. Right: Schematic representation of the 4 classes of states considered in screening. The large positive circle is the ionized adparticle. The dashed circle represents the lowest unfilled orbital. The wavy structure at the interface is a surface plasmon. The small negative circle is the "screening electron". As drawn <u>a</u> represents orbital, <u>b</u> no, <u>c</u> orbital and surface plasmon, and <u>d</u> surface plasmon screening

Before ionization, the unfilled orbital, with orbital energy ε_a, lies above the Fermi level and hence is unoccupied. Upon ionization, the Coulomb interaction U_{ac} between the hole and an electron which might occupy the level ε_a, may be sufficiently attractive to pull the level beneath the Fermi level, thus allowing for its self-consistent occupancy. The propensity for occupancy depends upon some hopping matrix element V_{ak} between the substrate and adparticle (or target atom if in the solid). Schematic displays of the various screening possibilities are also shown in Fig. 7, and described in the caption. Of course in reality all situations are some hybrid of both mechanisms with one or the other favored more or less, depending on specifics. Roughly the criteria depends upon the size of $|V_{ak}|^2/(\bar\varepsilon_{band} - (\varepsilon_a - U))$ vs. $\lambda^2/\hbar\omega_{sp} \simeq V_{image}$ where $\bar\varepsilon_{band}$ is some average band energy of the substrate electrons involved in the hopping. The first term is an estimate of the orbital screening energy whereas the second that of the surface plasmon. Typically both are in the $\sim$ 1-5 eV range.

LANG and WILLIAMS [9] have provided a useful rule of thumb for deciding upon one or the other mechanism. Based on density functional calculations for core hole energies associated with adsorbed atoms on jellium surfaces, they noticed that if the ionized atom has a closed valence shell (i.e., the screening electron must be promoted to a shell with a higher principle quantum number) then the adparticle screening is reasonably well simulated by the surface plasmon image charge screening. Experiments for I on Ag and Xe on several substrates support this conjecture. On the other hand, for open valence shells, screening is apt to be well modeled by the orbital mechanism. Since the screening energy is roughly the same in either case, experimental studies could focus on the positions of shakeup satellites, which would be quite different for plasmon vs. atomic-like excited states.

7. Adiabatic to Sudden Limits

Almost all of the modeling discussed in this chapter has depended upon the applicability of the sudden approximation. Thus it seems prudent to probe a bit into its validity and see what experimental implications one extreme or the other has. First we note that a time-dependent perturbation may appear adiabatic (slow) to a high-energy excitation but sudden to a low-energy (or slowly responding) one. Stated quantitatively, for a given excitation ϵ^* and a time-dependent interaction $V(t) = V_0 f(t)$, the adiabatic limit is approached as [1a]

$$\frac{df}{dt} << \epsilon^*/\hbar$$

or in terms of Fourier components, as

$$f(\omega^*) = \int dt \; e^{-i\omega^* t} \; f(t) << \frac{1}{\omega^*} \quad \text{(the sudden limit)}.$$

This is well illustrated in Fig. 1 for the case of the exponentially switched on potential. The right-hand side shows that $f(\omega)$ rapidly goes to zero for $\omega > \eta$, the rate at which the potential is introduced. The adiabatic or infinitely slow limit is reached as $\eta \to 0$, in which case $f(\omega) \to 0$ for all ω and $V(t)$ is thus incapable of creating any finite energy excitations. Consequently any adiabatic spectrum would contain only a single line, the so-called adiabatically-relaxed peak, which in terms of the harmonic oscillator scheme discussed in Section 5 would correspond to transitions from the ground state of one oscillator well to the ground state of another.

From the point of view of experimental verification, one should try to invent a situation in which the rate of hole-switching on can be controlled, in order to check out these ideas. Since a core-hole potential is equivalent to the absence of an electron, the rate at which this localized potential is switched on can be varied by varying the rate at which the excited electron exits the ionization region. This can be controlled using synchrotron radiation varying the excess kinetic energy of the photoionized electron continuously from threshold, where the exit velocity is zero and thus the switching is adiabatic, to "very large" where not only is the exit velocity large but also plasmon-mediated electron-hole interactions are small, both conditions being required in the sudden limit [1a,1c,29]. On this physical basis, theoretical estimates have been made concerning both pair-asymmetries in DONIACH-SUNJIC line shapes [1a,27] and intrinsic plasmon satellites [1c, 29], as a function of outgoing ionized electron kinetic energy which appear to be in accord with experimental observations [30].

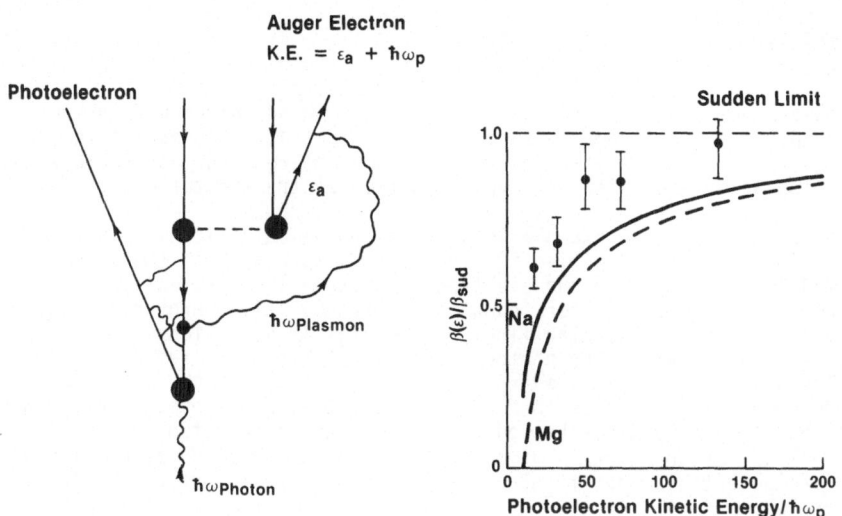

Fig. 8. Left: Diagramatic representation of the process responsible for Auger-plasmon-gain satellites. Right: Plasmon-gain-satellite intensity for KLL Auger transitions in Na and Mg, as a function of photoelectron energy, normalized to the intensity in the sudden limit. Calculated values from [1c] for Na and Mg are also shown [31]

   A particularly intriguing demonstration of this phenomenon has recently been reported by FUGGLE and coworkers [31] within the context of "plasmon-gain satellites" in Auger emission [32]. The basis of this experiment is shown on the left in Fig. 8. An x ray creates a photoelectron-core hole pair and so-called intrinsic plasmons associated with the transient localized potential. Subsequently, the core-hole Auger decays into a two hole-one electron state, where in lowest order, the kinetic energy of the Auger electron is characteristic of the particular core hole and is independent of the initial excitation process. However, a small possibility exists for the Auger electron to absorb a plasmon created in the initial photo-excitation process, in which case a "plasmon-gain satellite" would result; that is some Auger electrons would emerge with a kinetic energy $\hbar\omega_p$ in excess of the characteristic Auger energy. The intensity of this gain satellite should be proportional to the probability for intrinsic plasmon creation in the initial photoionization process. As already mentioned, the Poisson distribution follows from a sudden approximation and hence serves as an upper limit on the plasmon intensity. Reducing the photoelectron energy by lowering the ionizing photon energy should result in reduced satellite intensities, ultimately going to zero for threshold ionization. A semi-quantitative theory has been reported [1,c]. FUGGLE and coworkers measured the plasmon gain intensity, as a function of photoelectron kinetic energy for KLL Auger transitions in Na and Mg. The normalized results are shown in the right side of Fig. 8, together with theoretical predictions [1c]. Considering the simplifications in the theory and complications in the experiment, not only is the agreement gratifying but more importantly, the unambiguous support for the adiabatic to sudden transition ideas expressed in this section is reassuring.

8. Photoexcitation/Desorption Scenarios

We have now arrived at the point where many of the ideas introduced through-out this chapter can be assembled to form some illustrative scenarios in which initial electronic excitation leads to some interesting observable final state dynamics associated with the nuclear motion of the excited object.

The most straightforward example is that of photo-ionization in diatomic molecules. The rate of ionization is given by an expression of the form of (2), the modules squared of a one-electron $\underline{p} \cdot \underline{A}$ matrix element, multiplied by the relevant Franck-Condon factors consistent with overall energy conservation. Discrimination between particular final hole states is equivalent to taking $\varepsilon_{\alpha'} = \varepsilon_{hole} + KE$, where KE is the photoelectron kinetic energy, replacing the sum on α by a sum on KE for fixed ε_{hole}, and taking the Franck-Condon factors to be independent of KE. There then exists a one-to-one correspondence between the photoelectron energy distribution and the energy distribution of the relative nuclear motion which is given by (10). A number of different final or hole state potentials that might arise in photoionization of an AB diatomic molecule are shown in Fig. 9 [33].

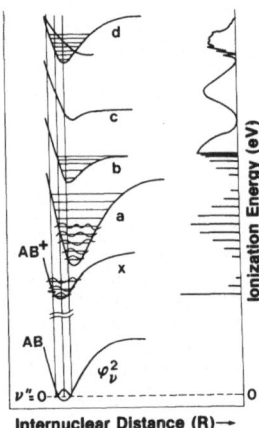

Fig. 9. Potential curves for the molecule AB in its ground state and the corresponding molecular ion AB$^+$ in several ionic states. The Franck-Condon region is between the vertical lines. Photoelectron bands resulting from the various transitions are shown schematically on the right ordinate [33]

The lowest lying AB$^+$ curve corresponds to ionization from an orbital not too involved in the AB bonding, hence little change in the internuclear separation. The projection of the AB vibrational ground-state wavefunction is thus largest on the ground state wavefunction of the ionic curve, and the nuclear energy distribution, (10), is the series of discrete delta functions of decreasing intensity shown on the right, characteristic of discrete to discrete transition: If the potential wells were harmonic, this distribution would be identical with the $\beta \ll 1$ Poisson distribution of (18), shown on the left in Fig. 2. Photoionization of a bonding orbital requires that the AB separation increases and the resulting vibrational excitation distribution is shown by curve a, similar to the $\beta \sim 1\text{-}5$ Poisson distribution, but still involving bound states only. Of potential relevance to DIET are the curves b-d, where not only does the original vibrational wavefunction overlap bound states of the hole potent-ial, but also dissociative continuum states. In fact an excitation event leading to curve c looks identical to a MGR process. Of course, the nuclear dynamics does not care what the final state of the electron is but only what is the potential surface resulting from that final state; thus distinctions between ionizing events and bonding to anti-bonding excitations are largely irrelevant, provided the potential curves are similar. It is for this reason that the present considerations of photoelectron spectroscopy are relevant to DIETs. The interesting case of predissociation is shown in curve d, in which initial excitation is to an electronic state with the potential well, hence the bound-to-bound vibrational transitions. However this electronic state is degenerate with another electronic state in which the nuclei are un-bound, but which states could be inaccessible via the direct photoexcitation. Transitions between these electronic states lead to pre-dissociation in which the fragment energy distributions show remnants of the discrete vibrational structure, effectively lifetime broadened [34].

Although in photoelectron spectroscopy, the energy distribution of the ejected electron is measured, one could also measure the photo-dissociated ion and atom energies, and the similarities with the experimental DIET procedure would be striking. A complication arises in the ion/atom measurements since the resulting distributions would be superpositions of all possible continuum distributions, independent of photoelectron or hole state. Thus in Fig. 9 the observed distribution would contain contributions from curves b, c, and d, from which it would be difficult to extract information unless the branching ratios overwhelmingly favored one alternative over all others. Fortunately, this problem can be overcome, at least in principle, by doing photoelectron-photo fragment coincidence measurements, which would then allow us to know which hole surface governed the time evolution and hence energy distribution of the dissociated fragment.

With these considerations in mind, it is easy to construct a possible DIET sequence in which most of what has been said here could lead to observable consequences. A hypothetical, but possible chain of events is shown in Fig. 10 in which initially a potentially desorbable atom is in the vibrational ground state of the well formed by the chemical bond between the atom and the surface. The initial excitation could be to a highly excited (perhaps auto-ionizing) neutral state, in which the potential well maximum is displaced outwards from the surface, labeled A* in Fig. 10. Excited neutrals are then ejected directly, some of which might subsequently radiate in the manner discussed by TOLK et al. in this volume. As considered so far, excitation and desorption appear to be via the MGR mechanism. However, for some adsorbate-substrate separations, the hypothetical A* state could be degenerate with a singly charged ionic state. Surface induced mixing between these states not only gives rise to a branching between excited neutrals and ions but also couples MGR and KF desorption processes. In this example, the $A^0 \rightarrow A*$ $\rightarrow A_1^+$ process can be thought of as not only MGR desorption on A* together with depletion of the A* state by coupling to A_1^+, but also KF initial excitation to A_1^+. Perhaps some "Fano-like" [36] behavior in total ion yields would result from interferences between the $A^0 \rightarrow A* \rightarrow A_1^+$ and $A^0 \rightarrow A_1^+$ paths for initial excitation. Once on the A_1^+ curve, return to A* is unlikely since Auger decay from one to two hole (A_1^+ to A^{++}) is probable since Auger life-times are generally shorter than the characteristic transit times for an atom through the few Å distance over which the Auger probability exponentially decreases to insignificance. This transition is represented by the line labeled Auger. One must know the potential curves in order to obtain vibrational and continuum nuclear wavefunctions needed for calculation of Franck-Condon factors. Since the states dictating the shape of curve $|1\rangle$ have a finite lifetime, the vibrational Franck-Condon factors between $|A^0\rangle$ and $|A^{++}\rangle$ or $|A^{++}, \varepsilon\rangle$ given that $|A_1^+\rangle$ was an intermediate state, would be of the form of (11) or (19). However still more possibilities must be considered. Auger

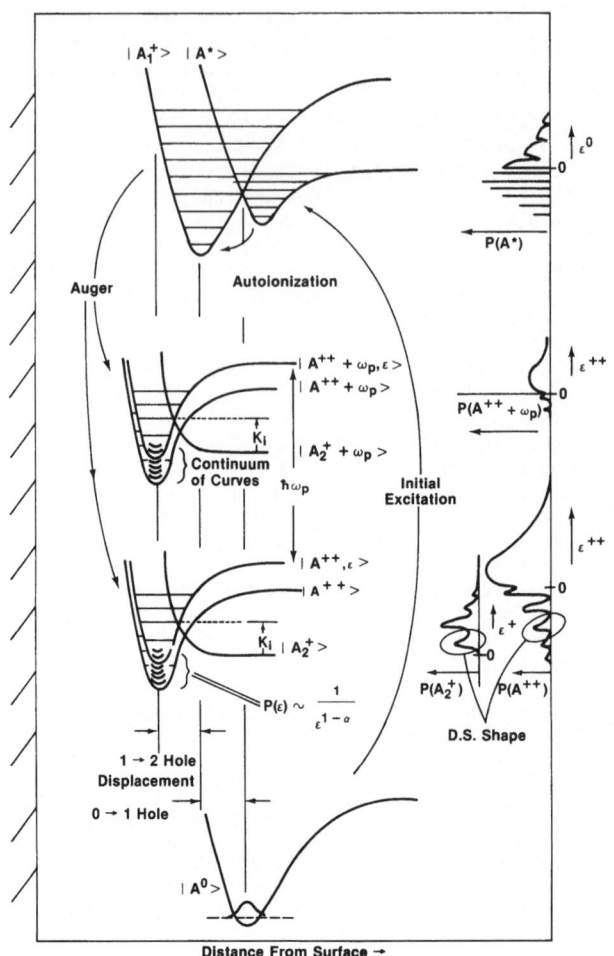

Distance From Surface →

<u>Fig. 10.</u> Hypothetical DIET scenario illustrating possible manifestations of many of the points discussed in this chapter. On the right are shown energy distributions for the various neutral and ionic products. Desorbed particles correspond to that portion of the energy distribution in which $\varepsilon > 0$

transitions from $|A_1{}^+>$ to $|A^{++}>$ involve switching on an additional hole charge and thus in the screening response, the possibility of surface plasmon satellites are present. A Poisson distribution of satellite curves, with intensities dictated by $\beta \simeq V_{image}/\hbar\omega_p \sim 0.1 - 0.5$, follow, one of which is shown as $|A^{++} + \omega_p >$ in Fig. 10. Further, in the one to two hole transition, pair creation leading to surface modified Doniach-Šunjić line shapes, (20) and (21), could result as well as either discrete vibrational or Gaussian broadening due to the ion-surface well, suggested by the continuum of 2 hole-surface attractive curves shown in Fig. 10 [13,35]. While the doubly charged ion vibrates in this well, still another predissociative transition could occur leading to final desorption as a singly charged ion on $|A_2{}^+ >$. The dynamical behavior while in this final well is exactly that of the time reversed dynamics

22

of sticking or adsorption involving substrate induced diabatic transitions [13] on a particle incident upon a surface. The energy distributions of the ions ejected from $|A^{++}>$ and $|A^{++} + \omega_p>$ would be complicated functions involving the coherence factors associated with the initial excitation and subsequent decay dynamics, the Poisson distributions and Doniach-Šunjić functions due to "p^3" excitations, as well as various predissociation and lifetime factors. This can be viewed as good or bad news. Bad news because the energy distributions contain so much information that they would be hard to disentangle using conventional experimental techniques. However good news when one realizes how rich in information pertinent to non-adiabatic surface dynamics these distributions are. Within the idealized world in which a theorist lives, the notion of high-resolution energy distributions of photo or auger electrons coincident with desorbed ions or atoms (experimental data which would be very helpful for disentangling this information) provides not only stimulating food for thought but also a DIET which, because of its high density of useful information, might be sufficiently rewarding to stimulate a not-so-idealized-world experimentalist to bridge the gap and enjoy the best of both worlds.

References

1. a) J.W. Gadzuk and M. Šunjić, Phys. Rev. B 12, 524 (1975);
   b) J. W. Gadzuk, Phys. Rev. B 14, 2267 (1976) and Surface Sci. 67, 77 (1977);   c) J.W. Gadzuk, J. Elec. Spect. 11, 355 (1977);   d) J.W. Gadzuk and S. Doniach, Surface Sci. 77, 427 (1978) and J.W. Gadzuk, Surface Sci. 86, 516 (1979);   e) J.W. Gadzuk in "Photoemission and the Electronic Properties of Surfaces", ed. by B. Feuerbacher, B. Fitton, and R. F. Willis, Wiley (1978);   f) J.W. Gadzuk, Kinam 2, 7 (1980).

2. G. Korzeniewski, E. Hood, and H. Metiu, J. Vac. Sci. Technol. 20, 594 (1982) and references therein.

3. J. Lin and T. F. George, J. Chem. Phys. 72, 2554 (1980); C. Jedrzejek, K.F. Freed, S. Efrima, and H. Metiu, Surf. Sci. 109, 191 (1981).

4. "Photoselective Chemistry" ed. by J. Jortner, R. D. Levine, and S.A. Rice, Adv. Chem. Phys. 47, (1981), for a good entry point.

5. D. Menzel and R. Gomer, J. Chem. Phys. 41, 3311 (1964); P.A. Redhead, Can. J. Phys. 42, 886 (1964); D. Menzel, J. Vac. Sci. Technol. 20, 538 (1982).

6. G. Wendin, "Breakdown of the One-Electron Pictures in Photoelectron Spectra", Structure and Bonding 45, Springer-Verlag, Berlin (1981).

7. P. H. Citrin, G. K. Wertheim, and Y. Baer, Phys. Rev. B 16, 4256 (1977).

8. D. C. Langreth, Phys. Rev. B 1, 471 (1970).

9. A. Kotani and Y. Toyozawa, J. Phys. Soc. Japan 37, 912 (1974); K. Schönhammer and O. Gunnarsson, Surf. Sci. 89, 575 (1979); A.R. Williams and N.D. Lang, Phys. Rev. Lett. 40, 954 (1978).

10. M. L. Knotek and P. J. Feibelman, Phys. Rev. Letters 40, 964 (1978); P. J. Feibelman, in "Inelastic Particle-Surface Collisions", ed. by E. Taglauer and W. Heiland, Springer, Berlin (1981).

11. T. F. O'Malley, Advan. At. Mol. Phys. 7, 223 (1971); J.C. Tully and R. K. Preston, J. Chem. Phys. 55, 562 (1971); J. W. Gadzuk, Surf. Sci. 118,180 (1982).

12. U. von Barth and G. Grossman, Phys. Rev. B 25, 5150 (1982).

13. J. W. Gadzuk and H. Metiu, Phys. Rev. B 22, 2603 (1980); H. Metiu and J. W. Gadzuk, J. Chem. Phys. 74, 2641 (1980).

14. G. A. Sawatzky, and T. D. Thomas, Phys. Rev. Letters 41, 1825 (1978); O. Gunnarsson and K. Schönhammer, Phys. Rev. B 22, 3710 (1980); S. M. Girvin and D. R. Penn, Phys. Rev. B 22, 4081 (1980).

15. For example see: A. D. Wilson and H. Friedman, Chem. Phys. 23, 105 (1977); W. Siebrand and M. Z. Zgierski, J. Chem. Phys. 71, 3561 (1979); K. Ohno, Chem. Phys. Letters 69, 491 (1980); Y. Fujimura, H. Kono, T. Nakajima, and S. H. Lin, J. Chem. Phys. 75, 99 (1981); P. M. Champion and R. C. Albrecht, Chem. Phys. Letters 82, 410 (1981).

16. E. Müller-Hartman, T. V. Ramakrishnan, and G. Toulouse, Solid State Comm. 9, 99 (1971); J.W. Gadzuk, Phys. Rev. B. 24, 1866 (1981); J.K. Norskov, J. Vac. Sci. Technol. 18, 420 (1981).

17. P. W. Langhoff, S. T. Epstein, and M. Karplus, Rev. Mod. Phys. 44, 602 (1972); H. D. Meyer, Chem. Phys. 61, 365 (1981).

18. S. Waldenstrom and K. Razi Naqvi, Chem. Phys. Letters 85, 581 (1982).

19. L. S. Cederbaum and W. Domcke, Adv. Chem. Phys. 36, 205 (1977).

20. C. O. Almbladh and P. Minnhagen, Phys. Rev. B 17, 929 (1978).

21. Y. Weissman and J. Jortner, Phys. Letters 83A, 55 (1981); E. B. Stechel and R. N. Schwartz, Chem. Phys. Letters 83, 350 (1981); J. Brickman and P. Russeger, J. Chem. Phys. 75, 5744 (1981).

22. E. Müller-Hartman, T. V. Ramakrishnan, and G. Toulouse, Phys. Rev. B 3, 1102 (1971).

23. J. W. Gadzuk, Phys. Rev. B. 24, 1651 (1981).

24. P. W. Anderson, Phys. Rev. Letters 18, 1049 (1967).

25. P. Noziéres and C. T. De Dominicis, Phys. Rev. 178, 1097 (1969).

26. P. Minnhagen, J. Phys. F 7, 2441 (1977); J. D. Dow and C. P. Flynn, J. Phys. C 13, 1341 (1980); K. Shung and D. C. Langueth, Phys. Rev. B 23, 1480 (1981).

27. S. Doniach and M. Šunjić, J. Phys. C 3, 285 (1970).

28. B. Gumhalter and D. M. Newns, Phys. Letters 53A, 137 (1975).

29. J. J. Chang and D. C. Langreth, Phys. Rev. B 5, 3512 (1972).

30. S. A. Flodstrum, R. Z. Bachrach, R. S. Bauer, J. C. Mc Menemin, and S. B. M. Hagstrom, J. Vac. Sci. Technol. 14, 303 (1977); F. J. Himpsel, D. E. Eastman, and E. E. Koch, Phys. Rev. Letters 44, 214 (1980); T. Jach and C. J. Powell, Phys. Rev. Letters 46, 953 (1981).

31. J. C. Fuggle, R. Lasser, O. Gunnarsson, and K. Schönhammer, Phys. Rev. Letters $\underline{44}$, 1090 (1980).

32. C. O. Almbladh, Il Nuovo Cimento $\underline{23B}$, 75 (1974).

33. J. W. Rabalais, "Principles of Ultraviolet Photoelectron Spectroscopy", Wiley, N.Y. (1977).

34. G. Herzberg, "Spectra of Diatomic Molecules", Van Nostrand, Princeton (1950).

35. J. W. Gadzuk, S. Holloway, C. Mariani, and K. Horn, Phys. Rev. Letters $\underline{48}$, 1288 (1982).

36. U. Fano, Phys. Rev. $\underline{124}$, 1866 (1961).

1.2 An Analysis of Electronic Desorption*

D.R. Jennison

Sandia National Laboratories, Albuquerque, NM 87185, USA

Abstract: An analysis of the desorption process in time is presented. Especially discussed are the roles of correlation in multi-particle excitations and of the strain induced localization of excitations in periodic systems.

A fundamental understanding of the mechanisms of bond scission due to electronic excitations is important not only to the surface spectroscopies of electron, photon and ion stimulated desorption, but also to such diverse interests as semiconductor device hardness, radiation biology, surface preparation for improved adhesion and polymer science. Considerable progress has recently been made in improving our understanding of such mechanisms for large molecules and surfaces. Since several authors in the present volume have reviewed the published models, I will attempt to present a general analysis of desorption.

The goals of this article are to organize our thinking by separating the various physical factors inherent in the total desorption process into elements which may be considered independently, and to indicate potentially profitable directions for future theoretical research.

It is useful to analyze the desorption process in time. Events occurring in times less than a few x 10^{-16} sec (e.g. the creation of the initial electron excitation and typical hole or electron hopping times) take place so rapidly that atomic motion may be ignored. In times of the order 10^{-15} sec, atomic motion may be affected by the presence of an electronic excitation and this coupling must be considered [1]. On time scales 10^{-14} sec - 10^{-15} sec, the excitation may decay by valence Auger processes. The time 10^{-15} sec is typical for the decay of deep-valence semiconductor holes in the bulk while a decay time of 10^{-14} sec is typical based on isotope effect analyses made using the Menzel-Gomer and Redhead models [2].

In Figure 1, an analysis of the desorption process is presented and the remainder of this article is devoted to explaining it. While oversimplified and incomplete, it is nevertheless hoped that the analysis will be found instructive and convenient as a foundation. The "initial electronic excitation" may only be a precursor to the "initial valence excitation" whose evolution in time we will analyze in Fig. 1. This is obviously true if the initial excitation involves a core or deep valence hole which may Auger decay (the Knotek-Feibelman mechanism [3]).

---

* This work was performed at Sandia National Laboratories supported by the U.S. Department of Energy under contract number DE-AC04-76DP000789.

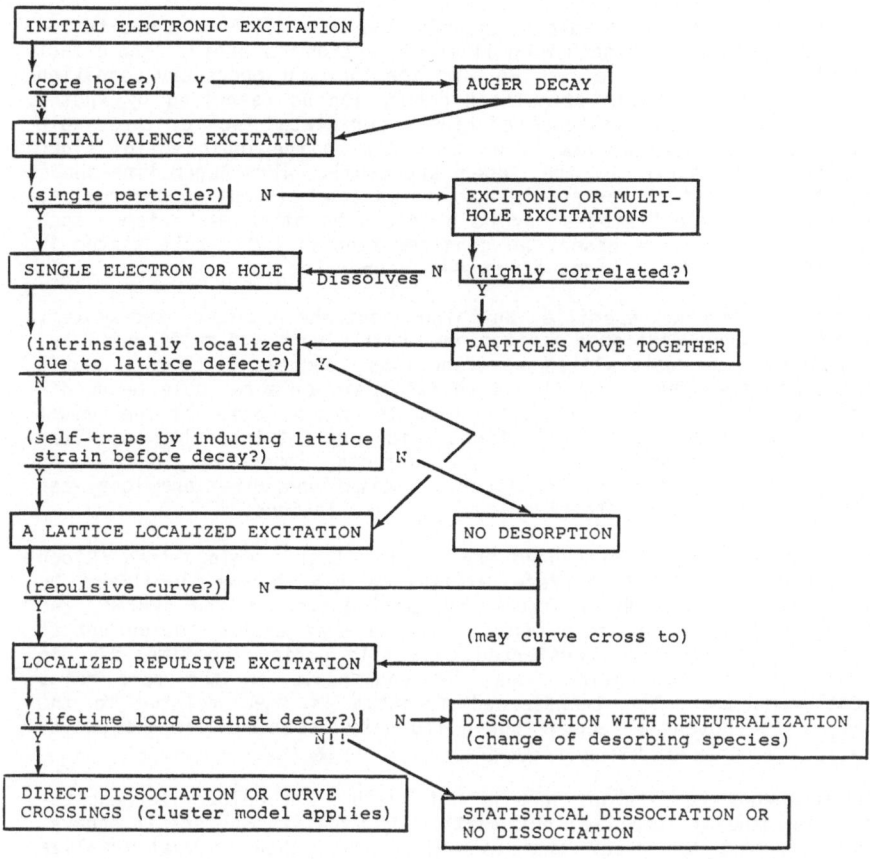

Fig. 1. The desorption process is analyzed. Y = yes, N = no

The "initial valence excitation" may be single or multi-particle in character. Examples of the former are single carriers in semiconductors or normal photoemission final states. Examples of the latter are shake photoemission states or Auger final states. Multi-hole valence excitations are of special interest because they have sufficient energy to produce positive ions (e.g. H^+), while single particle excitations frequently do not. However, in order to be relevant to desorption there must be a high degree of correlation between the holes[4] (or the electron and hole in the case of excitons). If the holes are not highly correlated, they will separate in times of the order of single hole hopping times, i.e. $\hbar/w$ where w is the valence bandwidth. Since these times are typically $\lesssim$ few x 10^{-16} sec, atomic motion will be unaffected by the short-lived multi-hole resonance and for purposes of desorption the multi-hole excitation may be thought of as having "dissolved" into a collection of independent single particle excitations. Recently, highly correlated multi-hole excitations have been observed in covalent molecules[5] and upon a silicon surface [6]. The physics of this phenomenon is now quantitatively understood [7].

27

We see that the essential role of correlation in a multi-hole excitation is to hold the excitation together until atomic motion may begin. The effect of the latter is discussed below. We also see that the necessary condition for the occurrence of Auger stimulated desorption as described by Knotek and Feibelman [3], is the existence of highly correlated two- or three-hole Auger final states (such as are known to occur in the ionic solids since these states are visible in the Auger spectrum). Thus Auger line-shape analysis is a useful tool for providing information relevant to desorption. However, the Auger final states may be produced by other mechanisms, such as shake photoemission processes, so consideration of multi-hole states is not restricted to exciting beams of core-hole energies.

Up to now, we have not specified anything about the nature of the system. If periodic, the above-mentioned excitations, both single and multi-particle, will be described by delocalized wavefunctions, possess dispersion and a local lifetime, $ℏ/w$, where w reflects an excitonic or a two-hole bandwidth in the case of highly correlated multi-particle excitations. If the system is not periodic, there may be excitations which are intrinsically localized due to an extrinsic feature, e.g. lattice defects, impurities, adatoms or admolecules. We will call all electronic excitations which are localized due to lattice properties, "lattice localized excitations."

At this conference, I introduced the concept that a delocalized (Bloch or molecular-orbital like) surface excitation may become localized by causing a local strain, which breaks the periodicity of the system, and leads to desorption. Intuition would indicate that atomic motion cannot be ignored if the momentum transferred to a site during the time, $ℏ/w$, in which a delocalized excitation hops, is comparable to the zero point vibrational momentum. The transferred momentum is then related to the slope, F, of the potential energy curve found by assuming that the excitation is localized (e.g. using a cluster model) by $ℏF/w$.

Of course, what is relevant here is not whether the delocalized excitation would eventually localize, but whether it localizes before it decays. (If we assume that the decay time is $\sim 10^{-14}$ sec, then it must localize in typically the order of a hundred hops.) Both situations are known from studies of bulk transport properties in which the strain-localized bulk carrier is referred to as a small polaron [8]. It is important to studies of desorption to find either solution for a given system: if delocalized excitations do not localize, observed desorption may be assigned to defect sites.

The similarity of the present case to the body of small polaron theoretical work is that a "feedback loop" exists[9]: the presence of the excitation causes strain which localizes the excitation and causes more strain. This leads to a non-linear Schrödinger equation whose solution is difficult. The dissimilarity of the present situation to conventional small polaron theory is that the potential energy curves of interest are repulsive and lead to dissociation [9]. It is gratifying, however, that rigorous small polaron theory, done by Emin[8], leads to the same criterion for localization obtained by the intuitive argument discussed above.

The criteria for strain induced localization may be summarized: it is desirable for the excitation to have a steep potential energy surface (large force, F) and a small bandwidth, w (long local lifetime before hopping). Highly correlated two-hole excitations display both characteristics since they are very repulsive and hop slowly due to their two-particle character. I have determined that it is very likely that such excitations on multi-hydride silicon surfaces readily localize by inducing Si-H strain [10]. Single hole excitations are currently being studied which, if they localize, would lead to neutral H production.

A "lattice localized excitation" permits a cluster analysis of desorption[11] providing that the decay of the excitation is not considered. If the decay must be considered a cluster model may be appropriate depending upon whether electrons which may enter or leave the cluster during the decay are accounted for. However, we have seen that the cluster model does not apply unless the excitation localizes, which involves considerations of correlation (in multi-particle cases), lattice periodicity and strain induction.

Cluster models, if appropriate, may show whether a localized excitation which is not on a repulsive curve may be an entry channel to desorption by curve crossing onto a repulsive curve (so-called electronic predissociation[11]). They may also form the basis for calculations of decay rates which may cause reneutralization or prevent dissociation. If decay returns the site to its ground state after appreciable kinetic energy has been transfered to the atoms, it is possible that dissociation may occur from the ground state curve (often called statistical dissociation) depending upon the rate of energy transfer from the local site to the remainder of the system.

Future theoretical work needs to continue to explore the nature of correlation in multi-hole excitations, to explore the occurrence of strain-induced localization, to compute decay rates for localized excitations and to study vibrational energy transfer on surfaces. Cluster models should also predict ion energy distribution curves, more of which should be measured. Neutral atom studies are also obviously very important, particularly near threshold where single particle excitations occur.

Since several theoretical efforts are currently underway, I expect considerable progress to be made by several groups by the time of the next DIET meeting.

I wish to thank D. Emin for discussions on self-trapping and H. H. Madden for critically reading the manuscript.

References

1. A hydrogen ion repelled by a bare Coulomb repulsion, e^2/R, will move 0.1 Å in 10^{-15} sec when starting from an R equal to the H-Si bond length.

2. For a review, see R. Gomer, this volume.

3. For reviews, see M. L. Knotek, this volume, and P. J. Feibelman, this volume.

4. D. Ramaker, this volume, and references therein.

5. D. R. Jennison, J. A. Kelber and R. R. Rye, Phys. Rev. B 25, 1384 (1982).

6. H. H. Madden, D. R. Jennison, M. M. Traum, G. Margaritondo, and N. G. Stoffel, Phys. Rev. B, 15 July 1982.

7. D. R. Jennison, J. Vac. Sci. Technol. 20, 548 (1982).

8. D. Emin, Adv. in Phys. 22, 57 (1973).

9. D. Emin, J. Vac. Sci. Technol., 22, Mar/Apr 1983.

10. D. R. Jennison, J. Vac. Sci. Technol., 22, Mar/Apr 1983.

11. C. F. Melius, R. H. Stulen, and J. O. Noell, Phys. Rev. Lett. 48, 1429 (1982).

1.3 Direct and Indirect Mechanisms of Stimulated Desorption

J.C. Tully

Bell Laboratories, Murray Hill, NJ 07974, USA

1. Introduction

The objective of this workshop has been to make progress toward a microscopic understanding of the mechanisms of bond breaking that lead to desorption. The stimulated desorption process can be pictured, loosely, as a sequence of four steps:

Step 1.   The initial electronic excitation (10^{-16}s).

Step 2.   A fast (10^{-15}s) redistribution of electronic energy.

Step 3.   A slower (10^{-14}-10^{-13}s) displacement of atomic positions, resulting in ejection.

Step 4.   A modification of the escaping species as it recedes from the surface (10^{-13}s).

Much of our knowledge of the desorption process is derived from examination of the ejected species; its identity, angular and velocity distributions, charge state, and electronic, vibrational and rotational states may all provide clues to the underlying desorption mechanism. However, these clues are indirect. At least in some cases, the modification of the desorbed species as it escapes (Step 4) can be substantial. Unless we understand this modification, we cannot reliably employ final state information to probe the interesting dynamics occurring in Steps 2 and 3.

To illustrate some of the problems, consider the ground and excited potential energy curves for the simple isolated molecule O_2, as shown in Fig. 1, taken from calculations of SAXON and LIU [1]. There are a very large number of excited potential energy curves, some repulsive and some attractive. If Figs. 1a and 1b were superimposed, the figure would appear almost black, with a myriad of intersections and avoided intersections. For the higher dimensional case of a molecule adsorbed on a surface, there are likely to be even more possibilities for transitions among potential energy surfaces. Furthermore, the potential surfaces will be broadened through coupling to excited states of the solid. Thus there is ample opportunity for electronic re-adjustment as the particle breaks away from the surface.

The situation may not be as murky as Fig. 1 would suggest, of course. The nonadiabatic and spin-orbit interactions responsible for inducing transitions among potential surfaces may be very weak for many pairs of states. Nevertheless, a multiplicity of probable pathways are still likely to exist. It is not correct to conclude from the absence of complicated structure in photoemission that only a few states are involved in desorption. The simplicity of electronic spectra arises from dipole selection rules, and from the short time scale of the excitation event. Dipole selection rules do not govern the subsequent evolution, and transitions which do not occur on the 10^{-16}s excitation time scale may well become important on the 10^{-14}-10^{-13}s time scale of desorption.

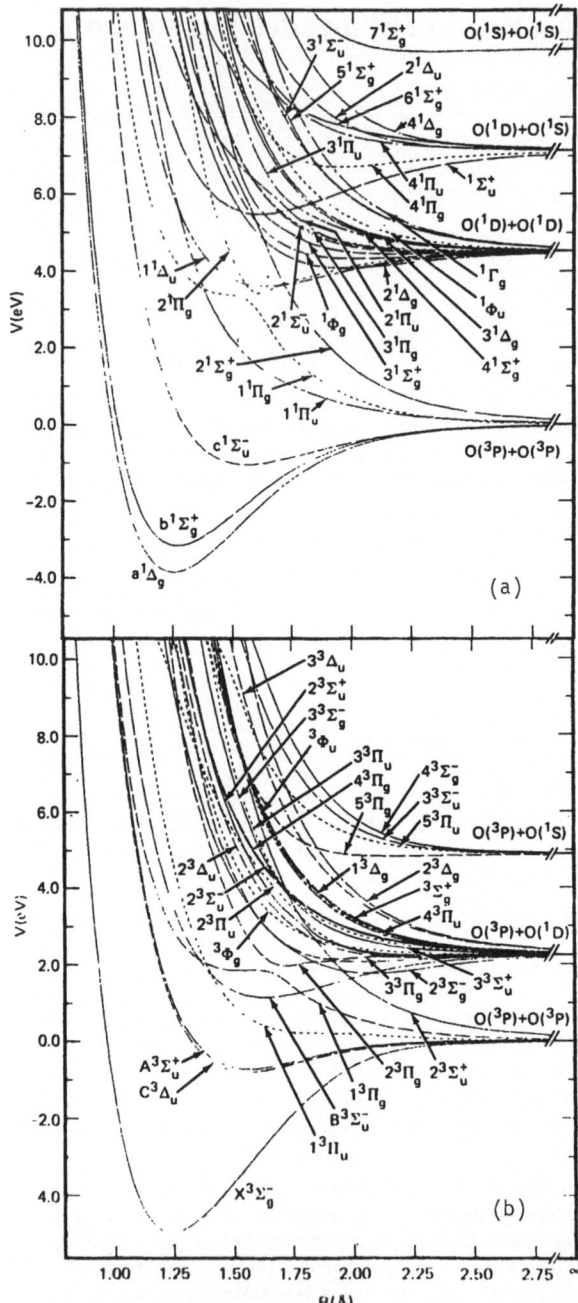

Fig. 1 Calculated potential energy curves for the excited states of the isolated O_2 molecule, from [1]. (a) Singlet states. (b) Triplet states. Reprinted with permission of the American Institute of Physics.

It is clear that the final electronic state of the ejected species is not necessarily a direct reflection of its electronic configuration when it was in contact with the solid. Observation of an ejected F^+ ion from an LiF crystal, e.g., does not imply a mechanism involving removal of two electrons during the initial excitation sequence, Steps 1 and 2. There may be ample opportunity to gain or lose electrons during Step 4 as the atom escapes. In fact, as suggested by Fig. 1, the integrity of excited electronic states lying 5 or 10 eV above the ground state is tenuous, at best. In many cases it may not be fruitful to attempt to describe stimulated desorption in terms of specific excited electronic states. Perhaps a more meaningful mechanistic understanding will emerge from consideration of concepts such as charge distribution and energy flow.

In the next section we will explore this direction, addressing in particular the role of energy flow during the desorption process. We will draw upon the related processes of gas-phase photodissociation and electronically enhanced defect reactions in solids. In Sec. 3 we will introduce a strategy for elucidating the mechanism of induced desorption that has sometimes proved useful in gas-phase inquiries, namely, the systematic search for deviations from statistical product distributions. This leads us finally to consider, in Sec. 4, the evolution of the electronic states of species as they escape from the surface (Step 4), since this evolution itself can introduce non-statistical behavior which might mistakenly be attributed to the desorption mechanism.

2. Energy Flow In Stimulated Desorption

Electron and photon stimulated desorption (ESD and PSD) are closely related to the gas phase processes of electron impact induced dissociation and photodissociation. There are many gas-phase examples where dissociation occurs via direct excitation to a repulsive potential energy surface [2], analogous to the MENZEL-GOMER-REDHEAD (MGR) mechanism of ESD [3]. At higher excitation energies, formation of a core hole may be the first step in molecular dissociation [4], as in the Knotek-Feibelman mechanism of stimulated desorption [5].

In both mechanisms, the dissociation process is *direct*; the molecule is excited to a repulsive electronic state (or to the repulsive portion of a bound state), and breaks apart on the timescale of one vibration, $\lesssim 10^{-13}$s. On the other hand, the most commonly observed dissociation mechanism for large polyatomic gas-phase molecules is an *indirect* one; the energized molecule hangs together for many vibrational periods before breaking apart. In the great majority of cases, the fragmentation of such large energized complexes occurs in a statistical manner; i.e., energy flows freely and randomly among vibrational degrees of freedom (the ergodic hypothesis) until a particular bond is sufficiently energized to cause it to rupture. The identity and internal (vibrational) state distributions of the fragments that are ultimately produced do not retain memory of the initial state of excitation of the molecule, except for conserved quantities such as total energy and angular momentum. Statistical models of reaction [6], notably the RRKM theory [7], have been very successful in describing these processes.

It is important to emphasize that the observation of a statistical distribution of product states does not necessarily imply rapid energy flow and randomization. A statistical distribution can result as well from an ensemble average over many individual events, each of which may have involved very specific energy flow.

One basic difference between the direct and indirect dissociation mechanisms is the nature of the potential energy surface upon which the nuclear motion evolves. Direct dissociation usually involves a repulsive (antibonding) potential energy surface. Fragmentation occurs, as in the MGR picture, because the electronic excitation has placed the molecule in a configuration of high potential energy. Indirect fragmentation need not involve a repulsive potential energy surface. Rather, the initial electronic excitation is converted, via non-radiative transition, to nuclear kinetic energy on a low-lying potential energy surface (frequently but not necessarily the ground state). Thus, at least for gasphase molecules, it is not necessary to invoke a repulsive potential energy surface to promote dissociation.

It is interesting to speculate whether electronically stimulated desorption from surfaces might sometimes proceed via nuclear kinetic energy rather than potential energy; i.e., without invoking excitation to a repulsive potential energy surface. In order for such a mechanism to be competitive, two conditions must be met. First, non-radiative conversion of electronic to vibrational energy must be very fast; i.e., it must occur before the electronic excitation has dissipated. Second, the local vibrational energy thus produced must, at least for a fraction of events, be able to promote bond breaking before it is dissipated to the bulk.

The first condition bears on a dilemma that has been referred to in several of the other papers of this volume. Electronic excitations are very quickly dissipated in solids. Thus, even for the conventional direct desorption mechanisms, it is difficult to understand how an excited repulsive state can exist for the $10^{-14} - 10^{-13}$s required for atomic motion to occur. Suggestions such as two-hole stabilization [8] and self-trapping [9] have been proposed, and may well play a role in some cases. Non-radiative transitions offer another possibility. Particularly in situations involving a high density of vibrational levels (large molecules or, possibly, surfaces), non-radiative transitions can occur almost instantaneously on the time scale of nuclear motion. For example, non-radiative transitions have been observed to occur in 10^{-15}s in organic molecules [10]. Thus it is quite possible that in some cases radiationless transitions may occur fast enough to compete with dissipation of electronic energy to the solid. This could offer a mechanism for populating excited electronic states which are "protected" from dissipation to the solid, analogous to the two-hole mechanism, and could then lead to bond-breaking on a more leisurely time scale. Non-radiative transitions also provide a mechanism for very rapidly converting substantial amounts of electronic energy into vibrational energy. The vibrational energy so produced would initially be localized in the vicinity of the localized electronic excitation since "acceptor" modes are only those which differ substantially between initial and final electronic state. It remains, then, to address the second condition stated above, can this localized vibrational energy produce bond-breaking?

This mechanism efficiently produces bond-breaking in isolated molecules because, with the exception of slow IR radiation, the vibrational energy is trapped in the molecule essentially forever. This is not the case for a localized surface site in a solid. Local vibrational energy will dissipate to the solid rapidly, perhaps as fast as 10^{-13}s. Thus, either a substantial fraction of the energy associated with the non-radiative transition must be initially deposited in the bond to be broken, or redistribution of this energy must occur fast. One can certainly imagine cases for which motion along the bond to be broken is the primary contributor to the nonadiabatic coupling. However, it is probably more generally true that the vibrational energy deposited by a nonradiative transition, although localized in the vicinity of the active bond, will not be in precisely the right linear combination of modes to lead directly to desorption. For such cases we must then consider the competition between flow of vibrational energy into the active bond vs. dissipation to the solid.

Very similar questions have arisen in regard to a certain class of defect reactions in semiconductors. It has been demonstrated experimentally that recombination of an electron and a hole can enhance substantially reactions such as diffusion of point defects [11]. The point defects act as recombination centers, and the energy of recombination is used to promote the reaction. In fact, it has been observed experimentally that the activation energy for enhanced diffusion is lower than that for non-enhanced diffusion by essentially the entire electron-hole recombination energy [11]. One possibility is that all of the recombination energy is deposited directly into the reaction coordinate mode. This appears very unlikely, however, and is inconsistent with the very low Arrhenius prefactors measured for enhanced reaction rate. WEEKS et al [12] have shown, alternatively, that the same activation energy is obtained if the energy of recombination is deposited randomly in the vicinity of the defect. Energy can then redistribute among local vibrations, occasionally producing reaction prior to dissipating to the solid. The relative rates of local redistribution vs. dissipation affect, to a first approximation, only the prefactor and not the activation energy.

This mechanism involving random deposition of vibrational energy via nonradiative electron-hole recombination appears consistent with all experimental studies to date of recombination enhanced defect reactions. The same mechanism might possibly be involved in stimulated desorption from surfaces in some cases. This might be particularly likely in cases of weakly bound adsorbates for which dissipation of vibrational energy to the solid might be somewhat slower than redistribution of energy within the molecule. A clue that a mechanism of this type may be operating might be found in the temperature dependence of stimulated desorption yields. Thermal energy will add to the energy deposited by nonradiative transition. Thus desorption yields should be Arrhenius-like, increasing with increasing temperatures, but with small prefactors.

3. Deviations From Statistical Behavior

As suggested in the previous section, and in several of the other papers in this volume, there is considerable uncertainty about the mechanisms of stimulated desorption, and it is clear that different mechanisms are operative under different conditions. A similar situation exists, of course, for gas-phase chemical processes. A strategy that has proved useful in gas-phase studies is to search for striking deviations from statistical behavior [13]. Distinctive product angular or velocity distributions, or characteristic patterns of energy disposal among the internal states of products can be vital clues to the reaction mechanism. However, one must have a standard against which to judge whether a result is distinctive or ordinary. The obvious standard is the result which would be obtained if products were populated statistically; i.e., if the probability of formation of a particular quantum state of a particular product were proportional to the number of possible ways of forming that state (or the volume of phase space corresponding to that state), subject to conservation of energy and angular momentum. Because reactions may be endo- or exo-thermic, or reaction intermediates may be formed with non-random angular momentum or energy, the statistical product distribution is usually not a simple Boltzmann distribution, and in fact may differ considerably. Sometimes product distributions which look at first glance to be unusual, turn out after analysis to be very nearly statistical.

In many situations one knows additional information about a reaction. For example, spin may be approximately conserved or nuclear motion may be assumed to evolve along a single adiabatic potential energy surface. This type of information can be imposed as added constraints on the statistical hypothesis. Thus a great many "statistical theories," with different constraints, can be invented. Transition-state theories [6], RRKM theory [7], phase-space theories [6] and many variations have proved useful for describing gas-phase processes. It is likely that several statistical models of desorption might also be constructed. Of course, even exact dynamics can be considered to be a statistical theory subject to the constraint imposed by the equations of motion (e.g., the principle of least action). Thus a statistical model is likely to be useful only if it is based on a minimum of very simple constraints.

The reason for developing statistical models of desorption is, of course, not because it is believed that a random, non-specific mechanism prevails. Rather, as stated above, it is to help identify specific or unusual phenomena when they occur. This strategy has proved quite useful in gas phase applications. In fact, a quantitative measure of the deviation from statistical behavior, called the "surprisal," S has been proposed [13]:

$$S = \ln (P/P_o)$$

where P is the observed probability of an event, and P_o is the "expected" probability as computed by a suitable statistical model. This quantification is probably superfluous, but the underlying strategy of seeking striking deviations from statistical behavior is likely to be a useful one toward unraveling the mechanisms of desorption.

There are two possible sources of non-statistical behavior in product state distributions. The first is the occurrence of a specific reaction pathway. The second is a non-statistical redistribution of states as the products separate. Thus the observation of non-statistical behavior does not necessarily signal a distinct mechanism.

Consider, e.g., a gas-phase process, the reaction of F atoms with substituted hydrocarbons, RX. Flourine atoms are very chemically active, so the reaction intermediate [F-R-X] is highly energized, and breaks apart to form RF and X. DURANA and MCDONALD [14] have studied the vibrational energy distribution of the product RF by infrared emission, for R = C_2H_3 and C_6H_5, and for X=H, CH_3, Cl and Br. They observed non-statistical translational and vibrational energy distributions for several of these reaction systems. However, isotropic product angular distributions were observed, showing that the [F-R-X] complex survives for at least a few rotational periods ($>5 \times 10^{-12}$s) in all cases. Those cases for which nonstatistical vibrational and/or translational distributions were observed are those for which a potential energy barrier exists in the exit channel. Thus it is believed that all of these reactions involve complete equilibration of energy among vibrational modes of the reaction intermediate prior to its dissociation. The mechanism is completely statistical. The non-statistical energy partitioning results entirely from the acceleration of the fragments by the exit channel barrier as they recede. This produces increased translational energy and decreased vibrational energy in the products, as observed.

4. Final State Effects in Desorption

Whether or not a statistical analysis such as that outlined in the previous section proves valuable, it will be important to understand how desorbed species are altered as they escape from the surface (Step 4 of the Introduction). For example, the translational energy of desorbed species can be modified substantially during the escape process. This has been demonstrated both experimentally and theoretically for atoms and molecules thermally desorbed from surfaces. If there is a potential barrier to surmount prior to desorption, then thermally desorbed particles may have much higher than thermal translational energies, and angular distributions that are focused toward the normal. If there are no potential energy barriers (other than the binding energy) to surmount, the mean translational energy of desorbed atoms may be considerably less than thermal [15].

The translational energy distributions of ionic or electronically excited species may be very strongly affected by neutralization or de-excitation processes during their escape. According to the accepted Hagstrum [16] picture of ion neutralization, faster particles have higher survival probabilities P_s, given by

$$P_s \propto \exp\left[-\frac{c}{v_1}\right],$$

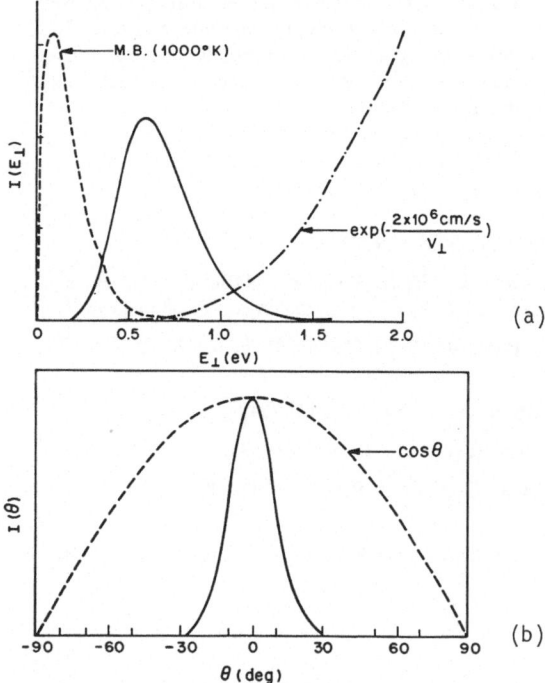

Fig. 2. (a) Kinetic energy distribution of ejected ions (solid curve), assuming 1000°K Maxwell-Boltzmann distribution initially leaves the surface (dashed curve), and ion survival probability given by dot-dashed curve. (b) Angular distribution of ejected ions (solid curve), using same assumptions as above. Dashed curve is statistical cos θ distribution

where v_1 is the component of velocity in the normal direction. Taking the constant c to be the typical value of 2×10^6 cm/s [17], the ion survival probability is proportional to the dot-dash curve of Fig. 2a. Multiplying this by a Maxwell-Boltzmann distribution at a temperature of 1000 K (dashed curve), produces the un-normalized velocity distribution of those ions which survive (solid curve). The translational energies of surviving ions are thus greatly increased because the slower ions (or excited states) are preferentially destroyed. Fig. 2b shows the angular distribution of surviving desorbed ions calculated for the same assumptions as Fig. 2a, with the additional assumption that neutralization is independent of the tangential component of velocity. Surviving ion distributions are focused considerably towards the surface normal (solid curve), compared to the statistical cos θ distribution (dashed curve).

MADEY and coworkers [18] have observed angular distributions of desorbed ions which exhibit very interesting (and non-statistical) patterns which cannot be explained by any simple exit channel effects. In fact, these distributions reveal direct information about the adsorbate binding geometry, and perhaps indirect information about the desorption mechanism. This type of very specific information about product state distributions promises to be very illuminating. Measurement of internal (vibrational and/or electronic) state distributions of desorbed species may also be very valuable. Finally, most

37

experimental studies of stimulated desorption have detected either ions or excited neutrals, whereas in most cases the major species desorbed are ground-state neutrals. Ion and excited state populations are much more likely to be substantially altered during escape. Measurement of ground-state neutrals may provide much more definitive information about the mechanism of stimulated desorption.

References

1. R. P. Saxon and B. Liu, J. Chem. Phys. *67*, 5432 (1977).

2. M. Dzvonik, S. Yang and R. Bersohn, J. Chem. Phys. *61*, 4408 (1974); R. K. Sparks; K. Shobatake, L. R. Carlson and Y. T. Lee, J. Chem. Phys. *75*, 3838 (1981).

3. D. Menzel and R. Gomer, J. Chem. Phys. *41*, 3311 (1964); P. E. Redhead, Can. J. Phys. *42*, 886 (1964).

4. T. A. Carlson and M. O. Krause, J. Chem. Phys. *56*, 3206 (1972).

5. M. L. Knotek and P. J. Feibelman, Phys. Rev. Lett. *40*, 964 (1978).

6. P. Pechukas, in Dynamics of Molecular Collisions, Part B, edited by W. H. Miller (Plenum, NY, 1976), p. 269.

7. W. L. Hase, in Dynamics of Molecular Collisions, Part B, edited by W. H. Miller (Plenum, NY, 1976), p. 121.

8. P. J. Feibelman, this volume.

9. D. Emin, Adv. Phys. *22*, 57 (1973).

10. R. M. Hochstrasser and C. Marzzacco, J. Chem. Phys. *49*, 971 (1968).

11. D. V. Lang and L. C. Kimerling, Phys. Rev. Lett. *33*, 489 (1974). L. C. Kimerling, Solid State Electr. *21*, 1391 (1978).

12. J. D. Weeks, J. C. Tully and L. C. Kimerling, Phys. Rev. *B12*, 3286 (1975).

13. R. D. Levine and R. B. Bernstein, in Dynamics of Molecular Collisions, Part B, edited by W. H. Miller (Plenum, NY, 1976), p. 323.

14. J. F. Durana and J. D. McDonald, J. Chem. Phys. *64*, 2518 (1976).

15. J. C. Tully, Surf. Sci. *111*, 461 (1981).

16. H. D. Hagstrum, in Inelastic Ion-Surface Collisions, edited by N. H. Tolk (Academic, NY, 1977), p. 1.

17. M. J. Vasile, Surf. Sci. *115*, L141 (1982).

18. T. E. Madey, J. J. Czyzewski and J. T. Yates, Jr., Surf. Sci. *49*, 465 (1975).

Part 2

Desorption Processes

2.1 Mechanisms of Electron-Stimulated Desorption

R. Gomer

Department of Chemistry and James Franck Institute, The University of Chicago
Chicago, IL 60637, USA

1. Introduction

The various phenomena associated with the desorption of neutral and ionic
species from surfaces under bombardment by electrons, generally in the 10-300
eV range, are collectively known as ESD (electron-stimulated desorption) or
EID (electron-impact desorption). Related effects, for instance, the rearrangement of adsorbates under electron impact or the ejection of weakly
bound species by direct momentum transfer from electron desorbed atoms or
molecules are also included. The excitation processes and the subsequent
evolutions of the substrate-adsorbate systems on excited state potential
hypersurfaces are of great intrinsic interest and can illuminate aspects of
bonding at surfaces hard to study by other means. ESD also serves as a probe
of different binding states, since desorption cross sections and products are
sensitive to the details of adsorption. Finally ESD and the closely related
photon stimulated desorption (PSD) are of technological importance, for
instance in plasma-wall interactions in fusion reactors of the Tokamak type.

Despite these facts ESD has been studied systematically only since 1964,
although some sporadic work goes back to 1950. PLUMLEE and SMITH [1] found
evolution of O^+ from Mo surfaces with an ion/electron yield of 10^{-6} at 1100
K; YOUNG [2] found O^+ evolution when oxidized Cu, Ni, Mo, Ta or Ti surfaces
were bombarded with electrons; MOORE [3] found O^+ evolution (only) from CO
adsorbed at room temperature on Mo ribbons, with thresholds of $\sim$18 eV. In
1964, REDHEAD [4] and MENZEL and GOMER [5] independently reported systematic
total depletion cross section measurements on a number of systems and formulated a semiclassical theory of electron impact, or electron stimulated
desorption, which has come to be known as the MGR (or any permutation of
these letters) model. It was recognized by these workers that ESD should
be very sensitive to the details of the binding mode of an adsorbate and
ESD was in fact used to explore different binding modes at a time when the
modern arsenal of surface techniques, UPS, XPS, electron-loss spectroscopy,
and so on was not yet available. Interest in ESD was considerably enhanced
by the discovery in 1974 by MADEY YATES and CZYZEWSKI [6] that the angular
distribution of ions desorbed by electron impact depended strongly on binding mode. They christened the phenomenon or technique ESDIAD. In 1978,
KNOTEK and FEIBELMAN [7] observed the formation of O^+ from TiO_2 and other
oxides by ESD and postulated an Auger mechanism for ion formation in ESD in
general. Their work rekindled interest in ESD with much of the attention
going to the primary excitation processes, particularly for ion generation,
and to the study of oxide and semiconductor surfaces. In 1979-80, a number
of workers started investigating photon stimulated desorption of ions, in
many cases utilizing newly available synchrotron radiation, and found striking
similarities to ESD, as would be expected.

The purpose of this article will be to give an overview of the basic features of the MGR model and to discuss briefly the various modifications of it suggested by BRENIG, ANTONIEWICZ and by KNOTEK and FEIBELMAN and others. The discussion will mostly deal with desorption from metal surfaces, the topic originally addressed by REDHEAD and by MENZEL and GOMER. Lack of space will prevent any discussion of PSD or of ESDIAD, which are covered in some detail by other articles in this volume.

2. The MGR Model

The basic argument of MGR is the following. Elementary considerations indicate that direct momentum transfer from $\sim$100 eV electrons to even the lightest adsorbates is wholly insufficient to lead to anything but slight vibrational excitation, so that desorption or rearrangement must be the result of electronic excitation of the adsorbate or the adsorbate-substrate bond. In this respect, ESD is identical to analogous processes in isolated atoms or molecules. In the latter, the primary excitation event is necessarily irreversible: If the atom or molecule is ionized, it will remain so, if it is excited to an antibonding state it will dissociate (or remain excited until radiatively deactivated).

For an atom or molecule adsorbed on a metal surface, however, ionization, or excitation of the substrate-adsorbate bond can be followed by neutralization or by a "bond-healing" transition because of the availability of conduction electrons in the metal. Thus, the overall ionization and desorption cross sections would be expected to be less than for comparable processes in isolated molecules even though the primary excitation cross sections should be comparable if not identical. Thus, the ESD cross section should have the form

$$\sigma = \sigma_e P , \qquad (1)$$

where σ_e is a primary excitation cross section and P an escape probability. The situation can be depicted schematically by potential energy diagrams, as in Fig. 1, which depicts excitation to an antibonding curve. The initial excitation is vertical, that is Franck-Condon-like, and is followed by descent of the adsorbate A along the repulsive curve, labelled $(M + A)^*$. Transition to the bonding curve M + A can best be represented by noting that the latter is merely the lowest member of an infinite manifold $\{M^* + A\}$ of curves which differ from M + A by some excitation in the metal M, which does not affect the bonding. In other words, a bond-healing transition from the antibonding state to the bonding one must be energy conserving and thus transfers the appropriate amount of energy to an electronic excitation in the metal. Since there is an infinite manifold of curves $M^* + A$, the transition can occur in principle anywhere along the trajectory but not all transitions will result in recapture. Figure 1 shows that there is a critical distance x_c beyond which a transition will still lead to desorption (along a copy of the ground state curve) because the adsorbate A has sufficient kinetic energy for escape. P can thus be written

$$P = \exp - \int_0^{t_c} dt/\tau = \int_{x_0}^{x_c} \frac{dx}{v(x)\tau(x)} , \qquad (2)$$

where $\tau(x)$ is the lifetime with respect to the transition, assumed to be distance dependent, and $v(x)$ the velocity of A along the repulsive curve. A detailed analysis of (2) assuming an exponential dependence of τ on distance

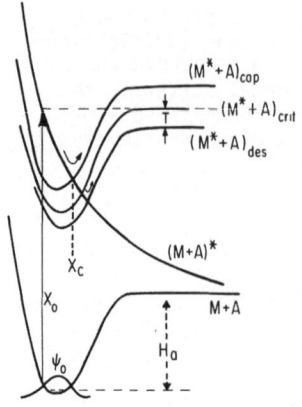

Fig.1. Schematic potential energy diagram for electron-stimulated desorption of neutral species according to the MGR model. The ground state curve is labelled M + A; excited copies, differing from it by excitations in the metal not affecting the bond are labelled (M* + A). The antibonding curve to which excitation is assumed is labelled (M + A)*. The subscripts on (M* + A) refer to recapture, critical curve for recapture, and desorption after transition, respectively. For vertical excitation at x_0 the critical transition distance for recapture is x_c. H_a heat of adsorption; T kinetic energy of escaping A after indicated transition. The vibrational ground state wave function ψ_0 is also sketched

from the surface, and an exponential form of the curve (M + A)* in the region of interest has been given by MENZEL and GOMER [5]. For present purposes, it suffices to assume that τ is constant over the range x_0 to x_c and that the repulsive curve can be taken to be linear over this small range. With these approximations P becomes

$$P = \exp - \Delta t/\tau , \tag{3}$$

where Δt is the time required for A to move from x_0 to x_c along the repulsive curve (M + A)*. Δt is given by

$$\Delta t = (2m\Delta x/S_r)^{\frac{1}{2}} , \tag{4}$$

where $\Delta x = x_c - x_0$ and S_r the slope of the potential curve (M + A)*, that is

$$S_r = \Delta E/\Delta x . \tag{5}$$

Eqs. (3) and (4) can be combined to give

$$P = \exp [-(2m\Delta x/S_r)^{\frac{1}{2}}/\tau] = \exp - cM^{\frac{1}{2}} , \tag{6}$$

where M is molecular weight and c a constant. The mass dependence, predicted in 1964, has since been found in a number of cases. It is worthwhile to get at least a crude idea of Δx; it could of course be determined accurately if the shapes of relevant potential curves were known. If the repulsive curve is taken to be linear and the ground state curves M* + A are approximated by triangular wells truncated on the right side it is simple to show that

$$\Delta x = \frac{a}{1 + S_r/S_c} , \tag{7}$$

where a is the half width of the triangular well and S_c the slope of the outgoing branch of this well, that is

$$S_c = H_a/a , \tag{8}$$

where H_a is the binding energy in the ground state. Since $S_r \gg S_c$ in general,

$$\Delta x \sim a(S_c/S_r) \ . \tag{9}$$

If we take $\underline{a}$ = 2Å, S_c/S_r = 5, we see that Δx = 0.5 Å. Obviously this will vary from case to case, but it is clear that Δx is of the order of 1 Å or less.

The discussion so far has considered only excitation to a repulsive but neutral state. Fig. 2 shows excitation to an ionic curve $M^- + A^+$, assumed to be so located that vertical excitations from the ground state curve put the system on the repulsive part of the ionic curve. The situation differs from the previous case in one important respect. Recapture will occur only for transitions at $x<x_c$, as before, but desorption as an <u>ion</u> requires that <u>no</u> transitions occur along the entire trajectory. Evidently this increases the effective width of the zone in which neutralization can occur; it is limited only by the increase in τ with distance from the surface, and the increasing velocity of the departing ion. The matrix elements for neutralization are probably much greater than those for transition from an antibonding to the bonding curve. Thus, at a given distance τ (neutralization) $<< \tau$ (antibonding-bonding) and one would expect cross sections for ion formation to be much smaller than for neutral desorption. Eq. (6) then also predicts a much greater isotope effect in ionic desorption. Both these expectations are confirmed by experiment.

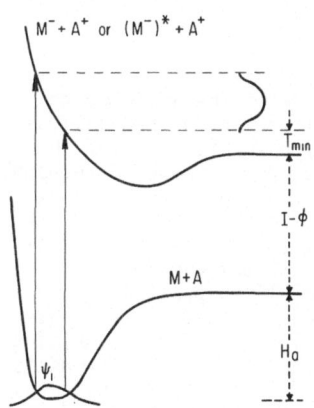

Fig.2. Schematic potential energy diagram for ionic desorption. The ground state curve is labelled M + A, the ionic curve $M^- + A^+$. I ionization potential of A; ϕ work function of M. The ion energy distribution, (unskewed by neutralization) is also shown, but excited copies of M + A are omitted for clarity

It is possible to exploit the mass dependence of P to dissect σ into σ_e and P, if cross sections can be determined for different isotopes. Combination of (1) and (6) gives

$$c = \frac{\ln(\sigma_1/\sigma_2)}{M_2^{\frac{1}{2}} - M_1^{\frac{1}{2}}} \ , \tag{10}$$

where the subscripts refer to the two isotopic species in question. Table 1 shows some typical cross section data, obtained in this way. These were all taken at $\sim$150 eV incident electron energy. Several things are noteworthy. Excitation cross sections are generally quite high, with the exception of that for O^+ formation for oxygen adsorbed on W(110). Quite generally, the P values of ionic desorption are much less than for neutral products, as

Table 1. Some cross section data

System	O/W(110) [8]		v-CO/W(110) [8]		K_r/W(110) [9]	
Species	O^+	O	CO^+	CO	K_r^+	K_r
σ [cm^2]	6×10^{-22}	$\sim 10^{-17}$	$\sim 10^{-21}$	5×10^{-17}	$< 10^{-25}$	3×10^{-18}
σ_e [cm^2]	2×10^{-19}	5×10^{-17}	5×10^{-16}	7.5×10^{-16}	2.3×10^{-16}	2.3×10^{-16}
P	4×10^{-3}	0.2	$\sim 1.5 \times 15^{-6}$	6×10^{-2}	$\sim 10^{-9}$	0.013

just discussed. σ_e for ions and neutrals in the same system generally differ, indicating that ion neutralization is not the main source of neutrals. The mechanism for Kr desorption is somewhat different from those treated so far and is discussed later in this article.

It is obvious from Fig.2 that the energy distribution of ions is a measure of the ground state vibrational wave function, although skewed by variations in neutralization probability, and it is possible to get a rough idea of the slope S_r from this distribution.

Nothing has been said so far about the actual nature of the antibonding curves, the details of the antibonding-bonding transition, or of neutralization in the case of ions. There are, of course, many possibilities for the antibonding-bonding transition. For instance, if the ground state corresponds to a localized electron pair bond, the antibonding state could be a triplet state and bond-healing a triplet-singlet transition. In the case of isolated molecules, this would be very slow, but there are obviously numerous ways of achieving such transitions via conduction electrons. In the case of re-neutralization, the transition could be resonance tunneling, possibly to an excited state of A, or an Auger transition to the ground state. In all these cases, the bookkeeping device of designating the final state after transition M* + A (or possibly M* + A*) is justified if the energy of Auger electrons is included as an excitation of the metal M. It is easy to show that regardless of which particular Auger or other transition occurs, the final energy of M* is the same. Obviously the states M* + A of given total energy are highly degenerate because of the degeneracy of M*.

It is easy to see from the model why ESD cross sections should be sensitive to details of binding. First x_e and thus Δx will depend on binding and second the matrix elements for the various transitions will also depend on geometry, number of neighbors, and so on. In all probability, similar factors enter, at least in part, into the anisotropies of the angular distributions of desorption products.

A slight generalization of the curves in Fig.1 also shows why electron induced rearrangements are possible. Excitation from one stable binding mode can occur to an intermediate excited state, followed by deexcitation into another binding state. To date, the rearrangements which have been observed consist mainly of the dissociation of diatomic molecules by electron impact. For instance, molecularly adsorbed CO on various surfaces can not only be desorbed as CO or CO^+ but also converted to beta-CO which consists of dissociated CO [10]. Similar behavior is observed with adsorbed N_2 on tungsten [11].

Although the MGR model was aimed at adsorption on metals, it is legitimate to ask what it would predict for other substrates. In a general way, this question has already been answered: Strong adsorption on semiconductors or insulators is similar to covalent bonding in small molecules, in the sense that mobile electrons for reneutralization or bond-healing are not available. This is particularly true of insulators and valence saturated oxides. Arguments why reneutralization is slow even for semiconductors have also been given recently by RAMAKER [12] and by FEIBELMAN [13]. These amount to saying that resonance tunneling is energetically forbidden because the Coulomb repulsion energy of two holes in the ad-bond cannot be converted into kinetic energy of holes in the valence band if the latter is too narrow. Qualitatively this is equivalent to saying that localized electrons stay localized and do not move to adjacent atoms or bonds from which electrons have been removed. Analogously there will be very little bond-healing after excitation to antibonding curves for neutral desorption. Thus, desorption cross sections should be essentially equal to excitation cross sections in non-metallic systems, and isotope effects should be minimal.

3. Modifications and Critiques

Let us turn next to various criticisms, modifications, or refinements which have been suggested. Most of these will be discussed in detail in other papers in this volume and only the main arguments will be given here. The first point to be made is that it is obviously a simplification to speak of a single antibonding or even a single ionic curve. There will be many such curves (not counting metal excitations irrelevant to the adsorbate-substrate bond). In the repulsive region, curves corresponding to different states may lie very close to each other and it is possible that well-defined transitions to a single excited state do not occur.

BRENIG [14] has questioned the validity of the recapture mechanisms proposed by MGR on the following grounds. Excitation to the upper state by vertical, i.e., Franck-Condon transitions lead to a distribution in x on the upper curve initially equal to that in the ground state; that is, the nuclear wave function is that of the ground state bonding curve. The matrix element for recapture or reneutralization has the form, in first approximation,

$$\langle \phi_g | V | \phi_e \rangle \ \langle \psi_g | \psi_e \rangle \ , \tag{11}$$

where ϕ_e and ϕ_g are electronic wave functions for the excited and ground state, e.g., functions of the variable $x_n - r_e$ where x_n is the coordinate describing the position of A and r_e the coordinate of an electron measured from the center of A. $V(x, r_e)$ is the electronic pertubation leading to transitions; ψ_e and ψ_g are the nuclear wave functions of A in the excited and ground states respectively. Since $\psi_e = \psi_g$, however, the only non-vanishing transition is that to $\psi_g = \psi_e$ since the $\{\psi\}$ form an orthonormal set. Consequently, BRENIG argues, recapture must occur, if at all, only into a ground state whose crossing with the excited state corresponds to the zero-zero vibrational transition. This argument is illustrated in Fig.3. In my opinion, there are several reasons why the orthogonality requirement is not necessarily fatal to transitions. First, V is a function not only of the electronic coordinates, but also of x_n, i.e., V decreases rapidly as x_n

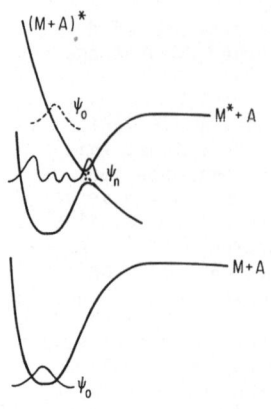

Fig.3. Potential energy diagram illustrating the arguments of BRENIG [14] regarding transition to M* + A. The ψ refer to vibrational wave functions in the ground state

increases. Thus, the separation into electronic and nuclear coordinates implied by expression (11) is invalid or alternately the Franck-Condon integral is truncated and therefore non-vanishing. Second, as the adsorbate moves along the excited state curve, the distribution in x rapidly evolves to one corresponding to the new Hamiltonian. Even if its form remained unchanged, the origin, i.e., the point of maximum amplitude moves to the right. For all these reasons, the orthogonality requirement is relaxed. These are, of course, very qualitative arguments. For a more detailed discussion, the reader is referred to the article by BRENIG in this volume, and to a forth-coming paper by FREED [15].

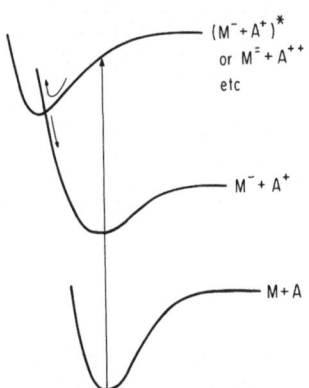

Fig.4. Potential energy diagram illustrating the argument of ANTONIEWICZ [16], which assumes that the ionic curve $M^- + A^+$ lies closer to the surface than the ground-state curve M + A. Ionic desorption is assumed to require excitation to a doubly ionized or otherwise more highly excited state, than $M^- + A^+$, followed by inward movement of A^{++}, transition to A^+, and eventual escape

An interesting objection to the MGR model of ion production has been raised by ANTONIEWICZ [16] who argues that the repulsive portion of ionic curves should lie closer to the surface than the repulsive portions of bonding curves because A^+ has one electron less than neutral A. Consequently, according to ANTONIEWICZ, ion generation in ESD requires something like the two-stage mechanism illustrated in Fig.4. Initial excitation occurs to the attractive branch of a curve corresponding to double ionization or to some other local excitation; this is followed by inward movement of A^{++} and transition to

$M^- + A^+$. The difficulty with this argument is that threshold energies for ion generation are too low for this model. (Thresholds will be discussed in some detail later in this article.) Further, there is no reason to assume that strong covalent bonding will not pull in the bonding curve to the point where vertical transitions from the ground state occur to the repulsive portion of the ionic curve.

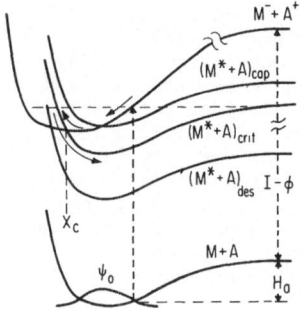

<u>Fig.5.</u> Potential energy diagram illustrating the mechanism of neutral desorption of weakly bound adsorbates via initial excitation to the ionic curve. Here neutralization occurs with 100% probability, but only those atoms which become neutralized at $x \leqslant x$ (<u>not</u> $x > x_c$ as in previous cases) escape

There is, however, one situation in which the ANTONIEWICZ argument is almost certainly correct, namely very weak adsorption, for instance that of inert gases on metals. ZHANG and GOMER have found that Xe [17] or Kr [9] adsorbed on W yield only neutral atoms in ESD. In the case of Kr, the cross section is sufficiently high to permit accurate threshold energy measurements [9]. The latter is just that required to form an ion; this could be determined accurately by photoelectron spectroscopy. The conclusion is almost inevitable that initial excitation leads to formation of an ion, which moves inward, makes a transition to the neutral weakly bonding curve and is then desorbed as a neutral atom. The situation is illustrated in Fig.5 which also shows why only a fraction of the neutralized atoms will escape, namely those which have made transitions to curves $M^* + A$ lying below the critical recapture curve $(M^* + A)_{crit}$, i.e., for which x at the transition is <u>less</u> than x_c. It was pointed out to me by Professor M.B. WEBB [18], who has seen ESD of inert gases (as a perhaps unwanted by-product of Leed Studies on Ag surfaces), that the relative positions of the ionic and neutral curves are important in determining escape cross sections. The reader can easily convince himself that x_c increases the closer to the surface the ionic curve lies, relative to the neutral one. The larger the value of x_c, the more likely it is that the neutralized atom will escape. The relative differences in the locations of ionic and neutral curves will be smallest for Xe and largest for He since the difference made by one electron is least in the former and greatest in the latter case. Thus, the expected desorption cross sections should be smallest for Xe and largest for He (which is not easily accessible to measurement of course). WEBB's values [18] on Ag and those of ZHANG and GOMER for Xe [17] and Kr [9] on W confirm this. The mass of an ion is also important, however, since the probability of neutralization at $x > x_c$ will increase as the ion velocity decreases, i.e., as its mass increases. This effect goes in the same direction as that just discussed. An isotope effect was in fact seen for Kr by ZHANG and GOMER [9] as the entries in Table 1 imply.

We turn next to the KNOTEK-FEIBELMAN mechanism [7] of initial excitation. The MGR model, as originally formulated, does not consider excitation processes in detail and contents itself with assuming that they will occur.

Furthermore, it considered only chemisorption on metals. KNOTEK and FEIBELMAN subjected the surface of an ionic oxide, TiO_2, to electron bombardment and found that O^+ ions were generated, despite the fact that oxygen is present in TiO_2 as O^-. They argued that the ejection of three electrons by conventional impact ionization was unlikely and proposed instead the interatomic Auger process shown schematically in Fig.6. Since the conduction band of Ti (present as Ti^{+4} in TiO_2) is empty, ejection of a core electron can only be followed by Auger filling via oxygen electrons. If two Auger electrons are ejected, O^- is converted to O^+. Since the O^+ ion so created has a Ti^{+4} ion as its nearest neighbor, there is strong Coulomb repulsion and O^+ will be ejected. Furthermore, in substances like TiO_2, unlike metals, reneutralization is inherently unlikely since there are no available electrons. Examination of O^+ yields as function of electron energy showed that the onset of O^+ generation corresponds to the energy required for Ti 3p ionization. It was argued by KF that electron energies just sufficient for O 2s ionization did not contribute appreciably because the process O $2s^1 2p^6 \longrightarrow$ O $2s^2 2p^3 + 2e^-$ was just barely exoenergetic, thus leaving only a very small amount of phase space to the Auger electrons. KF also argued that analogous processes should be expected in ion generation from chemisorbed layers on metals and also in the production of neutral O from metal surfaces.

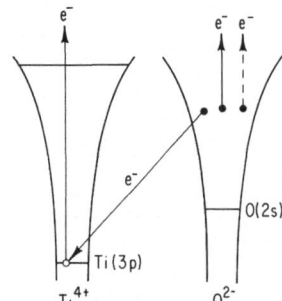

Fig.6. Schematic potential energy diagram illustrating the KNOTEK-FEIBELMAN [7] mechanism of O^+ generation from TiO_2 surfaces

The evidence for the KF excitation mechanism for ion generation in the case of oxides is convincing as well as eminently reasonable. In the case of oxygen and other substances adsorbed on metals, there is now evidence that interatomic Auger processes also contribute to ionization but are not solely responsible [19]. Ionization of O 2s, followed by intraatomic Auger processes may also make a contribution to O^+ (and O) desorption on metals; at least this has been proposed by KF [7]. The work of MENZEL and collaborators [20] also suggests that intramolecular Auger processes are important in ion generation from adsorbed molecules: For adsorbed CO, CO^+ production is enhanced by ejection of an electron from C 1s, while O^+ production is strongly enhanced by O 1s excitation.

It is worthwhile to consider these questions in more detail and to examine energy thresholds in chemisorption. Unfortunately, there is room for some ambiguity in the relation between the voltage applied between electron source and specimen and the electron kinetic energy available for ESD. Thermionic electron sources are normally employed, so that, disregarding the small thermal energy of electrons the electron kinetic energy T_e is

$$T_e = V_b + \phi_e \tag{12}$$

if the incident electron winds up at the Fermi energy ε_F of the specimen. V_b is the battery voltage between specimen and electron emitter and ϕ_e the work function of the latter. If the density of states at ε_F is small, or if the adsorbate sits at a distance from the surface, so that the incident electron cannot utilize the full drop in potential to ε_F at the scattering locus, T_e will be approximated by

$$T_e \simeq V_b + \phi_e - \phi_s , \tag{13}$$

where ϕ_s is the work function of the surface from which ESD takes place. It is easy to see from Fig.2 that for ions energy balance requires that

$$H_a + I - \phi_s + T_+ (min) = T_e (min) , \tag{14}$$

where T_+ is the ion kinetic energy and the designation (min) refers to quantities at threshold. T_e is generally assumed to be given by (12) rather than (13) but it is not obvious that this should always be the case. If for the same process, threshold energies for photon stimulated desorption are available an experimental answer can be obtained.

For neutrals, the situation is more complicated. Fig.1 shows that there is no meaningful minimum kinetic energy if there is escape after a bond-healing transition. In addition, internal excitations of the desorbing entity must also be taken into account. Although Fig.1 shows the curve $(M + A)^*$ deriving from M + A at infinite separation, the possibility that the predominant desorption path leads to some excited state, A^* cannot be excluded. In the case of CO desorbing from W(100) surfaces, SANDSTROM [21] has in fact found some evidence of this. To date, there are only very few measurements of threshold energies for neutral species and virtually no kinetic energy data at all. Consequently, only limited conclusions about mechanisms can be drawn from threshold energies of neutral desorption at present.

Table 2 lists some selected threshold and related data for some chemisorption and a physisorption system. For oxygen on W(110), the threshold for O^+ formation is roughly 7 eV higher than required for simple ionization, but approximately the energy required for transferring an oxygen 2s electron to the Fermi level, if the data of OPILA and GOMER [26] are correct. Thus, an intraatomic Auger process may well be operating at threshold. At higher energies, not shown in Table 2, there is definite evidence for contributions from interatomic processes as W core levels are ionized [24]. This has also been seen for O on other substrates [19]. The situation for F^+ production from F/W(100) [27], [28] seems to be different from that for O^+ just discussed. For F^+, the threshold is definitely too low to be explained by F 2s ionization, although it is again too high by ∿6 eV for simple hole formation. BAUER [27] interprets this to mean that T_e is given here by (13) rather than (12). However, the comparison of ESD and PSD results by TRAUM et al. [28] suggests that (12) is valid here and that it requires a few eV above threshold to lead to appreciable ionization. In the case of CO^+ from virgin-CO on W(110), there is a similar discrepancy between observed and calculated thresholds, but no obvious correlation with either CO or W levels. It is possible that some of these discrepancies do result from low densities of state at the Fermi level, or from small matrix elements connecting initial states to the Fermi level.

For neutral desorption, the situation is similarly unclear, except that threshold energies are clearly lower than for the production of corresponding ions. In the case of oxygen on W(110), the measurements are somewhat in-

Table 2. Some threshold data

	O/W(110) ## [eV]	F/W(100) [eV]	v-CO/W(110) [eV]	Kr/W(110) [eV]
H_a	~4 [22],[23]	~3 [27]	~1.5 [22]	~0.1 [31]
$I - \phi_S$	7.6	10.6 [27]	8.2 [10]	
T_+ (min)	~5. [24],[25]	1-2 [27],[28]	1 [24],[29]	
$H_a + I - \phi_S + T_+$ (min)	16.6	~15.	10.7	
T_e (min), ions #	~23.0 [24]	21 [27],[28]	18.5 [24]	
T_e (min), neut.#	16.5 [24]		8.5 [24]	7.1 [9]
Energies of relevant adsorbate levels *	2p ~-7 2s -23 [26]	2s -31[28]	$(5\sigma-1\pi)$ -7.3 [30] 4σ -10.8 [30]	$4p_{3/2}$ -6.9 [31]

* Relative to ε_F

Based on $T_e = V_b + \phi_e$

Symbols: H_a heat of adsorption; I ionization potential; ϕ_S substrate work function; T_+ (min) minimum ion kinetic energy; T_e (min) threshold electron energy; ε_F Fermi energy.

The results quoted here refer to the system at T≤100 K, where the state yielding both neutral 0 and 0$^+$ is not a minority state [8].

direct, since it has not yet been possible to observe neutral O desorption directly; they are based on Xe kickoff by desorbing oxygen [24]. For CO, the threshold energy is very much greater than the heat of adsorption, suggesting a high kinetic energy as well as the possibility of desorption of CO*. This has in fact been reported by SANDSTROM [21]. The only case which seems totally clear is that of Kr desorption [9], which has already been discussed. In summary, it seems that intra and interatomic Auger processes play some role in ion generation from chemisorbed layers <u>above</u> threshold but that they are not the sole contributors to excitation. No role for them seems clearly evident in neutral desorption from the experimental results so far available. One final point remains to be made. Since there definitely is an isotope effect in ionic desorption from metal surfaces, it seems clear that reneutralization processes occur. Thus, even where interatomic Auger processes contribute, electrons for reneutralization are locally available, as would be expected in a metal.

4. Conclusion

Where do we stand? In my perhaps biased opinion, the MGR model provides a reasonable perspective from which to view electron stimulated desorption from metals. It is quite clear, however, that actual situations are much more complicated than a naive interpretation of the model would suggest. There are unanswered questions connected with primary excitation processes both for neutral and ionic desorption, and also with the evolution of systems in excited states. The situation seems basically simpler for insulators and semiconductors since recapture processes are inherently less important and the primary excitations in some respects simpler. It is to be hoped that the intensive effort now being devoted to ESD and PSD will provide answers to some of these questions. In particular, it will be interesting to obtain threshold and energy distribution data for neutrals and it will be equally interesting to see theoretical advances along the several lines suggested by other articles in this book.

References

1. R.H. Plumlee and L.P. Smith, J. App. Phys. <u>21</u>, 811 (1950)

2. J.R. Young, J. App. Phys. <u>31</u>, 921 (1960)

3. G.E. Moore, J. App. Phys. <u>32</u>, 1241 (1961)

4. P.E. Redhead, Can. J. Phys. <u>42</u>, 886 (1964)

5. D. Menzel and R. Gomer, J. Chem. Phys. <u>41</u>, 3311 (1964)

6. J.J. Czyzewski, T.E. Madey and J.T. Yates, Phys. Rev. Lett. <u>32</u>, 777 (1974)

7. P.J. Feibelman and M.L. Knotek, Phys. Rev. <u>B18</u>, 6531 (1978)

8. C. Leung, Ch. Steinbrüchel and R. Gomer, App. Phys. <u>14</u>, 79 (1977)

9. Q.-J. Zhang and R. Gomer, to be published

10. C. Leung, M. Vass and R. Gomer, Surf. Sci. <u>66</u>, 67 (1977)

11. T.E. Madey and J.T. Yates, J. Vac. Sci. Tech. $\underline{8}$, 525 (1971)

12. D.E. Ramaker, C.T. White and J.S. Murday, J. Vac. Sci. Tech. $\underline{18}$, 748 (1981)

13. P.J. Feibelman, Surf. Sci. $\underline{102}$, L51 (1981)

14. W. Brenig, this volume

15. K.F. Freed, to be published

16. P. Antoniewicz, Phys. Rev. $\underline{B21}$, 3811 (1980)

17. Q.-J. Zhang and R. Gomer, Surf. Sci. $\underline{109}$, 567 (1981)

18. M.B. Webb, private communication

19. J. Kirschner, D. Menzel and P. Staib, Surf. Sci. $\underline{87}$, L267 (1979)

20. R. Franchy and D. Menzel, Phys. Rev. Lett. $\underline{43}$, 865 (1979)

21. I.G. Newsham and D.R. Sandstrom, J. Vac. Sci. Tech. $\underline{10}$, 39 (1973)

22. C. Kohrt and R. Gomer, J. Chem. Phys. $\underline{52}$, 3283 (1970)

23. T. Engel, H. Niehus and E. Bauer, Surf. Sci. $\underline{52}$, 237 (1975)

24. Q.-J. Zhang and R. Gomer, to be published

25. S.-L. Weng, Phys. Rev. $\underline{B23}$, 1699 (1981). This author lists considerably lower values of onset ion kinetic energy than [24]; the values of [24] are used in Table 2

26. R. Opila and R. Gomer, unpublished, using an He resonance lamp and $h\nu = 40.8$ eV. There is a slight possibility that intensity at ε_F -20 eV is masked by He (I) generated photoelectrons

27. Ch. Park, M. Kramer and E. Bauer, Surf. Sci. $\underline{109}$, L533 (1981)

28. D.P. Woodruff, M. Traum, H.H. Farrel, N.V. Smith, P.D. Johnson, D.A. King, R.L. Benbow and Z. Huryeh, Phys. Rev. $\underline{B21}$, 5642 (1980)

29. S.-L. Weng, Phys. Rev. $\underline{B23}$, 3788 (1981)

30. E.W. Plummer, B.J. Waclawski, T.V. Vorburger and C.E. Kuyatt, Prog. Surf. Sci. $\underline{7}$, 149 (1976)

31. R. Opila and R. Gomer, to be published

2.2 Mechanisms of Electronically Induced Desorption of Ions and Neutrals

D. Menzel

Physik-Department E20, Technische Universität München
D-8046 Garching, Fed. Rep. of Germany

Abstract

A summary is first given of the most important aspects of the experimental
data accumulated by older as well as recent work. An attempt is then made to
find the broadest common mechanistic basis compatible with these facts, and
from there to define more clearly the areas of knowledge deficit. The basic
view thus developed corresponds closely to the weathered MGR mechanism. As
in its original version, two important steps are assumed: FRANCK-CONDON excitation, and delocalization of the excitation (recapture) in competition
with desorption. The basic manybody nature of the primary excitation (as well
as the recapture step) is stressed, leading naturally to the possibility of
contributions from multiple excitations and from transitions consisting of
several steps, all in the FRANCK-CONDON regime (e.g., core ionization followed by Auger decay). It is important to realize that the two basic steps
influence each other in the sense that the nature of the (multiple) valence
excitation resulting after the F.C.-step has a profound influence on the
probability of the ensuing delocalization step so that primary processes with
intrinsically small cross sections compared to others (e.g., core vs. valence
ionization) still compete or win in the total cross section.

Emphasis is also laid on the neglected majority, neutral desorption. It is
shown that reneutralization of primary valence excitations does contribute
which is proof of strong nonadiabaticity of the evolution of excited states;
while at least the *relative* contribution of core-initiated processes compared
to valence initiation is much smaller than for ionic desorption. On the basis
of very recent results for angular distributions of neutral desorbates, a new
mechanism for neutral desorption is proposed which seems to contribute in
particular for weakly bound adsorbates: primary electronic excitation of the
adsorbate followed by relaxation with transfer of the excitation energy into
vibrations which can stay localized long enough to lead to desorption.

1. Introduction

Desorption processes caused by electronic transitions in adsorbate complexes
induced by electron or photon impact (commonly called EID or ESD, and PSD,
respectively) have been investigated for over 20 years now [1,2,3]. In the
past few years interest has been revived by the finding that core ionization
processes - despite their intrinsically smaller cross sections as compared
to primary valence processes - can contribute significantly or even overwhelmingly in the desorption of ions. It is the merit of KNOTEK and FEIBELMAN
[4] to realize this fact and strongly call attention to it. Their proposed
mechanism is of particular importance to maximal valence ionic compounds, but
subsequent work stimulated by their proposals has generalized this to covalent
adsorbates [5]. Unfortunately the impression is often found in the literature

53

that these processes follow basically new mechanisms (even a new general
name and acronym has been proposed: AID, Auger induced desorption) and are
mutually exclusive to older concepts. In this author's view, it is more helpful for an overall understanding to try and find the largest common denominator of all aspects to develop a general mechanism which is broad enough
to allow specialization in various respects; and, from there, to pinpoint
more clearly the differences of the various paths and the deficit in knowledge which so far prevents us, for each specific system, to predict the most
important contributions.

In the last part of this survey special attention is paid to desorption
of neutrals which - while being the vast majority under most circumstances -
have not been investigated in anything like comparable detail to ionic desorption. We will arrive at the conclusion that a large part of these processes can be understood in the same terms as desorption of ions. However,
recent findings suggest that another mechanism can contribute, at least for
weakly bound species, which could be termed electronically induced vibrational or quasithermal desorption, and for which direct vibrational counter-
parts may well exist.

2. Short Survey of Experimental Findings

Because of the existence of a number of reviews and surveys up to quite recent times [2,3,6,7], no detailed account is necessary here, but rather we
recapitulate only the most important findings which any mechanism must be
able to explain.

Products: Electronic excitation in surface complexes induced by electron
impact or photon absorption leads to liberation of ions (mostly positive,
but also negative [8]) and neutrals from surfaces as varied as adsorbate-
covered metals and semiconductors, to insulating ionic crystals.

Relative abundances: Under many circumstances neutrals are most abundant
by far. This is particularly so for primary valence excitations where usually
only a few percent (and down even to about 10^{-4} [9]) of the total desorbed
particles are ions. For core-initiated processes (see below) the ratio shifts
towards ions [10], and doubly ionized species are also observed.

Fragmentation and excitation: From molecular adsorbates, various fragments
(ionic and molecular) are observed, with a tendency towards small fragments
for ions (e.g., the most abundant ion from hydrogen-containing molecules is
usually H^+). In some cases, desorption of electronically excited neutrals
has been seen (CO^* from CO/Mo [11]; Na^* from NaCl [12]).

Absolute cross sections and their variations: Maximum cross sections in
the region of 100 - 1000 eV of primary energy vary widely, from gas-kinetic
cross sections down by more than 6 orders of magnitude. The fact that such
variations are present even for different binding states of the same particle on the same surface shows that they must be due not to variation of
primary excitation cross sections, but of secondary processes. This is also
borne out by the very strong isotope effects observed (up to a factor of
150 for H/D; see ref.[9]and references given therein).

Thresholds and energy dependences: Absolute thresholds for singly positive ion desorption are usually in the valence region, although they are
often not as low as one would expect from adiabatic considerations. Structure
at higher energies (which in some cases consists of very strong secondary

thresholds) is often, but not always, found at substrate and adsorbate core excitation energies [4,5,13-15]; in many cases the details of the excitation curves show that the probability of desorption is strongly enhanced by a shake-up excitation (valence excitation coupled to core [10,15] or valence [14,16] excitation). For ionic surfaces and adsorbates, the strongest features by far are cation core excitations without additional shake-up [4,13, 17]; this fact has been used for conclusions about the chemical nature of surface species [4,7] and for surface and species specific EXAFS [18].

Doubly positive ions are found in measurable quantities only above core thresholds (+ shake-up). For neutrals, absolute thresholds below those for ions are found, with strong secondary thresholds at ion thresholds. In between, indications of resonant behavior have been seen [19]. Negative ions have also been reported to have lower thresholds than positive ions [8].

Energy distributions: These have been investigated for singly positive ions only. They have been found to range from zero up to 12 eV, with maxima usually in the range of a few eV.

Angular Distributions: Singly positive ions come off as more or less focussed beams in most cases, with many systems yielding a beam of the order of about 15 to 30⁰ FWHM around the surface normal, but others with nonvertical and multiple beams. This forms the basis of the ESDIAD method [20] for determination of adsorbate orientation. It is interesting that so far no systems with isotropic distributions have been found.

For neutrals, very recent first results [21] have shown two types of angular distributions. In several chemisorption systems distributions comparable to those of the positive ions have been found, but in one system (N_2O from physisorbed $N_2O/Ru(001)$) a broad, though not isotropic distribution was measured.

3. Mechanism

For many years, these processes have been discussed in terms of the semi-classical MGR mechanism [22]. It assumes a primary FRANCK-CONDON transition to a repulsive (with respect to the surface bond) potential curve - ionic or neutral - followed by a recapture step which involves transfer of the excitation energy into the bulk, in competition with desorption. BRENIG [23] has developed a quantum-mechanical version of this mechanism. Recently, MELIUS et al. [16] have analyzed valence-initiated desorption of H^+ from H/Ni on the grounds of potential energy curves calculated for Ni-H clusters. KNOTEK and FEIBELMAN [4,7] have advanced a mechanism based on Auger decay of core holes (mostly interatomic in the case of ionic surfaces) which they view as an alternative so that the two mechanisms are mutually exclusive. It is the view of the present author that this is not so but that these two mechanisms can be easily unified because the original MGR mechanism is broad enough to encompass all the new aspects.

To see this it has first to be realized that the sequence of events postulated by KF - core ionization followed by Auger relaxation of the core hole - takes place on a very short time scale compared to nuclear motion so that it can be assumed to correspond to one FRANCK-CONDON transition. The main difference to the original scheme of MGR is that this FC sequence leads to a multiple valence excitation instead of a single one. It should be noted that contrary to erroneous assertions in recent literature and discussions, the MGR mechanism is not specified for, much less restricted to, primary one-electron excitations. The same is true a fortiori for the quantum mechanical extension by BRENIG. Examination of the original literature [22] shows that the valence bond description of the chemisorption bond which was used there naturally contains many-electron transitions and that it was made very clear that one-electron transitions constitute only the lowest-order approximation. Of course, at the time of development transitions of the type core ionization (+ shake-up) + Auger decay were not envisaged and their special efficiency was not seen. Furthermore, BRENIG's treatment has shown [23,24] that at least for the case of adsorbed H and O the lowest energy one-electron-like transition does not lead asymptotically to a free ion so that ionic desorption requires necessarily a double excitation. The recent measurements and calculations of MELIUS et al. [16] have led to the same conclusion. The assumption of primary one-electron transitions indeed made (implicitly or explicitly) by many authors in the past were made for simplicity, and preferential attention was paid to primary valence excitations because these appeared to lead to the largest primary cross sections; but there is nothing in the treatment that makes such a narrowing-down necessary.

We therefore propose to start from a very simple unifying picture which encompasses all features and leaves it open to examine in detail the remaining problems. Whether it is considered as a hybrid of MGR and KF, or a generalization of either one, the general features are the following.

We always have a *primary excitation step* which can be described as a FRANCK-CONDON transition to an excited valence state of the adsorbate complex. This may be a single step valence excitation (which in some cases may be describable without too much strain as a one-electron excitation, in others - maybe most - as a manybody excitation), or a two (or more) step process: core ionization followed by Auger decay (which itself could consist of an Auger cascade). In the sudden limit (i.e., at high primary energies relative to threshold), these subprocesses can be considered as separable steps; this treatment will break down close to threshold as then the relaxation following core ionization as well as the slow photoelectron emission (or that of the two slow final state electrons for ESD) can interfere with the Auger process. For an initial core-shake-up event the sequence may become even more complicated. However, in all cases the primary excitation (sequence) can be assumed to be decoupled from desorption itself by the different time scales. This means that the further evolution can be described by potential energy curves of possible excited states corresponding to the various final states of the primary excitation. In general, these will have multiple crossings and lead to various asymptotic states (and therefore products).

The second step which decides about desorption or recapture is contained in this evolution. It consists of the possibility of *delocalization* of the excitation before the leaving particle has reached the point of no return. The description of this process will have to be adapted to the characteristics of the final state of step 1 which is the initial state of step 2. If a one-electron excitation is an acceptable approximation of the latter, then the simplest version of the original MGR treatment is acceptable (one-electron tunneling). More sophisticated treatments will be necessary for more complex excitations. For two-hole states, treatments developed for Auger spectroscopy [25] should be applicable [26], although the different time scales of Auger electron emission and desorption have to be taken into account (a state which appears localized in Auger may well be strongly delocalized for desorption). Final states of core-satellite - Auger sequences will be even more complicated.

Manybody treatments leading to potential energy curves would express this absorptive part of the potential by curve broadening. Nonadiabaticity [27] will enter importantly into the final results.

This generalized scheme does not solve any of these problems yet, but it lets us pinpoint a number of questions derived from the experiments. While it is qualitatively understandable that excitations containing two or more holes tend to be more localized than one hole, the strong but very variable effect seen in different systems and for different levels is not understood yet. E.g., [10,15]: Why are increases at the C1s and O1s thresholds small or missing for CO^+ and moderate for O^+ desorption from adsorbed CO, while a huge effect is seen for N^+ from N_2/Ru at N1s; why is there no effect of substrate core ionization on ion desorption from covalent adsorbates in most, but not in all cases; why are core-satellite primary excitations so much more effective than "core only" processes in many cases but not in all: *cf.* O^+ from adsorbed CO at C1s (no) and O1s (yes); why is the 1^+ charge state so probable in core-initiated processes, i.e., why are neutral species relatively much less abundant - if they exist at all - in these as compared to valence-initiated processes. Obviously, detailed calculations for such systems and processes are out of reach of our present possibilities; so what is needed are simplifying models for these couplings between the two steps which, however, still contain the important features accounting for the observed differences.

Some such models have been proposed. For instance, instead of attempting to find the complicated potential curves corresponding to the evolution of multiply valence-ionized complexes, an explanation on the grounds of a "Coulombic explosion" [28,5] (disintegration caused simply by repulsion between the many charges accumulated) has been suggested. Such a description will neglect the possibility of stable doubly ionized species (taking out antibonding electrons even strengthens the bond) and all possible curve crossings. It may be inadequate for molecular systems, therefore, especially at low ionization, and will be best for well-defined charge distributions, i.e., for ionic systems (disintegration by reversal of the MADELUNG potential [4]), and for high charges. In fact, for very high charge accumulation it will be better than the picture of evolution along potential curves, as then the movement may become so rapid that the BORN-OPPENHEIMER decoupling of electronic and nuclear motion may break down. Another example is the emphasis laid upon the decrease of equilibrium distance for an ionic as compared to the ground state curve by ANTONIEWICZ [29]. Such an "orbital shrinkage" effect can indeed be expected for atomic adsorbates, but often not for molecular adsorbates; it will lead to the initial motion being *towards* the surface instead of *away* from it. Such models will be good if they encompass the main characteristics of a class of systems. Their limitations should be checked by viewing them in the context of the general scheme.

4. Neutral Desorption

The considerations of Sect. 3 apply to desorption of ions and neutrals alike although they have mainly been connected to ionic processes because of the available data. Within the scheme given, the desorption of neutrals can be understood in terms of either direct particle-hole excitation (simple molecular analogue: bonding - antibonding transition) or by primary ionization followed by reneutralization of the desorbing ion (which can be viewed as a curve-crossing event in the potential diagram). The threshold measurements mentioned in Sect. 2 have shown that both paths exist, with the second being stronger in the systems examined. Recent measurements by RUBIO et al. [19] which are compatible with earlier results of the present author [30] suggest a resonant behavior of the first contribution (particle-hole excitation) for CO/W. As has been discussed amply in the literature [27,31] the existence of the second path proves strong nonadiabaticity of the desorption event.

The open question as to why neutrals are so abundant after valence-initiated processes but so much less important (and maybe even absent) after core-initiation has been mentioned above. It should be noted that the observed variations in valence vs. core vs. core-satellite desorption cross sections mentioned above may also (partly) be caused by variations in the ion:neutral ratio.

All these processes have the common feature that they are not only electronically initiated but that the desorption is directly electronically controlled, i.e., follows an electronically excited curve. In principle one should expect that processes would be possible which are initiated electronically but proceed via vibrational excitation or even quasithermally. In the first case an internal electronic excitation can lead to internal vibrational excitation of an adsorbed molecule which can remain on the molecule even if the main electronic excitation is transferred into the metal. Vibrational coupling can then transfer part or all of the excitation to the desorption mode which will lead to desorption if the energy available suffices. A similar mechanism, but with direct excitation of the inner vibration with a laser, may be responsible in the reported cases of "resonant desorption" [32] from alkali halides. An alternative which is difficult to distinguish would consist of an electronic excitation of the adsorbate followed by its relaxation into vibrations of the layer or the substrate which, however, would not dissipate immediately but form a temporary local hot spot from which quasi-thermal desorption could occur. Such a mechanism is also being discussed for the mentioned resonant desorption. We believe there is evidence for similar processes in normal ESD. As mentioned in Sect. 2, neutral N_2O desorbing from chemisorbed N_2O on Ru has a peaked angular distribution, while neutral N_2O from physisorbed N_2O on the same surface has a much broader, though far from isotropic, angular distribution [21]. We believe that in our case the first mechanism ("predesorption" in analogy to predissociation) is the more likely one. It should be noted that direct electronic desorption of physisorbed species is very difficult to understand as the VAN DER WAALS bond would be expected to become *stronger* by excitation or ionization of the adparticle. GOMER [32] has suggested a "kick-off" mechanism mediated by chemisorbed particles, therefore. Such an explanation could also be advanced for our findings (but appears unlikely), and vice versa. It appears even possible that the neutral desorption effects observed below the ionic threshold for chemisorbed species [10,19,28] proceed via the vibrational predesorption described. Further experiments are necessary to resolve these questions.

We conclude that such paths have to be taken into account for neutrals desorption. They represent genuine DIET events, even though the final step is vibrational or quasithermal. They could even be incorporated into the general scheme laid out in Sect. 3 as again the desorption step is the final event competing with complete dispersion of the local excitation.

5. Conclusion

The field of DIET is in active development once again. It is believed that a better view of our understanding as well as of the remaining problems can be gained by first searching for the largest common denominator for all observations made and then differentiating the various processes. Such a frame has been proposed and some problem areas within this frame have been named.

For desorption of neutrals an additional mechanism has been described which differs from this frame by not following an electronically excited potential curve but going via a vibrational or quasithermal path.

Acknowledgements

I have greatly benefitted from the collaboration of P. Feulner, R. Jaeger, W. Riedl, and R. Treichler, and from many conversations with W. Brenig.

The mentioned experimental work of my group has been supported by the Deutsche Forschungsgemeinschaft through SFB 128.

References

1   J.R. Young, J. Appl. Phys. 31, 921 (1960)
    G.W. Moore, J. Appl. Phys. 32, 1241 (1961)
    P.A. Redhead, Vacuum 13, 253 (1963)
    D. Menzel and R. Gomer, J. Chem. Phys. 40, 1164 (1964)

2   D. Menzel, Surface Sci. 47, 370 (1975)
    dto., in "Interactions on Metal Surfaces", ed. R. Gomer, Springer, Berlin-Heidelberg-New York 1975, p. 101

3   D. Menzel, J. Vac. Sci. Technol. 20, 538 (1982)

4   M.L. Knotek and P.J. Feibelman, Phys. Rev. Lett. 40, 964 (1978);
    P.J. Feibelman and M.L. Knotek, Phys. Rev. B 18, 6531 (1978)

5   R. Franchy and D. Menzel, Phys. Rev. Lett. 43, 865 (1979)

6   T.E. Madey and J.T. Yates, Jr., J. Vac. Sci. Technol. 8, 525 (1971),
    and Surface Sci. 63, 203 (1977); T.E. Madey and R. Stockbauer,
    to be published

7   P.J. Feibelman, in "Inelastic Particle-Surface Collisions", E. Taglauer, W. Heiland, eds.; Springer, Berlin-Heidelberg-New York 1981, p. 104

8   Liu Zhen Xiang and D. Lichtman, Surface Sci. 114, 287 (1981), and work cited therein

9   This is found for β_2-H/W(100):
    W. Jeland and D. Menzel, Chem. Phys. Lett. 21, 178 (1973)

10 P. Feulner, R. Treichler and D. Menzel, Phys. Rev. B 24, 7427 (1981)

11 P.A. Redhead, Nuovo Cim. Suppl. 5, 586 (1967)

12 N.H. Tolk, L.C. Feldman, J.S. Kraus, R.J. Morris, M.M. Traum and J.C. Tully, Phys. Rev. Lett. 46, 134 (1981)

13 J. Kirschner, D. Menzel and P. Staib, Surface Sci. 87, L267 (1979);
    T.E. Madey, R.L. Stockbauer, J.F. van der Veen and D. E. Eastman,
    Phys. Rev. Lett. 45, 187 (1980)

14 T.E. Madey, R. Stockbauer, S.A. Flodstrom, J.F. van der Veen, F.J. Himpsel and D.E. Eastman, Phys. Rev. B 23, 6847 (1981)

15 R. Jaeger, J. Stöhr, R. Treichler and K. Baberschke, Phys. Rev. Lett.
    47, 1300 (1981); R. Jaeger, R. Treichler and J. Stöhr, Surface Sci. 117,
    533 (1982)

16 C.F. Melius, R.H. Stulen, and J.O. Noell, Phys. Rev. Lett. 48, 1429(1982)

17 R. Jaeger, J. Stöhr, J. Feldhaus, S. Brennan, and D. Menzel,
    Phys. Rev. B 23, 2102 (1981)

18 R. Jaeger, J. Feldhaus, J. Haase, J. Stöhr, Z. Hussain, D. Menzel, and
    D. Norman, Phys. Rev. Lett. 45, 1870 (1980)

19 J. Rubio, J.M. Lôpez-Sancho, and M.P. Lopez-Sancho,
    J. Vac. Sci. Technol. 20, 217 (1982)

20 T.E. Madey and J.T.Yates, Jr., Surface Sci. 63, 203 (1977);
    T.E. Madey, in "Inelastic Particle-Surface Collisions", E. Taglauer and
    W. Heiland, eds., Springer, Berlin-Heidelberg-New York 1981, p. 80

21 P. Feulner, W. Riedl and D. Menzel, to be published

22 D. Menzel and R. Gomer, J. Chem. Phys. 41, 3311 (1964);
    P.A. Redhead, Can. J. Phys. 42, 886 (1964)

23 W. Brenig, Z. Phys. B 23, 361 (1976)

24 W. Brenig, this volume

25 M. Cini, Solid State Commun. 24, 681 (1977);
    G.A. Sawatzki, Phys. Rev. Lett. 39, 504 (1977)

26 P.J. Feibelman, Surface Sci. 102, L51 (1981); D.E. Ramaker, C.T.White,
    and J.S. Murday, J. Vac. Sci. Technol. 18, 748 (1981)

27 W. Brenig, J. Phys. Soc. Japan 51, 1915 (1982);
    P. Schuck and W. Brenig, Z. Phys. B 46, 137 (1982)

28 T.A. Carlson and M.O. Krause, J.Chem.Phys. 56, 3206 (1972)

29 P.R. Antoniewicz, Phys. Rev. B 21, 3811 (1980)

30 D. Menzel, Ber. Bunsenges. phys. Chem. 72, 591 (1968)

31 W.L. Clinton, Surface Sci. 75, L796 (1978)

32 J. Heidberg, H. Stein, E. Riehl, and A.Nestmann, Z. Phys. Chem. NF 121,
    145 (1980); T. Chuang, J. Chem. Phys. 76, 3828 (1982)

33 C. Leung, Ch. Steinbrüchel, and R. Gomer, Appl. Phys. 14, 79 (1977);
    Q.J. Zhang and R. Gomer, Surface Sci. 109, 567 (1981)

2.3 Mechanisms of "Electronic" Desorption

P.J. Feibelman

Sandia National Laboratories, Albuquerque, NM 87185, USA

1. Introduction

This brief overview of what we know about mechanisms of electron-transition initiated desorption is mainly intended as an exhortation. I indicate areas where desorption can be useful as a surface analytic tool, now, and point out the kind of experiments that would provide important information for the further elucidation of the pathways by which desorption occurs.

I begin with a cautionary analogy: Consider the goals of laser selective chemistry. Here the idea is to pump particular molecular vibrations or electronic excitations with laser light, thereby to affect the product branching ratios of important chemical reactions. What one needs to know, for a given pumped excited state, is how long the absorbed energy remains localized in a particular bond or molecular sub-unit, and to what final products the initial state develops in time. Apart from the energy of the photons which is in the meV to few eV domain for laser chemistry, and in the 10 to 1000 eV range for desorption studies, the same problems dominate theoretical attempts to further our understanding of the mechanisms by which electron- and photon-stimulated desorption occur. The poignant question is then - given that laser chemistry could be infinitely more important to mankind than understanding desorption, why spend our limited resources on the latter problem? The answer, in my opinion, is that for the sake of using desorption as a surface analysis tool we require much less detailed information than in the realm of laser chemistry. If we are willing to ask simple questions, we stand a good chance of getting useful answers. Examples of simple questions that we know that desorption studies can answer, in certain cases, are:

1) By observing desorption yields, and desorbate angle- and energy distributions, what can we say about the atomic geometry of the unperturbed surface from which the desorption took place?

2) To what extent can we establish simple "laws of radiation damage?"

In what follows, I review a number of mechanisms for desorption that have been proposed in the literature over the last 18 years. In studying them I ask the reader to consider the extent to which they have made or might make it possible to answer simple questions of surface geometry or laws of radiation damage.

2. Catalog of Desorption Mechanisms

Essentially what we are after, in detailing a desorption mechanism, is a flow chart. In this workshop, we are limiting our attention to processes that are initiated by an electronic transition, which might be induced by an impinging electron or photon, or by a heavier particle. Following the initial event there is a sequence of electronic and/or ionic relaxation phenomena, at the end of which an ion or neutral particle emerges from the surface and may be detected.

A) Menzel-Gomer-Redhead (MGR) Model - The earliest attempts to understand desorption from surfaces focused on the question of why desorption cross sections are generally much smaller than the analogous dissocation cross sections for gas phase molecules. The model proposed to deal with this problem, introduced in 1964 by MENZEL and GOMER [1], and independently by REDHEAD [2], begins with a valence electron excitation. An electron, presumably from a bonding orbital between the desorbate species and the rest of the system is suddenly excited into a non- or an anti-bonding state. As a result of this Franck-Condon excitation, the desorbate species finds itself on a repulsive potential curve and begins to move away from the surface. As it moves away, there is a substantial possibility that either by resonant tunneling or by an Auger transition, the initially formed hole will be refilled. In this case desorption is no longer the final outcome or possibly desorption occurs, but the desorbate is a neutral and therefore escapes undetected. The main predictions of this MGR picture are that: i) desorption thresholds should be in the range of energies of bonding-antibonding transitions, a few to say, 20 eV, ii) there should be an isotope effect in desorption, since faster particles which have the same surface chemistry have a better chance of escaping before tunneling or Auger processes vitiate the effects of the initial excitation, and iii) the reason that desorption cross sections are low relative to those for apparently analogous gas-phase dissociation processes is that surfaces have available a large supply of electrons for reneutralization of the initially created hole.

From the point of view espoused in the introduction, the problems with the MGR picture are that since valence excitation spectra depend strongly on surface geometry, even if we knew that the MGR process were occurring it would be difficult to conclude anything about the initial surface atomic arrangement without detailed theoretical calculations. Moreover, the model suggests nothing in the way of laws of radiation damage. The model has provided an understanding of the smallness of desorption cross sections and of the existence of a desorption isotope effect [3]. However, it has not made desorption an important tool for surface analysis.

B) Antoniewicz's Model - An elaboration of the MGR desorption model has recently been suggested by ANTONIEWICZ [4]. Consider an oxygen atom adsorbed somewhat ionically on a metal surface. If an electron is removed from the O atom in the initial excitation event, it becomes much smaller. At the same time, if it was initially better than singly negative, the O will still be attracted to the surface by its image. Therefore, Antoniewicz suggests, the O will begin its post-ionization trajectory by moving toward the surface. As it does so, the electron capture probability increases, and the likelihood is that recapture will occur. When it does, however, the O will be much closer to the surface than at equilibrium. Therefore the O will find itself on a strongly repulsive potential curve, and will have a good chance of being desorbed, either as a neutral or as an ion, depending on further potential curve crossings on its trajectory. The problem with this scenario is not that it is unreasonable, but rather that one is at a loss to know how its reality is to be demonstrated. Antoniewicz's model does not help us know when desorption will and will not occur, nor does it provide insight as to how desorption can be used for surface analysis.

C) Equivalent atom desorption - This idea is related to Antoniewicz's, but has the advantage that it leads to a specific prediction. It has recently been suggested by A. R. WILLIAMS and N. D. LANG [5]. Consider a halogen atom, say a Cl, adsorbed on a metal. To a very good approximation it will be a Cl^-, and be bound to the surface by the image potential it induces. Now consider what happens following a <u>core</u> ionization event.

For as long as the core hole exists on the Cl, that atom will be "equivalent" to a neutral Ar. That is, its valence electrons will sense the core hole as an additional nuclear charge. But since the Cl was adsorbed as a Cl^-, it started out with Z+1 valence electrons. These Z+1 electrons see an effective nuclear charge of Z+1, and thus effectively the adsorbed halogen has been converted to a rare gas species. Now consider the forces on the excited adatom. Since the Cl^- was chemisorbed, it is much closer to the surface than the neighboring rare gas atom would be, in equilibrium, and therefore it is strongly repelled from the surface by exchange forces. At the same time, since it is neutral, the adatom feels no compensating attractive forces, and thus it is likely to be desorbed....if it can escape before the core hole decays. This latter point is important in that it leads to a specific prediction, namely that "equivalent atom desorption" should be strongly dependent on the core-hole lifetime. If it is a realistic mechanism for Cl, for example, it should be favored for an $L_{2,3}$ core hole whose width is ~ 0.07 eV compared to an L_1 hole whole width of ~ 1.2 eV, or a K hole, of width ~ 0.6 eV [6].

D) Pooley-Hersch desorption - In alkali halides there is considerable evidence (as Dr. Townsend shows in his presentation in this volume) for a mechanism of defect formation which begins with exciton creation and is followed by a nonradiative recombination that leads to significant ion transport. There is reason to believe that this process is also operative at surfaces, where it leads to desorption rather than F and H center formation and separation. The mechanism, attributed to POOLEY[7] and to HERSCH[8] operates as follows: A halogen electron is excited to a state in which it is still bound to the halogen, but in a large orbit. Because this electron is now absent for a large fraction of the time, the halogen relaxes toward one of its halogen neighbors, forming a so-called V_K center defect, a $(halogen)_2^-$ molecular ion. Some time later, the electron in the large orbit recombines with the hole left behind in the V_K center, and as a result the two halogen atoms that make up the center suddenly find themselves too close to one another. Their mutual repulsion sends one of them shooting off in the 110 direction (the direction in which the original relaxation had occurred), leading in the bulk case to the formation of a vacancy (F-center) interstitial (H-center) pair, or in the case that the excitation was near enough to the surface, an F center and a desorbing halogen atom. Several of the predictions of this model have been verified, including the observation of desorbing material emerging in 110 directions, and its dependence on the relative sizes of the alkali and halogen species. Until detection of desorbing neutrals is done routinely, it is questionable what use the desorption community will be able to make of the Pooley-Hersch idea. But at low energies, this mechanism must be considered.

E) Auger Stimulated Desorption - For ionically bonded surfaces, there is now considerable evidence[9-13] that the initial event in the desorption process is the Auger decay of a core hole. What is perhaps most important about the Auger initiated desorption process is that, for the class of systems where it applies, it leads to very definite predictions concerning when desorption, which is a form of surface radiation damage, will and will not occur [14], and it also makes it possible to use desorption as a surface analytic tool.

The Auger desorption mechanism is operative when the stoichiometry of the surface is that of a "maximal valence" or closed shell compound, such as TiO_2, V_2O_5, or WO_3. In these cases, the metal atoms are nominally devoid of valence electrons, and therefore, if a shallow core hole is created in a surface atom, that hole will Auger decay by stripping 2 or more electrons from a surface anion, in Auger decay. Since in equilibrium each anion is surrounded by cations, if enough charge is stripped to make it positive, the former anion will be ejected from the surface by the Madelung potential. This mechanism explains the common observation of positively charged species of atoms desorbing that exist as anions in ionic solids, and it also explains the fact that desorption thresholds occur at core level ionization potentials.

What is interesting about the Auger desorption mechanism is that it makes very specific predictions about what should and should not be observed. To begin, one expects desorption thresholds to occur at the core ionization potentials of the metal atoms to which the desorbate species was bonded. This immediately implies the significance of Auger induced desorption as a probe of bonding site. For example, suppose that one wants to determine where the H atoms sit on a molybdena-alumina hydrode-sufurization catalyst. One can easily resolve the Al core threshold at ~ 74 eV from the Mo threshold at ~ 34 eV, and that of O at about 21 eV. Thus one can determine which of the three species the H's reside on. A second important feature of the Auger model of desorption is its sensitivity to metal atom valence. In sub-maximal valence compounds valence electrons do reside on the metal atoms, and consequently in the Auger decay of a core hole, less charge is removed from the anions, and the liklihood of changing an anion to a cation that will desorb is greatly reduced. This phenomenon was demonstrated by KNOTEK[9] for various sub-oxides such as Ti_2O_3 and V_2O_3, and implies that one can monitor surface stoichiometry via desorption measurements.

Recently another important application of the fact that desorption can be initiated by core-hole ionization has been demonstrated. JAEGER et al.[15] have measured surface EXAFS for O adsorption on Mo (001), by using the photodesorption of O ions as a measure of Mo core ionization. They are thus able to draw conclusions concerning the site at which the O atoms are adsorbed. Although the data of Ref.[15] are somewhat controversial at this point, [16] there is no doubt that PSD SEXAFS measurements will be very important in surface geometry determination. Such measurements have the advantages that there is no background of positive ions, and that they are extremely surface specific. Neither of these statements is true for SEXAFS experiments in which electron currents are measured. On the other hand, as is discussed below, it is frequently the case that desorption cross sections are much higher for surface species that form a small minority of surface atoms. In the absence of information revealing just which atoms are the ones that have a chance of desorbing, the SEXAFS experiments will be on a subset of surface atoms that may or may not be the "important" ones for a particular application.

Although it is obviously of interest to unravel further details of the various mechanisms for desorption, I think it is important to stress that for ionically bonded surfaces, for which the Auger model appears to be correct, the opportunities for surface analysis via photon and electron stimulated desorption are great. I would hope to see work of this nature go on in parallel with efforts to uncover equally simple mechanisms for desorption in other cases. Surprisingly little use of the surface analytic power of the Auger idea has been made since its publication four years ago. Concerning the other mechanisms that I have discussed here, the major challenge is to demonstrate that there is a range of systems for which they are operative, and then show how they can be used to solve important surface problems.

3. What Might We Learn From Further Study of Desorption Mechanisms?

This section is devoted to three topics of major interest for the near term, in untangling the mechanisms of desorption and learning how to make use of desorption data.

A)   What Surface Species Desorb? - In early studies of desorption of O from W[17,18] and from Mo [2], it was recognized that different surface species could have drastically different desorption cross sections. Moreover, the systems whose desorption yields have been measured have generally been distinct from those systems whose atomic geometry has been analyzed, whether by LEED, or by one of the spectroscopies.

The consequence is that there are few cases of a well-characterized surface system, where we know just what species is desorbing from what site. What would be very valuable, would be to establish a few such proto-type desorption systems. With such data base it would be much easier to test new ideas about desorption definitively. Two recent examples come to mind, that might make useful prototypes. In work that is as yet unpublished, BARKER and ESTRUP[19] have studied H desorption from Mo (100) and compared measured yields vs. H exposure with measured LEED beam intensities. H deposition is known to cause a 2x2 reconstruction of the Mo(100) surface, which results in the appearance of a 1/2, 1/2 LEED beam. What Barker and Estrup find is that at 140 K, the intensity of this LEED beam and the H yield behave extremely similarly as a function of H exposure, showing a peak at roughly .2L. This result plausibly indicates that the H's that are desorbing are the ones which are responsible for the appearance of the 1/2, 1/2 beam. An analysis of the LEED data, perhaps together with a LEELS study of the H vibrations should make it possible to determine the geometry of the reconstructed bonding sites. This then would be a very valuable system to study in desorption. Another example of desorption from a system of reasonably well known geometry occurs in the beautiful angular desorption measurements of NETZER and MADEY [20]. They showed, in agreement with photoelectron diffraction results [21], that trace amounts of O on a Ni(111) surface cause adsorbed NH_3 or H_2O to order. Thus H desorption from either of these desorption systems would be a useful test-ing ground for desorption models. It is worth noting that neither of the systems mentioned is an ionic one. The surface geometry of ionic materials is far from well understood.

B)   What Good Are Desorbate Energy Distributions? - The Franck-Condon pic-ture of desorption after an electronic transition suggests that by measuring ion energy distributions one can gain information about the shape of the repulsive potential curve along which the ion desorbed. This assumes that the binding potential curve appropriate to the unperturbed state is known, e.q., by measurements of binding energy and vibration spectra. However, no one has done this.

The first reason is that it is not only the potential energy curves that determine an IED, but also the reneutralization and recapture probabilities. The slower ions are removed from the IED by these poorly quantified processes. A more important problem, however, is that discussed in the last paragraph: In the absence of a model of the bonding site from which an ion is desorbed, it is not easy to draw conclusions from a measured IED. Once we have desorption data from well-characterized systems, I think IED's will become much more interesting and possibly even useful.

C)   What Principles Govern Desorption of Covalently Bonded Surface Species? - The Auger model of desorption, useful as it potentially is, does not apply to covalently or metallically bonded surfaces. Recently, there has been a great deal of interest in finding out how to describe desorption from such systems, and some simple principles have emerged from calculations of JENNISON[22] and of RAMAKER [23], which can be stated in terms of elementary Hubbard model ideas [24-26].

The first principle is that desorption of singly bonded species is favored over multiply bonded ones. This is not the case for ionic systems. In that case, an anion's charge changes sign as a result of an Auger transition, and it is then repelled from the surface by the electrostatic force due to its cation neighbors. The more cations there are the stronger the force. Thus multiple coordination favors desorption for ionically bonded surfaces. Consider the opposite extreme, causing a C atom to desorb from a graphite surface. Each C is bonded to three neighbors by covalent bonds. Removing the bonding electrons between a C and one of its neighbors is unlikely to give rise to a net repulsive force. On the other hand, consider an H atom attached to a dangling bond on diamond(111). If one or both bonding electrons are removed from the C-H bond, and if, as discussed below, other electrons can somehow be prevented from flowing back into it, there is no longer anything holding the H on the surface, and it seems reasonable to expect that it will desorb.

The idea that single bonding favors desorption is supported, when one considers the question of how fast an initially created hole is refilled by electrons from the rest of the solid. Generally one expects that the more paths that are available for the electrons to get to the site of the hole, the faster it will be filled. Thus if the hole is on a single bonded atom or molecular sub-unit, there is a bottleneck of sorts as regards its neutralization, which provides more time for the atom or sub-unit to escape from the surface.

The second important idea, in the understanding of desorption from covalent surfaces can be stated pithily as "two holes are better than one." This is true not only because if two bonding electrons are removed from a bond, then that bond no longer exists, but also because the reneutralization of a two hole state can be much slower than that of a one-hole configuration [27]. To explain this phenomenon, we use two ideas, one geometric, and one from the Hubbard model, that are related at a deeper level. The geometric idea is: if two holes are created on an atom, the decreased screening of the nuclear charge causes the atom to shrink considerably in size. Therefore the overlap integrals that govern the rate at which electrons can hop back on to the atom are reduced in magnitude, and the atom

66

has more time to escape. The Hubbard model idea is the basis of CINI and SAWATZKI'S explanation of the atomic-like Auger spectra of closed d-shell metals, Cu being the prime example [24-26]. Two holes created on the same site repel each other electrostatically. Suppose we call the repulsion energy U. Because they feel a mutual repulsion, the holes will attempt to move away from one another, converting U into kinetic energy. But the amount of kinetic energy that a hole can gain is limited by the finite band width W in the solid. Thus if U is greater than 2W the holes will not be able to escape each other. They will be forced to hop together at a rate W^2/U, which can be much smaller than the single hole hopping rate W.

Incidentally, the idea that single coordination is favored merges with the idea that two holes are more easily localized than one. Consider for example a CH_3 group singly bonded through the C to a dangling bond on a diamond(111) surface. The argument involves the question of what is the appropriate value of W for the various atoms. A simple tight binding model (nearest-neighbor hopping) tells us that W=Zt, where Z is the number of neighbors that have a resonant energy level, and t is the interatomic hopping integral. Now the C of the methyl group only has one C neighbor, and therefore, W=t for it. On the other hand the surface C to which it is bonded has 4 C-neighbors, and therefore W=4t for that atom. As a consequence, it is perfectly possible that U will be greater than 2t but less than 8t, in which case two holes created on the methyl carbon would be confined to that sub-unit, but two holes created on a surface carbon atom could rapidly escape from that atom and from each other. This picture has been nicely confirmed in the experiments of KELBER and KNOTEK [28], reported in this volume.

The final idea governing desorption from covalent or metallic systems is that a closed shell configuration is desirable. This idea is related to the concept of initial state screening [29], and is originally due to HOUSTON [30]. Let us return to CINI and SAWATZKI'S ideas concerning atomic-like Auger spectra, and compare what we expect for Cu as against, say, Ti. In the latter case the d-shell is far from being filled. If a core hole is created in Ti the first thing that will happen is that a screening electron will be drawn from the neighboring atoms to the site of the hole. Thus before the hole Auger decays, the Ti atom on which it resides will have Z + 1 valence electrons. It will be an "effective" V atom. After the Auger decay, if two electrons are removed, the Ti atom will have Z - 1 valence electrons.

Where is the other hole? It has diffused far from the scene of the Auger event to follow. Thus in the case of a system like Ti, the result of an Auger transition is that at time $t=0^+$, right after the decay of the core hole, the Ti atom will contain in its valence shell, two holes and a screening electron. The screening electron greatly reduces the hole-hole interaction, U, if it has the same principal quantum number as the holes. This is not possible in Cu, where the d band is full. In that case, the only states available for a screening electron are very large, and thus incapable of screening two d holes from one another. The result is that in Cu the effective U is big while in Ti it is small. This explains why the Cu Auger line is atomic-like and that of Ti simply resembles the self-convolution of the d-band density of states [31].

Now let us return to desorption. Here we wish two holes to remain localized so that the bond between desorbate and surface cannot be reestablished before desorption can occur. If the holes are created in a closed shell they cannot be screened from one another. Thus the effective U will be large, and they are likely to be localized. For an open shell, this is apt not to be the case. This then explains why desorption is likely to be favored for molecular subunits that are singly bonded to a surface, in which the electronic configuration is that of a closed shell [28].

Of course, all that I have presented here concerns what might be thought of as necessary conditions for desorption. The sufficient condition, that there be a repulsive force strong enough to cause the molecular unit to emerge to a detector is the subject of active current research.

4. Directions

My purpose in this presentation has been to stimulate efforts to make use of what we already know about desorption in the case of ionically bonded surfaces, and to point out ways in which further research can make desorption studies a very important analytic tool. One of the chief experimental goals is to establish prototype systems whose surface structure is characterized. Desorption studies on such systems would be extrordinarily valuable, in that measured cross sections would be meaningful, and predicted angle- and energy distributions could be tested. One system that might be very interesting to investigate is CO from Cu(100). The geometry of this sytem is very well known, and because the CO is only weakly bonded in this case, it might be easier to identify the states that lead to desorption, and to untangle the effects of reneutralization.

Acknowledgements. It is a pleasure to acknowledge helpful conversations with R. R. Rye, J. E. Houston, and D. R. Jennison.

This work performed at Sandia National Laboratories, supported by the U.S. Department of Energy under contract number DE-AC04-76DP00789.

References

1. D. Menzel and R. Gomer, J. Chem. Phys. 41, 3311 (1964).

2. P. A. Redhead, Can. J. Phys. 43, 886 (1964).

3. T. E. Madey, J. T. Yates, Jr., D. A. King, and C. J. Uhlaner, J. Chem. Phys. 52, 5215 (1970); C. Leung, Ch. Steinbruchel, and R. Gomer, J. Appl. Phys. 14, 79 (1977).

4. P. R. Antoniewicz, Phys. Rev. B21, 3811 (1980).

5. A R. Williams and N. D. Lang, unpublished.

6. These core hole widths reflect the fact that the $L_{2,3}$ hole cannot decay via a Koster-Cronig transition while the K and L_1 hole can. The lifetime values are taken from: K. D.. Sevier, Low Energy Electron Spectrometry, J. Wiley & Sons, New York (1972).

7. D. Pooley, Proc. Phys. Soc. (London) 87. 245, (1966).

8. H. N. Hersch, Phys. Rev. 148, 928, (1966).

9. M. L. Knotek and P. J. Feibelman, Phys. Rev. Lett. 40, 964 (1978).

10. P. J. Feibelman and M. L. Knotek, Phys. Rev. B18, 6531 (1978).

11. M. L. Knotek, V. O. Jones and V. Rehn, Phys. Rev. Letter, 43, 300 (1979).

12. T. E. Madey, R. Stockbauer, G. F. van der Veen and D. E. Eastman, Phys. Rev. Lett. 45, 187 (1980).

13. D. P. Woodruff, M. M. Traum, H. H. Farrell, N. V. Smith, P. D. Johnson, D. A. King, R. L. Benbow, and Z. Hurych, Phys. Rev. B21, 5642, (1980).

14. M. L. Knotek and P. J. Feibelman, Surf. Sci. 90, 78 (1979).

15. R. Jaeger, J. Feldhaus, J. Haase, J. Stohr, Z. Hussain, D. Menzel, and D. Norman, Phys. Rev. Lett. 45, 2160 (1980).

16. M. L. Knotek, private communication.

17. T. E. Madey, and J. T. Yates, Jr. Surf. Sci. 11, 327 (1968).

18. D. A. King, T. E. Madey, and J. T. Yates, Jr., J. Chem. Soc., Faraday Trans. 68, 1347 (1972).

19. R. J. Barker and P. J. Estrup, unpublished.

20. F. Netzer and T. E. Madey, Phys. Rev. Lett. 47, 928 (1981).

21. W. M. Kang, C. H. Li, S. Y. Tong., C. W. Seabury, K. Jacobi, T. N. Rhodin, R. J. Purtell, and R. P. Merrill, Phys. Rev. Lett. 47, 931 (1981).

22. D. R. Jennison, J. Vac. Sci. and Tech. 20, 548 (1982).

23. D. Ramaker, this volume.

24. M. Cini, Sol. St. Comm. 24, 681 (1977); 20, 605 (1976).

25. E. Antonides, E. C. Janse, and G. A. Sawatzky, Phys. Rev. B 15, 1669 (1977); G. A. Sawatzky, Phys. Rev. Lett. 39, 504 (1977).

26. P. J. Feibelman and E. J. McGuire, Phys. Rev. B 15, 3575 (1977).

27. P. J. Feibelman, Surf. Sci. 102, L51 (1981).

28. J. A. Kelber and M. L. Knotek, this volume; see also D. R. Jennison, J. A. Kelber and R. R. Rye, Phys. Rev. B 25, 384 (1982).

29. R. Laesser and J. C. Fuggle, Phys. Rev. B22, 2637(1980); D. R. Jennison Phys. Rev. B 21, 430 (1980).

30. J. E. Houston, unpublished.

31. M. L. Knotek and J. E. Houston, Phys. Rev. B 15, 4580 (1977); D. M. Zehner, J. R. Noonan and H. H. Madden, J. Vac. Sci. and Tech. 20, 859 (1982).

2.4 Models for Desorption in Covalent Systems

D.E. Ramaker

Naval Research Laboratory, Washington, DC 20375, USA

1. Introduction

The electron- and photon-stimulated desorption (ESD/PSD) of
ions and neutrals from various surfaces has been shown to be a
sensitive probe for the study of surface-adsorbate interactions
and bonding [1], and even for the determination of such physical
properties as the chemisorption bond angle and bond site (e.g.,
from ESD ion angular distributions, ESDIAD [2]). Despite this
success, until recently little understanding of the ESD/PSD
mechanism has been apparent for covalent systems. For example,
the desorption cross section is seen to be highly dependent on
adsorption site [3] and surface coverage [4,5], but this is not
well understood. However, recent comparative investigations [6]
between molecular dissociation in the gas phase and chemisorbed
on the surface have been very helpful in understanding the
desorption mechanisms in molecularly chemisorbed systems (e.g., CO
and H_2O on metals). Other work utilizing Auger and photoelectron
data along with theoretical calculations has provided similar
insight into the desorption mechanisms involved in extended
covalent solids (such as SiO_2 and Si_3N_4 [7]), and for chemisorbed
H (e.g. H/Si, Ni, Pd, and W [8-11]). The results of these investigations in covalent systems will be discussed here in the
context of models previously proposed for desorption from
surfaces, as well as for molecular dissociation and defect
formation in solids.

Ionic desorption from the surface is usually discussed in
the context of two different models. The first of these is known
as the MENZEL-GOMER-REDHEAD (MGR) model [12]. In this model, the
primary process is a FRANCK-CONDON valence excitation or
ionization to a repulsive neutral or ionic state (i.e., leaving a
one hole-one electron (1h1e) or one hole (1h) excited state)
which results in dissociation or desorption. This view is
supported by desorption thresholds in the valence region, and by
the dominance of the neutral yield over the ion yield [1e]. An
important secondary step in the MGR model is recapture of the
leaving particle resulting from transfer of the excitation
energy into the bulk (e.g., reneutralization or hole hopping).
The competition between the recapture and desorption processes
suggests that the observed desorption yield should depend on the
ion mass. A large isotope effect has indeed been observed, and
has been used (perhaps erroneously, see below) for further
support of the MGR model [1e].

The second model is the KNOTEK-FEIBELMANN (KF) model [13] which is particularly applicable to highly ionic systems (i.e., maximal valency systems). This model for ion desorption indicates that the ionization of a core level is a primary process. The subsequent Auger decay of the core hole creates a two-hole (2h) positive anion at an initially negative ion site. The expulsion of the positive ion results from the reversal of the Madelung potential. The Auger process has been demonstrated to be of importance in desorption of some covalent systems, but its role is not always clear [1,14-23]. In covalent systems, the ESD/PSD thresholds sometimes correspond to Auger yield thresholds, but often times not [1,15,19-23].

Models similar to those summarized above for desorption from the surface have been proposed much earlier for other electronic to nuclear energy transfer mechanisms, such as for molecular dissociation and defect formation in ionic solids. These models are categorized in Table 1 according to the nature of the excited state initiating the process. The year the model was first proposed is given to emphasize both the initial development of the concepts for molecular dissociation, followed by application to solids and surfaces, and the progression to an increasing number of ''particles'' (holes and electrons) in the more recent models. The similarity of the concepts in the widely different systems allows for an interesting and detailed comparison between dissociation in gas phase and chemisorbed molecules, defect formation in the bulk, and desorption from the surface.

In ionic solids, such as the alkali halides, the POOLEY-HERSCH [24] mechanism assumes a one-electron transition to a self-trapped excitonic state (a 1h1e state) involving two neighboring anions [25]. This is followed by a radiationless transition back to the ground electronic state (i.e., a recombination of the electron hole pair) via a potential energy curve crossing. The curve crossing with the ground state occurs, however, at a halogen-halogen distance far removed from the ground state equilibrium distance, and hence the two halogen atoms gain considerable momentum. A subsequent 110 replacement sequence separates the ion from the vacancy. The interstial halogen atoms sometimes diffuse to the surface where they may desorb by thermal evaporation [26,27].

The quasi-equilibrium theory (QET) or statistical model [28,29] is also included in Table 1. It assumes that the process of internal conversion is so fast, that all electronic excited states undergo radiationless transitions via curve crossings to the ground state, before any significant nuclear motion occurs. This mechanism is not unlike the POOLEY-HERSCH mechanism discussed above, in that a radiationless transition to the ground electronic state follows the initial excitation. However, the primary process in the POOLEY-HERSCH mechanism is the excitation of the excitonic state. For large molecules, the QET theory is not concerned with the initial excitation, although it is tacitly assumed to be one electron in nature [29]. The probability for a dissociation process in the QET theory is determined purely on a statistical basis, utilizing specific reaction sequences with estimated activation energies [29]. The validity of the QET theory is evident for some large molecules; however, it does not

Table 1. Models for electronic to nuclear energy transfer in molecules, solids, and at surfaces

MODEL	MOLECULES Dissociation	SOLIDS Defect formation	SURFACES Desorption
1h or 1h1e	Small molecules: FRANCK-CONDON (1930's) Large molecules: Statistical theory- QET (1952)	POOLEY-HERSCH (1960)	MENZEL-GOMER REDHEAD (1964)
2h	Coulomb explosion (1954)	VARLEY (1954)	Ionic Systems: KNOTEK- FEIBELMAN (1978) Covalent Systems: Auger plus localization (recent work)
2h1e	''Multielectron excitations'' (1970's)		Condensed and chemisorbed molecules (recent work)

properly predict the dissociation of others [30]. Its validity has not been examined for atoms or small molecules chemisorbed on a surface.

The concept of molecular dissociation, as a result of a one-electron FRANCK-CONDON transition to a repulsive state, has existed since the early days of quantum mechanics and the advent of vibrational spectroscopy [3]. The most significant concept, in all of the 1h or 1h1e models, is the transfer of electronic excitation energy to nuclear kinetic energy via transit of a repulsive section of an excited or ground state potential curve. Alternative excited state decay mechanisms, such as photon emission, exist in all chemical systems, including simple molecules. However, in extended covalent systems or for molecules chemisorbed on a metal surface, additional fast decay mechanisms, such as reneutralization or hole hopping (i.e., interaction of the molecule or cluster with the extended system) complicate this simple picture; indeed it may render the 1h or 1h1e states ineffective for the desorption or dissociation process.

Recent work [6] emphasizes the importance of competitive decay phenomena in all models, the 1h, 2h, and 2h1e models. A knowledge of the relative decay rates for one or two hole hopping, resonant decay, Auger decay, and desorption are essential to an understanding of the desorption thresholds, as well as differences in the dissociation yields of molecules in the gas, solid, and chemisorbed phases. It follows therefore that an isotope effect is not necessarily indicative of the MGR model, as mentioned above, since all three models are expected to exhibit such an effect. It is true, however, that the 1h MGR mechanism might be expected to exhibit the largest isotope effect because the 1h states tend to have the shortest lifetimes (see Sec.6).

Turning now to the 2h models in Table 1, the dissociation of a doubly charged benzene molecule into two singly charged fragments was indicated already in 1937 by the appearance of satellite peaks of ions with high kinetic energy [32] (also referred to as ''excess'' kinetic energy [33]). This process was not systematically studied, however, until 1954 [34], and was termed a ''Coulombic explosion'' first in 1966 by CARLSON and WHITE [35]. Its occurrence in small molecules, such as N_2, O_2, CO, NO, CO_2, CH_4, C_3H_6, etc., has been verified more recently by coincident ion time of flight spectroscopy. This shows that fragment ions of high kinetic energy are created coincidently with momentum in opposite directions, thus satisfying energy and momentum conservation laws [36-39]. Evidence also exists for the Coulomb explosion of microclusters with different kinds of bonding, i.e., metal clusters (Pb_n), ionic clusters (($NaI)_n$), and vander Waals clusters (Xe_n) [40]. The 2h excited states have been shown to be produced by a direct process (two-electron ionization or ionization plus shakeoff [41, 42]), in which case the threshold energy lies in the deep valence region, or via the Auger process initiated by ionization of a core hole [43-45].

VARLEY [46] first proposed that F center creation in alkali halides results from multiple ionization of the halogen anions (i.e., 2h states). The positive halogen ion moves to an interstitial position to decrease the repulsive lattice potential at the initial halogen lattice site. Although some arguments against the VARLEY mechanism have been presented [47,48], it remains a viable mechanism for defect formation at core level excitation energies [49] and for damage in semiconductors [50].

Results from electron-electron and electron-ion coincidence studies [51-52] and from theoretical calculations [53-56], on small molecules such a N_2, CO, and H_2O, indicate that 2h1e or ''multielectron excitations'' may also lead to dissociation. These states appear as satellites in photoemission spectra [53-56]. They derive their intensity either from correlation mixing with the one-electron states in the deep valence region, or from core hole shakeup. Although they are of relatively minor importance in the gas phase, because of the small branching ratios, recent work [6] (to be presented in Sec. 4) indicates they are most significant for dissociation of chemisorbed molecules.

Very recent work on PSD of ions from CeO_2 [57] indicates 2h1e states may also lead to desorption in ionic systems.

However, it is not clear that the excited electron plays a significant role in the desorption process in this case. The close correlation between the H^+ ion yield and photon adsorption coefficient indicates that the two holes may be sufficient to initiate the H^+ desorption, with the excited electron acting only as a spectator or perhaps even resonantly decaying to the continuum before the desorption sequence gets underway. In contrast, results in Sec. 4 indicate that in the molecularly chemisorbed systems, the excited electron is critical to the desorption sequence. In the latter case, if the excited electron decays to the continuum, the desorption sequence is aborted.

2. Atomic Desorption from Covalent Solids

Recently, both experimental and theoretical ESD studies have been performed on the covalently bonded solids, SiO_2 and Si_3N_4. Utilizing electron energies below 100 eV, KNOTEK and HOUSTON [58] [58] reported O^+, H^+, OH^+, and F^+ ion yields from samples obtained by electron activated stepwise oxidation or nitridation of a Si(111) surface. TRAUM [59] has examined the ESD ion yield at higher electron impact energies for SiO_2. RAMAKER, WHITE and MURDAY [7] have made the first attempt at quantitative comparison between theoretical and experimental desorption cross sections in these solids.

Probably the most significant finding from the experimental studies is the extent of surface hydrogen contamination in SiO_2 coming from the initial Si crystal, and the high levels of H^+ and OH^+ ion desorption arising from this contamination. Surfaces oxidized with activated O_2, where no OH is present, were found to be much more stable; indeed below 100 eV, the relative O^+ yield for clean SiO_2 is found to be rather small [58]. However, above the Si 2p core level around 103 eV, the O^+ yield apparently becomes very large [59]. In SiN_x, no N^+ or N containing fragments were detected [58]. The SiN_x was found to be stable under an electron beam with energies from 0 to 2000 eV [60].

The above experimental results are consistent with an Auger-induced 2h desorption mechanism. Although the O(2s) level in SiO_2 around 32 eV does produce a small O^+ desorption threshold, the O(2s) level does not generally have sufficient energy to give rise to the Auger decay, which results in localized valence holes necessary for subsequent O^+ desorption [58,7]. The absence of N^+ desorption in SiN_x, and the large O^+ desorption in SiO_2 above 100 eV, can be explained by examining the alternative decay mechanisms of the 2h state.

Critical to the 2h desorption models is the localization of the two holes, a condition necessary to provide the Coulomb repulsion for expulsion of the ion. Although a localization criterion was already proposed by VARLEY [46] in his ionic model, more general criteria have been proposed recently in the context of Auger spectroscopy. According to the CINI-SAWATZKY (CS) theory [61] or configuration interaction (CI) theory [62], localization of the valence holes results only if the effective hole-hole repulsion U^e is greater than some appropriate covalent

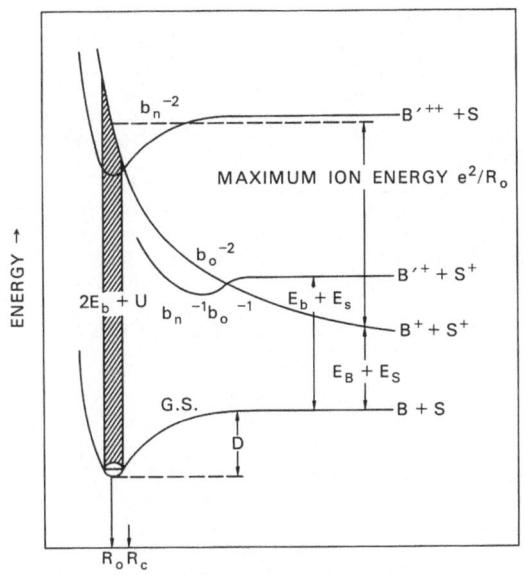

interaction V (i.e., $U^e > V$). Core-valence-valence (CVV) Auger lineshapes provide direct experimental evidence for localization or delocalization of the valence holes [61,62].

The alternative decay mechanisms of the 2h states can best be considered in the context of Fig.1 [7]. For definiteness, consider a surface atom S covalently bonded via bonding orbital b_0 to a single bulk atom B with bond energy D. The Auger process resulting from a core hole in S or B can leave the system initially in the repulsive state b_0^{-2}, with excitation energy $2E_b + U^e$, where E_b is the one-electron binding energy of b_0 and U^e is the Coulomb interaction energy resulting from two holes in b_0. Due to this repulsion, the atom S is pushed off the surface initially in state b_0^{-2}. If the holes decay before a critical distance R_c, S is recaptured; otherwise S gains sufficient kinetic energy to be desorbed. Beyond R_c, S can be neutralized leading to the desorption of neutrals (escape along b_n^{-2} where b_n is a bulk bond orbital). The maximum kinetic energy of the outgoing ions is e^2/R_0 (for escape along b_0^{-2}), although the maximum ion intensity will occur at smaller energies if, e.g., escape along $b_n^{-1}b_0^{-1}$ dominates. The most obvious means for decay of b_0^{-2} involves one hole hopping into the bulk without appreciable nuclear motion. However, this process is blocked if $U^e > V$ as indicated. It may not become important until larger R values are reached (Fig.1) when $U^e < V$. Two-hole hopping off the surface bond orbital can occur immediately, but it generally occurs with smaller probability.

To quantify the lifetime aspect of the model, RAMAKER, WHITE, and MURDAY [7] considered a HUBBARD-like single-band Hamiltonian motivated throught the bond orbital approximation.

Assuming that $U^e \ll \gamma$, and that the bulk band is described by a Bethe lattice of bandwidth γ, the probability P of finding two holes, created in the surface bond orbital at time t = 0, both in the orbital at a later time t is given by $P \sim \exp(-2\alpha\gamma t/\hbar)$, if $\gamma t < \hbar$. The factor α ($\sim 1/2$) arises because b_0 is a surface bond orbital. This expression must be replaced by $\beta(\hbar/\gamma t)^6$ when $\gamma t \gg \hbar$, where $\beta = (4\alpha)^4/[\pi^2(1-\alpha)^8]$. If $U^e > \gamma$, the two holes are essentially bound together, in which case $P \sim \exp(-\gamma'\alpha t/\hbar)$ so long as $\gamma't < \hbar$ and $P \sim \beta^{1/2}(\hbar/\gamma't)^3$ when $\gamma't \gg \hbar$, where γ' is the bound two-hole bandwidth. Second-order perturbation theory can be used to estimate γ'; this gives $\gamma' = \gamma^2/KU^e$, where K is the coordination number of the lattice.

The desorption or damage cross section σ_D can be estimated from the expression [7]:

$$\sigma_D(\text{theor.}) = \Sigma_{\text{core}} \quad \sigma_I(\text{core}) \quad f_D(\text{core}) \quad P(t_c), \qquad (1)$$

which involves two other important factors besides $P(t_c)$ (t_c is the critical desorption time for the atom to move beyond R_c. σ_I (core) is a core ionization cross section, which may include a backscattering factor in the solid, and can be obtained either from theory or experiment [63]. f_D is the fraction of core ionization events which results in two holes localized in a bonding orbital via the Auger process. Its magnitude can generally be determined from the Auger lineshape; this indicates the importance of AES to estimating ESD cross sections [7].

Both the CI theory and the experimental $L_{23}VV$ Auger lineshape [62] indicate that two valence holes remain localized in a Si(sp)-O(2p)-Si(sp) bond orbital (sp is an sp^3 hybrid orbital on the Si atom) consistent with $U^e > \gamma$ (i.e., 11eV compared with 4eV). In Si$_3$N$_4$, the corresponding bond orbital is the trigonal planar N(2p)-(Si(sp))$_3$ orbital [64]. The larger bond orbital causes a decreased U^e; the less polar N-Si(sp) bond causes an increased γ. Hence in Si$_3$N$_4$, $U^e < \gamma$, and the holes delocalize in times short compared to desorption.

Utilizing (1), the electron beam damage cross section for SiO$_2$ has been estimated to be $\sim 10^{-21}$ cm^2 [7]; experimentally it has been determined to be of the same magnitude [65]. The importance of hole localization in SiO$_2$ when $U^e > \gamma$, can be realized from the result ($\sim 10^{-39}$ cm^2) obtained when one assumes $U^e \ll \gamma$, which is more appropriate for Si$_3$N$_4$. In SiO$_2$, f_D for the ionization thresholds Si(2s,2p),O(2s) and O(1s) have been estimated to be 0.13, 0, and 0.03 respectively [7]. These results illustrate an important point not always fully appreciated [1]; each ionization or Auger yield threshold should not necessarily have a corresponding desorption threshold.

ESD and PSD studies on similar oxide systems have been reported. Recently ESD results on rf sputtered SnO$_2$ films were published [66]. Large O$^+$ desorption yields were observed under 2000 eV electron excitation, and were attributed to a POOLEY HERSCH type mechanism. However, this data is not inconsistent with the 2h Auger-induced mechanism. The O$^+$ desorption is a accompanied by a corresponding increase in film conductivity attributed to the oxygen vacancies. O^+ desorption has also been seen recently via PSD from the more ionic oxides (either bulk or chemisorbed O), such as TiO_2 [17, 67], WO_3[68] and MoO_2[69], but here the KF mechanism is more clearly valid.

3. H^+ Desorption

Several ESD/PSD studies on chemisorbed H have been reported recently; these include H on Si, Ni, Pd, and W [8-11]. In these systems, the KF maximal valency condition is clearly not present (i.e., the bonding is more covalent), nevertheless, the localized 2h excited state model is still favored. Analysis of the Si $L_{23}VV$ Auger lineshape for H/Si [8] indicated a localized 2h contribution with 23 eV binding energy consistent with the lowest H^+ ion threshold. This suggests the localized 2h state is populated directly via a two-electron ionization or shakeoff process. Recent cluster calculations modeling H/Ni indicate that only the 2h or 2h2e excited states are sufficiently repulsive to initiate H^+ desorption from Ni [9]. A strong correlation has been reported between the PSD H^+ ion yield from H/Pd (100) and the intensity of the hydrogen induced band in a constant initial state (CIS) photoelectron spectrum [10]. Although this may seem to imply a 1h desorption model (such as the MGR model), it does not rule out a 2h model, since direct shakeoff might also have a cross section proportional to the one electron transition matrix element. Finally, an extremely small ESD H^+ ion yield was obtained from H/W (110) [11]. This has been attributed to H atoms bonding only to highly coordinated sites. Whatever the reason, apparently in this case the 2h excited state can undergo a fast decay via one hole hopping (i.e., U^e<V).

4. Molecular Dissociation in the Gas, Condensed, and Chemisorbed Phases

Recent electron-electron (e, 2e) and electron-ion (e, e+ion) coincidence measurements [51,52], photodissociation data [70], and theoretical calculations [53-56], for the CO and H_2O gases, have provided quantitative information on ionization branching ratios, fragmentation ratios, and UPS satellite structure. Even more recent PSD data on condensed and chemisorbed CO and H_2O [15, 20-23] have allowed for a detailed comparative investigation and analysis of the data [6]. The results of these investigations are summarized here.

In CO(g), the 1h $5\sigma^{-1}$, $4\sigma^{-1}$, and $1\pi^{-1}$ states do not lead to dissociation. Three different excitation channels are responsible for the dissociation, they may be characterized as the $5\sigma^{-1}1\pi^{-1}2\pi$ state (1.0 C^+), the $5\sigma^{-2}6\sigma$ state (.54 C^+,.46 O^+) and the ''$3\sigma^{-1}$'' state (.39 C^+,.61 O^+). The fragmentation ratios for the three channels are indicated in parentheses [51]. The states are characterized by the dominant configuration contributing to the many electron CI wave-function in the ground state Franck-Condon region [53,54]. The $3\sigma^{-1}$ state, although labeled as a 1h state, undergoes very heavy mixing with other 2h1e states so that its character is not clear. The 2h1e states responsible for dissociation derive their excitation intensity also by mixing

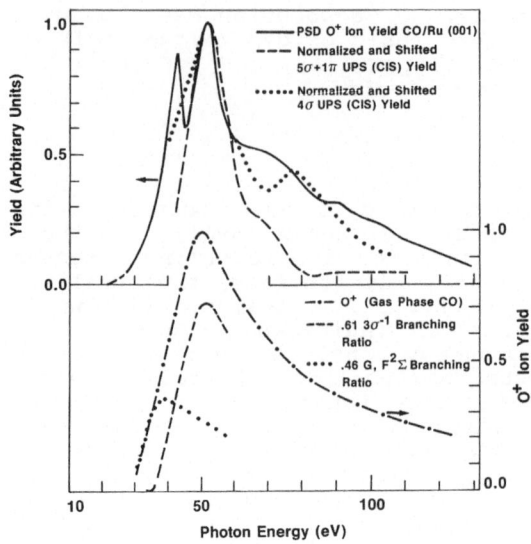

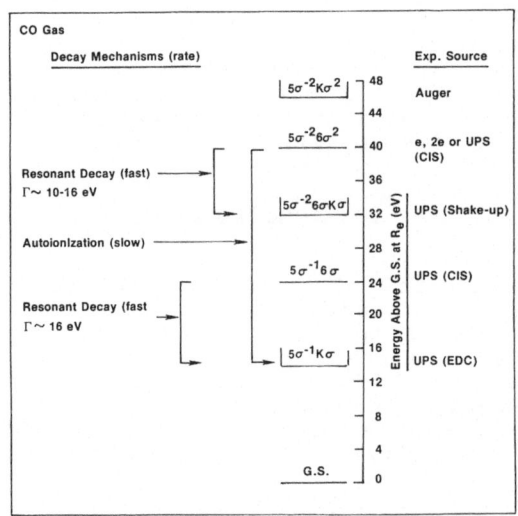

Fig.2. Top: Comparison of t PSD O$^+$ ion yield for CO/Ru (001) with the (5σ+1π) and 4 UPS(CIS) yield as obtained from [15]. The (5σ+1π) and UPS yields have been normalized and aligned with the O$^+$ ion yield at 50 eV. Bottom: The O$^+$ ion yield from photodissociation of CO gas [70]. Also shown is the ''3σ-1'' and the 5σ-26σ (G,F ^{2}Σ) branching ratios multiplied by the fragmentation ratios for O$^+$ formation as determined by (e,2e) and (e,e+ion) measurements [51]

Fig.3. Illustration of the possible decay mechanisms for the states involving the 5σ and 6σ orbitals in CO gas [6]. The energy given for each state involving the kσ orbital corresponds to the threshold energy. The experimental data source of each energy level or threshold is indicated at the right

with the 3σ-^{1} state. The O$^+$ ion yield is compared with the normalized 5σ-26σ G^{2}Σand 3σ-^{1} branching ratios in Fig.2.

An examination of the orbital character and excitation energies provides an understanding of these data. The 4σ, 5σ, 1π, and 2π orbitals are either non-bonding, or weakly bonding or antibonding. The 3σ and 6σ orbitals are strongly bonding and antibonding respectively. Ionization from a 3σ orbital or electron excitation into the 6σ orbital should significantly weaken the CO bond, and subsequently initiate dissociation via an MGR type mechanism. However, the 6σ orbital is so strongly antibonding, it lies above the $k\sigma$ threshold allowing for a fast (10^{16}/sec) 6σ to $k\sigma$ resonant decay. This introduces some additional decay channels which may abort the dissociation process, such as indicated in Fig. 3 for the $5\sigma^{-1}6\sigma$ state [6]. However, the additional 5σ hole in the $5\sigma^{-2}6\sigma$ state stabilizes the 6σ orbital, i.e., the 6σ orbital becomes a discrete excitonic like state in the presence of two 5σ holes. This is evident in Fig. 3 by observing that the excitation energy for the $5\sigma^{-2}6\sigma$ state is well below that of the $5\sigma^{-2}$ 2h state. The $5\sigma^{-2}$ state excitation energy is so large because of a large two-hole repulsion energy U^e (~20eV). Thus, a large hole-hole repulsion in the $5\sigma^{-2}$ state (suggestive of the 2h Auger model), or more correctly a large electron-hole attraction in the $5\sigma^{-2}6\sigma$ state, is necessary to prevent the decay and allow for the dissociation.

Fig. 2 compares the O^+ ion yield obtained by PSD from CO/Ru (001) [15] with that obtained by (e, 2e) and photodissociation from CO(g). The narrower $5\sigma^{-2}6\sigma$ peak has been attributed to the metal d_π to CO(2π) charge transfer which occurs on the surface, particularly in the presence of a core or valence hole [6]. This is most evident from the C or O KVV Auger lineshape, which indicates that U^e for the $5\sigma^{-2}$ state is effectively zero on the metal surface [70]. The large reduction in U^e drops the $5\sigma^{-2}$ excitation energy below the $5\sigma^{-2}6\sigma$ energy allowing for the resonant 6σ to $k\sigma$ decay. This aborts the desorption, since with $U^e \sim 0$ the $5\sigma^{-2}$ state will certainly not Coulomb explode. The $5\sigma^{-2}6\sigma^2$ resonance must undergo a relatively slow (10^{14} sec^{-1}) autoionization process ($5\sigma^{-2}6\sigma^2$ to $5\sigma^{-1}k\sigma$) or a double resonant decay ($5\sigma^{-2}6\sigma^2$ to $5\sigma^{-2} k\sigma^2$) to abort the desorption; consequently, this excitation has an excellent chance of leading to desorption. Therefore, in CO(g) the $5\sigma^{-2}6\sigma k\sigma$ excitation (including the $5\sigma^{-2}6\sigma^2$ resonance) dissociates; in CO/Ru only the relatively narrow $5\sigma^{-2} 6\sigma^2$ resonance leads to dissociation.

The similarity of the ($5\sigma + 1\pi$) and 4σ UPS (CIS) lineshapes for CO/Ru, and the O^+ ion yield peak around 50 eV (Fig. 2), strongly suggests this peak arises from the ''$3\sigma^{-1}$'' excitation; therefore, the 50 eV peak has the same source in both CO(g) and CO/Ru [6]. The structure above 60 eV in the PSD O^+ yield for CO/Ru has been attributed to the $5\sigma^{-2}7\sigma$ excitation and other 2hle states involving the 4σ and 1π orbitals [6].

Figure 4 shows the PSD O^+ yield from CO/Ni(100) at excitation energies near the O K level [20]. The excitation cross sections of the $1\sigma^{-1}(2\pi, 6\sigma)$, $1\sigma^{-1} 5\sigma^{-1}6\sigma$, and $1\sigma^{-1}3\sigma^{-1}$ states are also indicated [6]. Although the $1\sigma^{-1}$ ($2\pi, 6\sigma$) excitation is expected to dominate the CO(g) dissociation spectrum, for CO/Ni it is barely visible. This can be understood upon examination of Fig. 5, which illustrates the various decay mechanisms of the levels involving the C and O K levels. In general, resonant decay is faster (10^{16} sec^{-1}) than Auger decay (10^{14} sec^{-1}), which is faster than the desorption process (10^{13} sec^{-1}). O^+ desorption results only if the final electronic state has a 6σ electron or a 3σ hole, since the 2h mechanism is not effective with $U^e \sim 0$ on the surface [6].

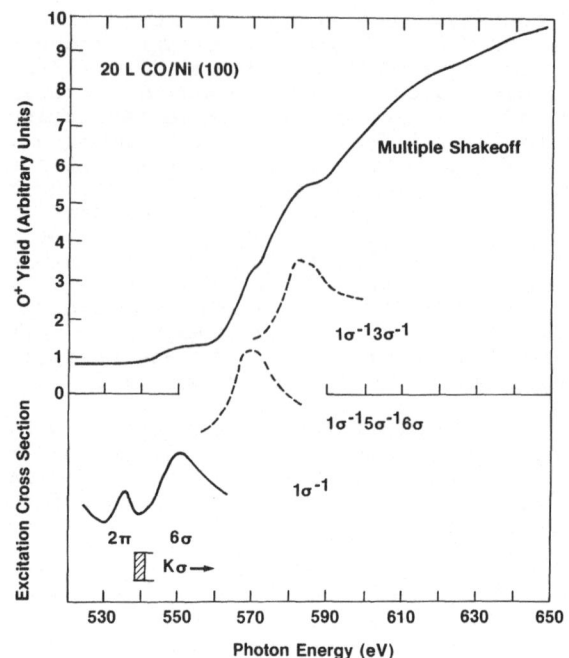

Fig. 4. Comparison of the O^+ PSD ion yield for 20 L exposure of CO/Ni(100) with the 1σ excitation cross section as determined by the O KVV Auger yield [20]. The $1\sigma^{-1}5\sigma^{-1}6\sigma$ and $1\sigma^{-1}3\sigma^{-1}$ excitation cross sections are also schematically indicated (dotted lines are essentially the $1\sigma^{-1}6\sigma$ excitation shifted to the proper energy). The ion yield above 600eV is attributed to multiple shake-off [6]

Very little if any O^+ desorption occurs near the C K level. The $2\sigma^{-1}5\sigma^{-1}6\sigma$ state can resonantly decay and does not lead to desorption. The $1\sigma^{-1}5\sigma^{-1}6\sigma$ state cannot resonantly decay, hence it produces the first major desorption threshold above the O K level. This difference between the corresponding O and C K levels results from differences in screening, i.e., the $2\sigma^{-1}5\sigma^{-1}$ holes are effectively screened by the 2π screening charge (located primarily on the C atom), the $1\sigma^{-1}5\sigma^{-1}$ holes are not screened as effectively. A significant $1\sigma^{-1}3\sigma^{-1}$ shakeoff excitation also produces O^+ desorption as indicated in Fig. 4 [20,6]; the $2\sigma^{-1}3\sigma^{-1}$ shakeoff intensity near the C K level is apparently much smaller.

Finally, it is of interest to note that the core level O^+ desorption yield increases monotonically with coverage while the valence level O^+ yield reveals a broad maximum with coverage [4,5]. The decreasing valence level O^+ yield near full coverage has been attributed to two-hole CO-CO hopping, or delocalization of the $5\sigma^{-2}6\sigma^2$ state, which aborts the desorption sequence [6]. This two-hole hopping process depends strongly on the 5σ bandwidth and hence on the coverage. After the Auger decay of the core levels, 3h1e or 3h states result which cannot easily decay by two-hole hopping. Consequently, the core level O^+ yield does not show the same behavior, rather it merely increases with coverage.

The situation with H_2O is similar to CO, but with some significant differences. The strongly antibonding $4a_1$ orbital in

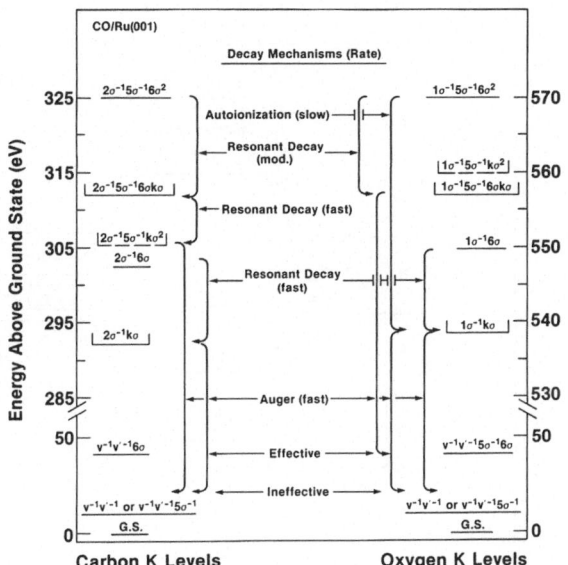

Fig.5. Illustration of some of the possible decay mechanism for various C and O K level 1h, 1h1e, and 2h1e excitations for CO/Ru(001) [6]. The energy given for each continuum threshold is relative to the vacuum. Experimental energies are used where possible; the $2\sigma^{-1}5\sigma^{-1}k\sigma^2$ and $1\sigma^{-1}5\sigma^{-1}k\sigma^2$ energies are rough estimates. The final valence level states are denoted as ''effective'' or ''ineffective'' for desorption as determined by the presence or absence of a 6σ electron. Relative rates of decay are indicated (e.g., fast, moderate, slow)

H_2O plays the same role as the 6σ orbital in CO; however, the $4a_1$ orbital lies below the ionization threshold so that resonant decay of the $4a_1$ electron is not important in H_2O. The deep valence $2a_1$ level is comparable to the 3σ level in CO. Both undergo a large mixing with the 2h1e states so that the $2a_1$ or 3σ orbitals are not strictly one-electron in character [53-56]. The $2a_1$ level is significantly broadened by hydrogen bonding between neighboring H_2O molecules in the condensed and molecularly chemisorbed systems [22].

The dissociation spectrum for $H_2O(g)$ is given in Fig.6 [52]. O^+ and H^+ from the ''$2a_1$'' state is clearly evident above 32 eV. The H^+ contribution just below 32 eV and the O^+ around 25 eV have been attributed to the $1b_1^{-2}4a_1$ and $1b_1^{-1}3a_1^{-1}4a_1$ states respectively [6]. The $1b_2^{-1}$ state clearly leads to OH^+ and H^+ ions. This has been attributed to predissociation as a result of potential curve crossings with other states [72]. The correlation diagram in Fig. 7 shows the $1b_2^{-1}$ $^2B_2$ state crossing with the 2h1e states, $1b_1^{-1}$ $3a_1^{-1}$ $4a_1$ $^4B_1$ and the $1b_1^{-1}$ $3a_1^{-1}$ $4a_1$ $^2B_1$, and correlating to OH^+ and H^+ ions respectively [73]. A third crossing with a $^4A_2$ state provides a channel for O^+ production, but none is seen experimentally, probably because the crossing occurs too far from the Franck-Condon region [72].

Comparison of the UPS spectrum for $H_2O(g)$, $H_2O(s)$, $(H_2O)m/GaAs$ and $(H_2O)d/Ti$ shows little differences in the orbital binding energies, except for the $(H_2O)d/Ti$ spectrum where significant shifts to lower binding energies are found for the $1b_2$, $3a_1$, and $1b_1$ orbitals, and a merging of the $3a_1$ and $1b_1$ energies is apparent [6]. This is consistent with the orbital structure

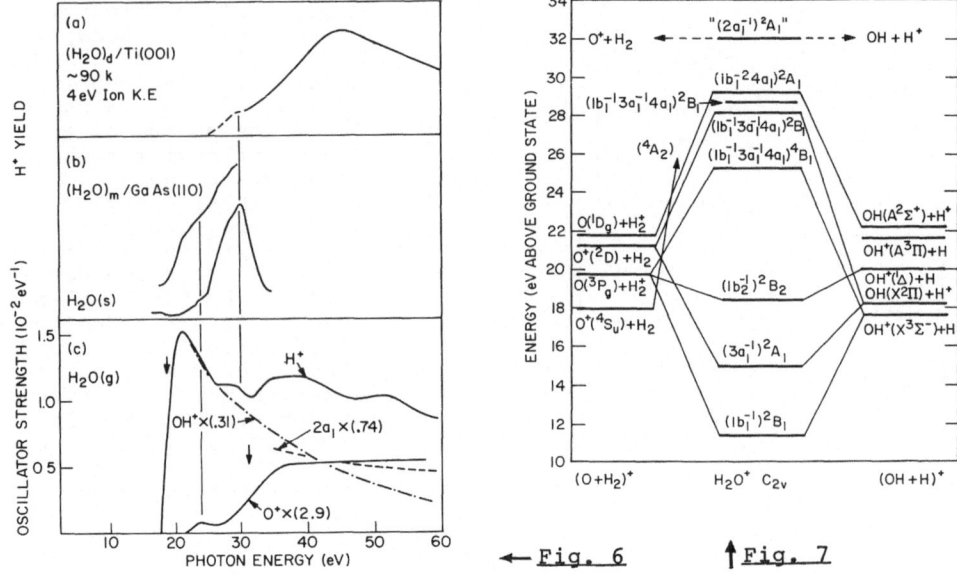

← Fig. 6   ↑Fig. 7

Fig. 6. (a)   The PSD H^+ ion yield for H_2O on Ti(001) at 90 K as reported in [21].   The H_2O apparently dissociates on the Ti surface at this temperature.   (b) Comparison of the PSD H^+ion yield for molecular H_2O on GaAs (110) and condensed (solid) H_2O as reported in [23] and [22] respectively.   (c) Oscillator strengths for ion production (H^+, OH^+, O^+) as determined by (e, 2e) and (e, e+ion) data [52].   The OH^+ and O^+ oscillator strengths have been normalized by experimentally determined fragmentation ratios, H^+/OH^+ and H^+/O^+ respectively, for easy comparison.   The normalized (H^+/total ions) $2a_1$ branching ratio is also indicated

Fig. 7.   Correlation diagram for pertinent states of H_2O as obtained from [72].   The states which correlate to the $2a_1^{-1}$ ^2A state is not certain, however, the fragmentation data suggests it correlates to $O^+ + H_2$ and $OH + H^+$ like states

of the OH fragment, i.e., H_2O dissociates on Ti at 90K [21]. H_2O molecularly chemisorbs on GaAs, thus ''m'' and ''d'' denote molecular and dissociated H_2O respectively.

Comparison of the PSD yields for these H_2O systems reveals some dramatic differences with the $H_2O(g)$ data [6].   The $1b_2^{-1}$ contributions are missing, apparently because the slow ($\sim 10^9$ sec^{-1}) predissociation process is aborted by the faster (10^{15} sec^{-1}) hole hopping decay mechanism.   The $2a_1^{-1}$ contributions are also absent in $H_2O(s)$ and $(H_2O)m/GaAs$.   This has been attributed to hydrogen bonding, which broadens the ''$2a_1^{-1}$'' peak aiding the hole hopping decay mechanism [6].   At less than monolayer coverages of OH/Ti, the hydrogen bonding does not occur.

5. Molecular Desorption from Metals

The experimental data for molecular desorption is much less extensive than for atomic desorption. However, ESD N_2^+ and CO^+ ion yields have been reported from N_2 and CO/Ru, and recently even neutral N_2 and CO desorption yields have been published [1e, 19]. A detailed investigation of the CO and CO^+ thresholds, and comparison with satellite peaks in the UPS spectrum for CO/Ru, has again pointed to the importance of the 1h1e excited states [6]. Table 2 summarizes the conclusions reached from these investigations.

It is well known that the bonding of the CO with the metal involves an interaction and charge donation to the metal via the 5σ orbital, and a backbonding with the 2π orbital. The 2π interaction with the metal leads to $2\pi_b$ and $2\pi_a$ (bonding and antibonding) orbital combinations, with the Fermi level lying between these two orbitals. It has been suggested that an electron in the $2\pi_a$ orbital or a hole in the 5σ or $2\pi_b$ orbital will significantly weaken the M-CO bond, and consequently result in CO or CO^+ desorption via a 1h or 1h1e mechanism [6]. The 3-6 eV CO threshold has been attributed to the $2\pi_b^{-1}2\pi_a$ state, but a strong mixing with 2h1e states probably assists in increasing the lifetime of this state. The remaining thresholds (e.g., CO at 14 eV and CO^+ at 14, 22-27, and just above the C K level) have been attributed to 2h1e states involving the 5σ, $2\pi_b$, and $2\pi_a$ orbitals. No evidence for molecular desorption as a result of a 2h excitation has been found for CO/Ru; however, in other non-metallic systems, such as for H_2O on TiO_2 and Al_2O_3, a 2h Auger mechanism is apparently consistent with the observed PSD OH^+ yields [18].

Recent ESD of H^+, CH_3^+, and H_2^+ from condensed phase neopentane, tetramethylsilane, and other branched alkanes suggest that ''multielectron (hole) states'' may also be involved here [74]. The excitations which produce the desorption appear to be largely localized on single methyl groups indicating, that at least for the desorption process, the branched alkane molecule simulates a small molecule chemisorbed on a surface (e.g., CH_3 ''chemisorbed'' on $C(CH_3)_3$ [75]). Thus similar 2h1e mechanisms may also be active in the dissociation of some large covalent molecules. More work must be completed in this area before definite conclusions can be made.

6. Summary and Conclusions

A summary of the systems examined in this review, and the states responsible for the observed dissociation or desorption, are given in Table 2. A word of caution about these assignments is needed here. They are the dominant configuration in a many electron CI wavefunction near the ground state Franck-Condon (FC) region. Although one configuration may be dominant in the FC region, another quite different configuration may be dominant at larger internuclear separations. Also, the dominant configuration in any region may contribute as low as 30% to 50% of the total charge density [53-56]. Nevertheless, these assignments

Table 2. Summary of PSD/ESD active excited states for the covalent systems reviewed in this work

System (Desorbed species)	$1h^a$ or 1 hle	$2h^{b,d}$	$nhle^{c,d}$ $nh2e$	Comments
SiO_2 (O^+)		$(b{-}^1b{-}^1)$		$U^e > V$
Si_3N_4				no ESD, $U^e < V$
H/Si,Ni,Pd(H^+)		$b{-}^1b{-}^1$		$n{-}^1n{-}^1, n{-}^1b{-}^1$ not effective
H/W(H^+)				little desorption
CO(g), CO(s) (C^+,O^+)	"$3\sigma{-}^1$"	$(v{-}^1v{-}^1)$	$5\sigma{-}^2 6\sigma, 5\sigma{-}^2 6\sigma^2$ $5\sigma{-}^1 1\pi{-}^1 2\pi$	$4\sigma{-}^1, 1\pi{-}^1, 5\sigma{-}^1$ not effective
CO/M (O^+)	"$3\sigma{-}^1$"	$(v{-}^1 3\sigma{-}^1)$	$5\sigma{-}^2 6\sigma^2$ $(v{-}^1v{-}^1 5\sigma{-}^1 6\sigma)$ $(v{-}^1v{-}^1 3\sigma{-}^1)$	$v{-}^1v{-}^1$ not effective $U^e \sim 0$
(CO)	"$2\pi_b{-}^1 2\pi_a$"		$5\sigma{-}^1 2\pi_b{-}^1 2\pi$	thres. at 3–6 & 14 eV
(CO^+)			$5\sigma{-}^1 1\pi{-}^1 2\pi_a$ $5\sigma{-}^1 2\pi_b{-}^1 2\pi_a$	thres. at 14 & 22–27 eV
			$(v{-}^1v{-}^1 5\sigma{-}^2 2\pi_a^2)$	thres. above C K level
			$(v{-}^1v{-}^1 1\pi{-}^2 2\pi_a^2)$	
H_2O(g) (H^+, OH^+, O^+)	$1b_2{-}^1$ "$2a_1{-}^1$"	$(v{-}^1v{-}^1)$	$1b_1{-}^2 4a_1$ $1b_1{-}^1 3a_1{-}^1 4a_1$	$1b_2{-}^1$ by predissociation
H_2O(s), H O/M (H^+)		$(v{-}^1v{-}^1)$	$1b_1{-}^2 4a_1$ $1b_1{-}^1 3a_1{-}^1 4a_1$	H bonding imp.
OH/Ti(H^+)	"$2\sigma{-}^1$"		$1\pi{-}^2 4\sigma$	H bonding not imp.

[a] Quotation marks below indicate "spectroscopic" state, however 2hle states mix in heavily. The role played by this mixing is not clear.

[b] n,b, and v indicate non-bonding, bonding, and valence orbitals respectively.

[c] nh indicates 2 or more holes.

[d] States produced primarily by core hole Auger decay are in parenthesis; all other states are excited at valence level energies.

serve a useful purpose in identifying the initial excitation and alternative fast ($>10^{14}$ sec^{-1}) decay mechanisms. Many 1h1e ionic states exist in the deep valence region of the CO and H_2O molecules. Fortunately, only a small number mix in heavily with the 1h valence states (e.g., only those of the same symmetry and having significant overlap) and hence have significant excitation intensity [53-56]. This allows specific assignments to be made in these cases. The role of subsequent curve crossings with other 2h1e states is not entirely clear [9].

Several conclusions can be made based upon the assignments in Table 2. They are listed here.

1) At core excitation energies, the Auger-induced 2h states are the most important for desorption if $U^e>V$. If $U^e<V$, the 2h states are ineffective and the 2h1e states become more important.

2) At valence excitation energies, the 2h states are excited with much lower intensity, so that in most instances the 2h1e states have comparable or even greater intensity. In either case, the 2h1e states generally have lower excitation energies than the 2h states (especially true for large U^e), so that the initial desorption thresholds on the surface generally result from 2h1e excitations.

3) In general, the lifetime, τ, of the 2h1e states are greater or of the order of the 2h states, with the 1h states having much shorter lifetimes. This is particularly true if $U^e>V$. If $U^e<V$, the 2h states have lifetimes of the order of the 1h states. This can be summarized as follows:

$U^e > V$: $\tau(2h1e) \sim \tau(2h) \gg \tau(1h)$

$U^e < V$: $\tau(2h1e) \gg \tau(2h) \sim \tau(1h)$.

4) The above relationships explain why the 1h states are rarely effective for desorption on the surface. However, the deep valence 1h states, such as the 3σ state in CO, may be sufficiently localized and narrow (i.e., τ (3σ) sufficiently large) that it can initiate desorption. In both CO and H_2O however, the role of the strong mixing with the 2h1e states is not clear. Further work on other systems is needed here.

5) Important alternative decay mechanisms of the excited states include resonant decay (RD), valence one hole and two hole hopping (this may occur via a valence VVV Auger process), and core hole Auger decay (AD). Core hole Auger decay generally initiates desorption (creates a 2h state), the other decay mechanisms abort the desorption process. Typical decay rates, R, may be summarized as follows: $R(RD) \sim 10^{16}$ sec^{-1}, $R(1h) \sim 10^{15}$ sec^{-1}, $R(2h) \sim 10^{14}$ sec^{-1} ($U^e>V$), $R(AD) \sim 10^{14}$ sec^{-1}. Typical desorption rates are $10^{13} - 10^{14}$ sec^{-1}.

6) Covalent interaction effects can dramatically alter the decay rates and hence the ion yields. Bond polarity effects in SiO_2 and Si_3N_4, adsorbate-substrate effects in the form of metal to CO screening charge transfer, adsorbate-adsorbate interaction effects for CO/Ru, and hydrogen bonding effects in H_2O(s) and H_2O/GaAs were indicated in the previous sections. These effects will be more critically examined elsewhere [6]

References

1. Early and recent review articles include:
   a. T.E. Madey and J.T. Yates, Jr., J. Vac. Sci. Technol. $\underline{8}$, 525 (1971).
   b. D. Menzel, in Interactions on Metal Surfaces, edited by R. Gomer (Springer, Berlin, 1975), p. 124; Surf. Sci. $\underline{47}$, 370 (1975).
   c. D. Lichtman and Y. Shapira, in Chemistry and Physics of Solid Surfaces, edited by R. Vanselow (CRC, Boca Raton, 1978), p. 397; M.J. Drinkwine and D. Lichtman, Prog. Surf. Sci. $\underline{8}$, 123 (1977).
   d. T.E. Madey and J.T. Yates, Jr., Surf. Sci. $\underline{63}$, 203 (1977).
   e. D. Menzel, J. Vac. Sci. Technol $\underline{20}$, 538 (1982).
2. See for example: T.E. Madey, in Inelastic Particle-Surface Collision, ed. by E. Taglauer and W. Heiland (Springer-Verlag, Heidelberg, 1981), p. 80; T.E. Madey, J.T. Yates, A.M. Bradshaw, and F.M. Hoffman, Surf. Sci. $\underline{89}$, 370 (1979); H. Niehus, Surf. Sci. $\underline{92}$, 88 (1980); R. Jaeger and D.Menzel, Surf. Sci. $\underline{93}$, 71 (1980).
3. T.E. Madey, Surf. Sci. $\underline{94}$, 483 (1980).
4. T.E. Madey, Surf. Sci. $\underline{79}$, 579 (1979); F.P. Netzer and T.E. Madey, J. Chem. Phys. $\underline{76}$, 710 (1981).
5. J.E. Houston and T.E. Madey, to be published in Phys. Rev. B.
6. D.E. Ramaker, to be published.
7. D.E. Ramaker, C.T. White, and J.S. Murday, J. Vac. Sci. Technol. $\underline{18}$, 748 (1981); Phys. Lett. $\underline{89A}$, 211 (1982).
8. H.H. Madden, D.R. Jennison, M.M. Traum, G. Margaritondo, and N.G. Stoffel, to be published.
9. C.F. Melius, R.H. Stulen, and J.O. Noell, Phys. Rev. Lett. $\underline{48}$, 1429 (1982); J. Vac. Sci. Technol. $\underline{20}$, 559 (1982).
10. R.H. Stulen, T.E. Felter, R.A. Rosenberg, M.L. Knotek, G. Loubriel, and C.C. Parks, Phys. Rev. B $\underline{25}$, 6530 (1982); R.H. Stulen, J. Vac. Sci. Technol. $\underline{20}$, 846 (1982).
11. S.L. Weng, Phys. Rev. B $\underline{25}$, 6188 (1982).
12. D. Menzel and R. Gomer, J. Chem. Phys. $\underline{41}$, 3311 (1964), P.A. Redhead, Can. J. Phys. $\underline{42}$, 886 (1964).
13. M.L. Knotek and P.J. Feibelman, Phys. Rev. Lett. $\underline{40}$, 964 (1978); Surf. Sci. $\underline{90}$, 78 (1979); Phys. Rev. $\underline{B18}$, 6531 (1978); P.J. Feibelman, Surf. Sci. $\underline{102}$, L51 (1981).
14. R. Franchy and D. Menzel, Phys. Rev. Lett. $\underline{43}$, 865 (1979).
15. T.E. Madey, R. Stockbauer, S.A. Flodstrom, J.F. Vander Veen, F.J. Himpsel and D.E. Eastment, Phy. Rev. $\underline{B23}$, 6847 (1981).
16. R.J. Baird, R.C. Ku, and P. Wynblatt, Surf. Sci. $\underline{97}$, 346 (1980); J.L.Hock, J.H. Craig, and D. Lichtman, Surf. Sci. $\underline{87}$, 31 (1979); 85, $\underline{101}$ (1979); J.H. Craig and J.L.Hock, Surf. Sci. $\underline{103}$, L81 (1981); $\underline{102}$, 89 (1981); $\underline{100}$, L435 (1980).
17. M.L. Knotek, Surf. Sci. $\underline{101}$, 334 (1980); Surf. Sci. $\underline{97}$, L27 (1980); M.L. Knotek, V.O. Jones, V. Rehn, Surf. Sci. $\underline{102}$, 566 (1981); Phys. Rev. Lett. $\underline{43}$, 300 (1979).
18. M.L. Knotek and J.E. Houston, J. Vac. Sci. Tech. $\underline{20}$, 544 (1982); Preprint.

19. P. Feulner, R. Treichler, and D. Menzel, Phys. Rev. B24, 7427 (1981).
20. R. Jager, J. Stohr, R. Treichler, and K. Baberscke, Phys. Rev. Lett. 47, 1300 (1981); R. Jaeger, R. Treichler, and J. Stohr, to be published in Surf. Sci.
21. R. Stockbauer, D.M. Hanson, S.A. Flodstrom, and T.E. Madey, to be published in Phys. Rev. B.
22. R.A. Rosenberg, V.Rehn, V.O. Jones, A.K. Green, C.C. Parks, G. Loubriel, and R.H. Stulen, Chem. Phys. Lett. 80, 488 (1981).
23. G. Thornton, R.A. Rosenberg, V. Rehn, A.K. Green, and C.C. Parks, Solid State Commun. 40, 131 (1981).
24. D. Pooley, Proc. Phys. Soc. 87, 245 (1966); 257 (1966); H.N. Hensch, Phys. Rev. 148, 928 (1966).
25. K. Kagawa, J. Phys. Soc. Japan 41, 507 (1976).
26. H. Overeijnder, M. Szymonski, A. Haring, and A.E. DeVries, Radiation Effects 36, 63 (1978).
27. L.S. Cota Araiza and B.D. Powell, Surf. Sci. 51, 504 (1975).
28. H.M. Rosenstock, M.B. Wallenstein , A.L. Wahrhaftig, and H. Eyring, Prac. Natl. Acad. Sci. U.S. 38, 667 (1952).
29. M.L. Vestal, J. Chem. Phys. 43, 1356 (1965); and references therein.
30. S. Ikuta, K. Yoshihara, and T. Shiokawa, Mass. Spect. 22, 233 (1974); 22, 239 (1974); 23, 61 (1975).
31. See any physical chemistry textbook, e.g., P.W. Atkins, Physical Chemistry, 2nd ed. (Freeman Co., San Francisco, 1982), p. 611.
32. P. Kusch, A. Hustrulid, and J.T. Tate, Phys. Rev. 52, 843 (1937); 54, 1037 (1938).
33. J. Olmsted III, K. Street, and A.S. Newton, J. Chem. Phys. 40, 2114 (1964).
34. F.L. Mohler, V.A. Dibeler, and R.M. Reese, J. Chem. Phys. 22, 394 (1954).
35. T.A. Carlson and R.M. White, J. Chem. Phys. 44, 4510 (1966).
36. B. Brehm and G. De Frenes, Int. J. Mass. Spect. Ion Phys 26, 251 (1978).
37. A.K. Edwards and R.M. Wood, J. Chem. Phys. 76, 2938 (1982).
38. R.G. Hirsch, R.J. Van Brunt, and W.D. Whitehead, Int. J. Mass Spect. Ion Phys. 17, 335 (1975).
39. K.E. McColloh, T.E. Sharp, and H.M. Rosenstock, J. Chem. Phys. 42, 3500 (1965).
40. K. Sattler, J. Muhlback, O. Echt, P. Pfau, and E. Recknagel, Phys. Rev. Lett. 47, 160 (1981).
41. F.H. Dorman and J.D. Morrison, J. Chem. Phys. 35, 575 (1961).
42. J.E. Monahan and H.E. Stanton, J. Chem. Phys 37, 2654 (1962).
43. T.A. Carlson and M.O. Krause, J. Chem. Phys. 56, 3206 (1972).
44. R.J. Van Brunt, F.W. Powell, R.G. Hirsch, and W.D. Whitehead, J. Chem. Phys. 57, 3120 (1972).
45. R.B. Kay, Ph. E. Van der Leeuw and M.J. Van der Wiel, J. Phys. B10, 2521 (1977); M.J. Van der Wiel and Th. M. El-Sherbini, Physica 59, 453 (1972).

46. J.H.O. Varley, Nature $\underline{174}$, 886 (1954); J. Phys. Chem. Solids $\underline{23}$, 985 (1962).

47. J.Sharma and R. Smoluchowski, Phys. Rev. $\underline{137A}$, 259 (1965); B.A. Cruz-Vidal and H.J. Gomberg, J. Phys. Chem. Solids $\underline{31}$, 1273 (1970).

48. A.A. Vorobev and V.M. Listisyn, Soviet Phys. J. $\underline{20}$, 1051 (1977).

49. J.W. Corbett, Surf. Sci. $\underline{90}$, 205 (1979).

50. V.S. Vavilov, A.E. Kiv, and O.R. Niyazova, Phys. Status Solidi $\underline{32A}$, 11 (1975); H.J. Pabst, Radiation Effects $\underline{25}$, 279 (1975).

51. A. Hamnett, W. Stoll and C.E. Brion, J. Electron. Spectrosc. Related Phenom. $\underline{8}$, 367 (1976); G.R. Wight, M.J. Van der Wiel, and C.E. Brion, J. Phys. $\underline{B9}$, 675 (1975).

52. K.H. Tan, C.E. Brion, Ph. E. Van der Leeuw, and M.J. Van der Wiel, Chem. Phys. $\underline{29}$, 299 (1978).

53. P.W. Langhoff, S.R. Langhoff, T.N. Rescigno, J. Schirmer, L.S. Cedarbaum, W. Domcke, and W. Von Niessen, Chem. Phys. $\underline{58}$, 71 (1981).

54. P.S. Bagus and E.K. Viinikka, Phys. Rev. $\underline{A15}$, 1486 (1977).

55. J.C. Leclerc, J.A. Horsley, and J.C. Lorquet, Chem. Phys. $\underline{4}$, 337 (1974).

56. R. Arneberg, J. Muller, and R. Manne, Chem. Phys. $\underline{64}$, 249 (1982).

57. B.E. Koel, G.M. Loubriel, M.S. Knotek, R.H. Stulen, R.A. Rosenberg, and C.C. Parks, Phys. Rev. $\underline{B25}$, 5551 (1982).

58. M.L. Knotek and J.E. Houston, J. Vac. Sci. Technol. $\underline{20}$, 544 (1982), also to be published.

59. M.M. Traum, private communication.

60. Y.E. Strausser and J.S. Johannessen, NBS Special Publ. 400-23, ARPA/NBS Workshop IV, Surface Analysis for Silicon Devices, held at NBS, Gaithersburg, MD, April 23-24, 1975.

61. M. Cini, Solid State Commun. $\underline{20}$, 605 (1976); Phys. Rev. $\underline{B17}$, 2788 (1978); Phys. Rev. $\underline{87}$, 483 (1979); G.A. Sawatzky, Phys. Rev. Lett. $\underline{39}$, 504 (1977).

62. D.E. Ramaker, Phys. Rev. $\underline{B21}$, 4608 (1980); B.I. Dunlap, F.L. Hutson, and D.E. Ramaker, J. Vac. Sci. Technol. $\underline{18}$, 556 (1981).

63. E.J. McGuire, Phys. Rev. $\underline{A16}$, 73 (1977); J.I. Vrakking and F. Meyer, Surf. Sci. $\underline{47}$, 50 (1975).

64. J. Robertson, Phil. Mag. $\underline{44}$, 215 (1981).

65. The experimental value indicated is the total damage cross section as reported in [7]. It includes all oxygen atom desorption (0^-, 0, and possible OH, OH^+, etc.) from thermally grown SiO_2 samples under a 2 Kev electron beam, and as such is not expected to be dramatically affected by H contamination (i.e., the value indicated is adequate for the order of magnitude comparisons made here).

66. Y. Shapira, J. Appl. Phys. $\underline{52}$, 5696 (1981).

67. D.M. Hanson, R.Stockbauer, and T.E. Madey, Phys. Rev. $\underline{B24}$, 5513 (1981).

68. T.E. Madey, R.L. Stockbauer, J.F. Van der Vean, and D.E. Eastman, Phys. Rev. Lett. $\underline{45}$, 187 (1980).

69. R. Jaeger, J. Stohr, J. Feldhaus, S. Brennan, and D. Menzel, Phys. Rev. $\underline{B23}$, 2102 (1981); R. Jaeger, J. Feldhaus, J. Haase, J. Stohr, Z. Hussain, D. Menzel, and D. Norman, Phys. Rev. Lett. $\underline{45}$, 1870 (1980).

70. J.C. Fuggle, E. Umbach, P. Feulner, and D. Menzel, Surf. Sci. 64, 69 (1977); J. Elect. Spect. Related Phenon. 10, 15 (1977); ibid and R. Kakoschke, J. Elect. Spect. Related Phenom. 26, 111 (1982).
71. T. Masuoka and J.A.R. Samson, J. Chem. Phys. 74, 1093 (1981).
72. A.J. Lorquet and J.C. Lorquet, Chem. Phys. 4, 353 (1974).
73. H.H. Harris and J.J. Leventhal, J. Chem. Phys. 64, 3185 (1976).
74. J.A. Kelber and M.L. Knotek, to be published.
75. D.R. Jennison, J.A. Kelber, and R.R. Rye, Phys. Rev. B25, 1384 (1982).

2.5 The Role of the Excited State in DIET Electronic Structure and Evolution in Time

W. Brenig

Physik-Department E20, Technische Universität München
D-8046 Garching, Fed. Rep. of Germany

1. Introduction

There are at least three conditions which an excited state of
an adsorbate has to fulfill in order to lead to desorption:

1. It has to have the appropriate asymptotic behavior at large
distance from the surface. For instance, if one is interested
in the desorption of (positive) ions the adsorbate has to
approach the corresponding (positive) ionic state at large separation. This does not mean, in general, that the charge has
to be localized at the adsorbate already at the equilibrium distance. In fact, in most cases the charge will be spread out
over the entire "surface molecule" participating in the chemical
bond involving the adsorbate and a few nearby substrate atoms.

2. The ionic potential $V_i(r)$ has to be sufficiently repulsive.
This does not only mean that the potential difference $V_i(R_0)$-
$V_i(\infty)$ at equilibrium and infinite separation has to be positive
so that the ion can come off with positive kinetic energy, but
also that the slope $V_i'(R_0)$ is sufficiently steep so that the ion
leaves the surface quickly enough before neutralization.

3. The decay width $\gamma(R)$ of the excited state for decays back into bound states has to be sufficiently small. This condition
can be quantified in terms of the decay width $2\gamma(R)$ of the excited (ionic) state in the MENZEL-GOMER-REDHEAD[1]model and the
corresponding probability $P(t)$ of the excited state still being
ionic at time t. In a simple classical description $P(t)$ is given
by

$$P(t) = \exp \left(-2 \int_0^t \gamma(t') \, dt'\right), \qquad (1.1)$$

where $\gamma(t) = \gamma(R(t))$ is the decay width at time t along the
classical path $R(t)$ of the desorbing ion. We shall see later
on that (1.1) is strongly modified by quantum effects. In the
classical description the probability for decay into bound
states is given by $1 - P(t_c) = p_b$ where t_c is the time the
desorbing ions have reached the critical distance R_c below which
the adsorbates after reneutralization do not have enough kinetic
energy to leave the surface (see Fig. 1).

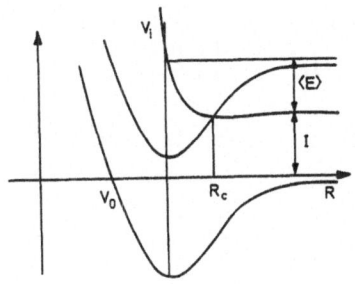

Fig. 1. If an ion reneutralizes at R_c, i.e., undergoes a transition to the potential curve of a neutral (parallel to the ground state curve) its kinetic energy is just enough to desorb in a neutral state with zero kinetic energy at infinity. Any transition at $R>R_c$ allows neutral desorption with positive kinetic energy, transitions from $R<R_c$ lead to recapture into bound states

2. The Role of Excited States with two Bonding Holes

Consider a light nonmetallic adsorbate such as H and O chemisorbed on a transition metal. More generally speaking, consider an adsorbate with valence levels below the center of the s-p and d band of a transition metal. Then the single particle bonding states of the "surface molecule" will have more weight at the adsorbate whereas the antibonding states will be more concentrated at the substrate (see Fig. 2).

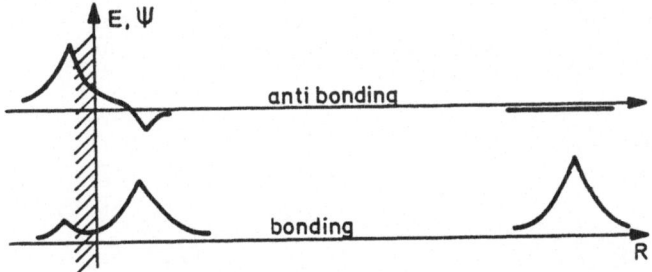

Fig. 2. Schematic view of the single particle bonding and antibonding state wave functions of the surface molecule. At infinite separation the bonding state is localized completely at the adsorbate

The leading contribution in a configuration interaction expansion of the ground-state wave function will have the two bonding orbitals occupied. They therefore will both have to be evacuated in order to have an excited state which has no charge at the adsorbate asymptotically. Two holes such as these cannot be produced directly in a single particle excitation by a photon absorption or an inelastic electron scattering. There have to be ground-state correlations (i.e., admixtures with at least one bonding hole in the C.I. expansion of the ground state). Or else in terms of a FEYNMAN diagram (see Fig. 3): The primary excitation leads to a one-hole configuration and then the electron-electron interaction in the excited many body system finally produces the two-hole state. The difference of this mecha-

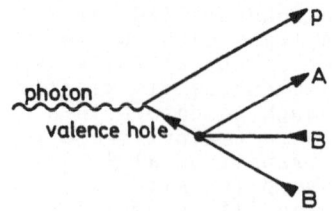

Fig. 3. Primary excitation (in this case via a photon) and secondary one via the e-e-interaction of the two-hole state

nism from the one proposed by KNOTEK and FEIBELMAN [2] for the excitation of the ionic state adsorbates on ionic crystals is the occurrence of a valence- instead of a core hole-state. In terms of creation operators of antibonding states A^+ and annihilation operators of bonding states B the ionic state $|i>$ leading to the desorption of a positive ion can be written as

$$|i> \quad = \text{const } A^+_\uparrow B_\uparrow B_\downarrow |0> \; . \tag{2.1}$$

This state with effectively one positive charge localized within the surface molecule is not an eigenstate of the Hamiltonian H. The coupling of the localized states A, B to the nearby continuum states of the s-p and d bands leads to a decay of $|i>$ and neutralization of the surface molecule. The decay width γ as well as the excitation energy V_i of (2.1) can be obtained from the pole of the GREEN's function

$$G(z) = \; <i| \; \frac{1}{H-z} \; |i> \tag{2.2}$$

located at

$$z_{+,-} \quad = E_0 + V_i \pm i\gamma \; . \tag{2.3}$$

Instead of (2.2) one may also consider the simpler one where $|i>$ is replaced by $a|0>$, where a is the annihilation operator of the valence state at the adsorbate. Because of the ground-state correlations and the nonvanishing overlap between a and A, the overlap between $|i>$ and $a|0>$ is nonzero, too. At the same time $a|0>$ also overlaps with the single bonding state $B^+_\uparrow B_\downarrow B_\uparrow |0>$ at the surface molecule. The corresponding Green's function therefore has two poles instead of one as in (2.3). As long as these are sufficiently well separated this does not lead to any problems. In fact, it has the advantage of allowing a direct comparison of the bonding and antibonding level.

$$G_a(z) \quad = \; <0|a^+ \; \frac{1}{H-z} \; a|0> \; . \tag{2.4}$$

Fig. 4 shows results [3] for the local density of states at the adsorbate

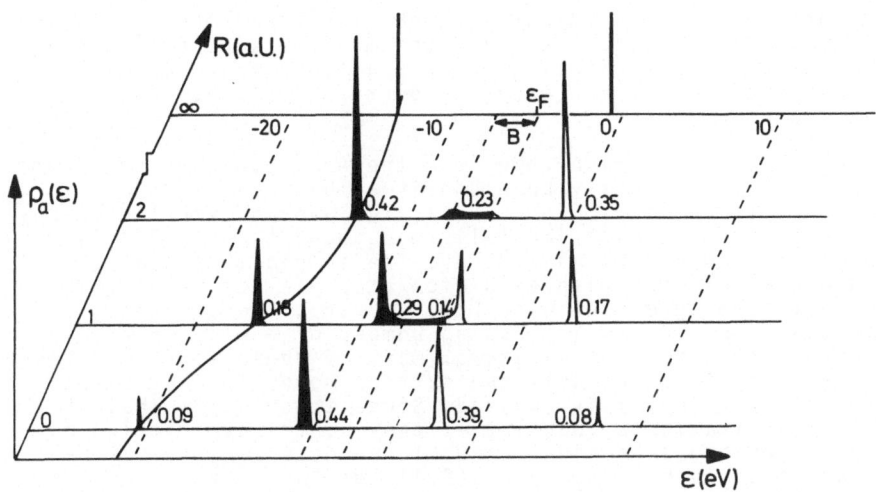

Fig. 4. The local density of states $\rho_a(\varepsilon,R)$ at the adsorbate as a function of the distance from the surface. The hole part given by (2.4,5) shaded in black

$$\rho_a(\varepsilon) = \text{Im } G_a(\varepsilon + i0)/\pi . \qquad (2.5)$$

Actually, this figure includes in addition to the hole contributions (2.5) the particle contributions, which are obtained by interchanging a and a^+ in (2.4). These results clearly exhibit what is to be expected from our qualitative discussion above: There are two peaks in the hole density of states (i.e., for ε below the Fermi level). One of these peaks corresponds to the single bonding hole state. It is located within the d-band hole states region where the adsorbate induced peaks occur which are seen in photo emission. The other peak occurs below the d band. It corresponds to the double bonding hole state. Clearly this state asymptotically approaches the single hole state at the adsorbate.

Figure 4 shows that at the same time that the double bonding hole state also fulfills the condition 2 mentioned in the introduction. While the position of the "photo emission peak" is only weakly dependent on the adsorbate distance from the surface, the "desorption peak" shows a strong dependence. Since the bonding-antibonding splitting is essentially given by twice the hopping integral $V_{am}(R)$ between the adsorbate and the nearest metal valence state, the R-dependence of the potential $V_i(R)$ is essentially the same as the one of $V_{am}(R)$.

At the equilibrium distance $R \sim 0.3$ A° the excitation energy of the desorption peak is several eV above the asymptotic limit leading to a kinetic energy of the desorbing ions of the right order of magnitude. If one adds to this kinetic energy the

binding energy of the adsorbate (∼3 eV for H/Ni) the ionization
energy of the free H (13,6 eV) and assumes the liberated electron in the primary excitation process at the vacuum level,
one predicts a threshold energy for photo-stimulated desorption
of a little over 20 eV.

These predictions which came from the Anderson model of chemisorption [3] have recently been confirmed by experiment and in
CI cluster calculations [4] for H/Ni. The average kinetic energy
of desorbing protons experimentally was found to be <E> ∼ 5 eV
and the threshold energy was ∼23 eV. The simplest cluster consisting of one Ni-atom with an H-atom atop yielded results in
semiquantitative agreement with Fig. 4. For bigger clusters
one finds, of course, an increasing number of energy levels
with increasing number of Ni atoms. Most of these levels correspond to states where not only the surface molecule, but also
the rest of the cluster is excited. Since these states have
only small excitation probabilities they contribute little to
the desorption cross section. It would be necessary to calculate quantities such as (2.2) or (2.4) for the bigger clusters
in order to single out those levels from the level diagrams
which are relevant for desorption and in order to compare the
results with, e.g., (2.5).

3. Spectroscopic parameters of the excited state

The imaginary part γ of the pole (2.3) would show up as the
width of the peaks in the local density of states. The calculations within the Anderson model in principle allow the determination of a decay width γ of the two levels. Since for the
results of Fig. 4 only the d-band states of the metal were taken into account only the photo emission level which has an
energy within the d band acquires a nonzero width. The levels
outside the d band in Fig. 4 are broadened arbitrarily replacing the infinitesimal i0 in (2.5) by a small nonzero imaginary part. If the coupling of the desorption level with the s-
p band is taken into account its width becomes nonzero, too.
In fact it can become quite large. We are going to come back
to this point later on.

The (electronic) width γ of the level should not be confused
with the width, say, ΔE of the energy distribution of the desorbing particles (see Fig. 5). This is only a consequence of
the spread in potential energy of the initial state Φ_0 due to
its spatial extension. A simple semiclassical estimate for the
width ΔE of a wave packet Φ_0 of spatial width ΔR in a potential with slope $V_i'(R_0)$ for instance would yield

$$\Delta E = V_i' (R_0) \quad \Delta R \ . \tag{3.1}$$

The potential curves of [3] and [4] have a slope of
$V_i'(R_0) \sim 7$ eV/Å . Thus for a width of $\Delta R \sim 0.2$ Å of the vibrational ground state Φ_0 (3.1) one would predict a width
$\Delta E \sim 1.5$ eV which is too small by a factor of 3 compared to the
experimental results of Fig. 5.

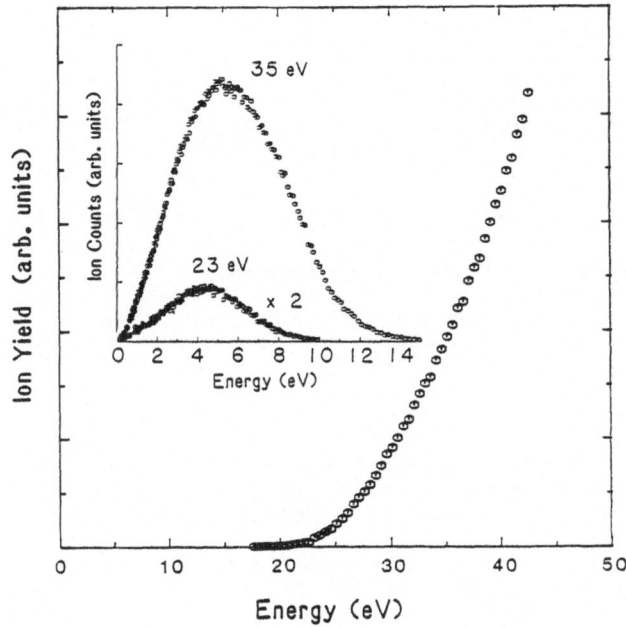

Fig. 5. Relative ESD yield of protons as a function of incident
electron energy for a Ni <111> surface exposed to hydrogen at
low temperature. The inset shows the kinetic energy distribution of H ions resulting from excitation with either 35 eV or
23 eV electrons (data corresponding to 23 eV incident energy
has been multiplied by two), (from [4])

Concluding this section, we may say that from the experimental point of view there are three data (average kinetic energy
<E>, width ΔE and the "reduction factor" $P(\infty)$ see (1.1)) from
which one can extract results for the quantities $V_i(R)$ and
$\gamma(R)$ according to the scheme

$$\langle E \rangle \;\; \to \;\; V_i(R_0)$$

$$\Delta E \;\; \to \;\; V_i'(R_0) \tag{3.2}$$

$$P(\infty) \;\; \to \;\; \gamma(R_0) \;.$$

Simple model calculations yield rather accurate values for
$V_i(R_0)$, the slope of the ionic potential in these calculations
has the right order of magnitude while the value of $\gamma(R_0)$ so
far could not be calculated very well. The variation of the
width with R is even harder to determine. On the other hand,
this variation is needed for the evaluation of (1.1) in order
to determine $\gamma(R_0)$ from the experiment. In the next section
we are going to describe the procedure indicated in (3.2) in
some more detail including the quantum effects in the evaluation of the desorbing ionic state.

4. Quantum Effects in the Ionic Motion

It was shown in [5] that the full pole energy (2.3) acts as a
generalized Born-Oppenheimer potential describing the motion
as well as the decay of the excited state according to the
Schrödinger eq.

$$(T + V_i + i\gamma) \, \Phi (t) = i \, d \, \Phi(t)/dt , \qquad (4.1)$$

where T is the kinetic energy of the ionic motion.

For the probability $P(t) = <\Phi(t)|\Phi(t)>$ an expression
such as (1.1) can be derived where now $\gamma(t)$ is given by

$$\gamma(t) = <\Phi(t)|\gamma| \, \Phi(t)>/P(t) . \qquad (4.2)$$

If the variation of the complex potential $V_i - i\gamma$ within the
spatial width ΔR of the wave packet can be neglected, a semi-
classical description can be used in which the center of gravi-
ty $R(t) = <\Phi(t)|R|\Phi(t)>/P(t)$ evolves according to Newton's
equation of motion. A simple look at the experimental data of
Fig. 5, however, teaches us that this is not the case. A measure
for the variation of the potential within the width ΔR is given
by the width ΔE of the energy distribution in Fig. 5 (~ 5 eV)
which is approximately equal to the value $V_i(R_0) = <E>$ of the
potential at the center of the wave packet Φ_0 at t=0.

Figure 6 exhibits the initial state Φ_0 in comparison to
several other structures and length scales.

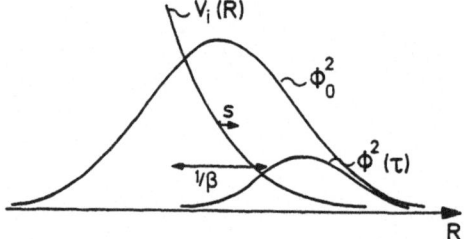

Fig. 6. The initial wave packet $\Phi_0(R)$ in comparison to the
potential $V_i(R)$, the wave packet $\Phi(\tau)$ after the life time τ,
the decay length $1/\beta$ of the potential and the distance
$s = V_i'(R_0)\tau^2/2m$ the ion travels in its lifetime according to
classical physics

The width of the initial wave packet can be determined from
the vibrational frequency (~ 0.15 eV) of the adsorbate. In the
analysis of the data according to the scheme (3.2) an exponential dependence of the potential on R $V_i = V_0 \exp \beta(R_0 - R)$ and
$\gamma = W_0 \exp \alpha(R_0 - R)$ on R was assumed.

Figure 6 shows drastically that the notion of a classical mass point moving along its potential curve does not hold. In fact the smallness of s compared to the width of the wave packet says that the particle in zeroth order does not move at all - it just decays.

To first order there is a motion but of a completely nonclassical nature which may be called 'asymmetric decay': Due to the strong variation of the imaginary part of the potential the tail of the wave packet is eaten up faster than the nose. This leads to a shift of the center of gravity of the particle without any momentum. Indeed one finds for the effective velocity of the center of gravity using (4.1)

$$dR(t) / dt = p(t)/m + 2 <(R-R(t))\gamma> , \qquad (4.3)$$

where $p(t) = <p> = <\Phi(t)|p|\Phi(t)>/P(t)$ is the expectation value of the momentum. Expanding $\gamma=\gamma(R)=\gamma(R(t))+\gamma'(R(t))(R-R(t))+...$ one finds

$$dR(t)/dt = p(t)/m + 2\gamma'(R(t))(\Delta R)^2 + ... \qquad (4.4)$$

Similarly one finds for the effective acceleration

$$dp(t)/dt = - V_i(R(t)) + \gamma'(R(t))<\Delta R\Delta p + \Delta p\Delta R>. \qquad (4.5)$$

Notice that the quantum corrections to the classical equations of motion due to the spatial variation of the real part of the potential (which we have neglected in (4.4) and (4.5)) would involve second derivatives of V_i while the corrections due to the imaginary part γ involve only the first derivative. The asymmetric decay leading to the second term on the rhs of (4.4) is a large effect which can completely dominate the behavior of the wave packet during the initial stages of desorption.

Figure 7 shows the effective velocity of the center of gravity according to (4.4) for some typical values of the relevant parameters. They were calculated [5] from an approximate solution of the Schrödinger equation (4.1).

If an exponential dependence of V_i and γ on R is assumed, one has four parameters V_0, W_0, α and β but only three data according to the scheme (3.2). In order to remove the remaining ambiguity we have assumed a relation $\alpha= 2\beta$. The argument behind this assumption is that on the one hand the potential V_i has essentially the same R-dependence as the hopping integral V_{am}. On the other hand, in a simple 'golden rule' calculation the width γ will be proportional to the square of this integral, hence $\gamma \propto V_i^2$.

It was pointed out already above in connection with (3.1) that theoretical values for β turn out to be too small: From the analysis of data [6] for H/W and O/W one finds surprisingly small values of $1/\beta$ between 0.05 and 0.2 A. The assumption

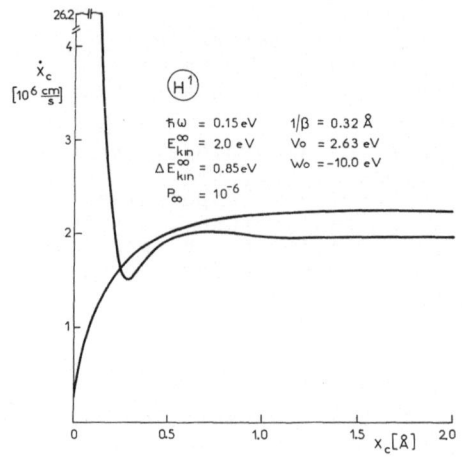

Fig. 7. The effective velocity of a desorbing ion as a function of distance from the surface according to (4.4) as compared to the classical result (taken from [6])

$\alpha = 2\beta$ then forces one to the introduction of decay lengths $1/\alpha$ for the width γ which seems unreasonably small. We have therefore also tried the somewhat "weaker" assumption $\alpha = \beta$.

Unfortunately not only the quantum effects but also their dependencies on the spatial variation of $\gamma(R)$ are large. As can be seen from the table below, the most important quantum effect is a drastic increase in the magnitude of W_0 needed to fit the same data on ion energy distributions and reduction factors $P(\infty)$: The inclusion of quantum effects leads to values which are between 10 and 20 times bigger than the corresponding classical values.

Table 1. Spectroscopic parameters of excited ionic states deduced from experimental data with and without quantum effects

	V_0 eV	W_0 Ev	$1/\beta$ Å	V_0^{class} eV	W_0^{class} eV	$1/\beta_{class}$ Å
H/W ($\alpha=2\beta$)	1.0	35.0	0.19	2.0	3.4	0.17
H/W ($\alpha=\beta$)	7.3	29.0	0.09	2.0	2.2	0.17
O/W ($\alpha=2\beta$)	37.0	60.0	0.05	8.8	3.0	0.14

For values of W_0 as large as shown in the table one might wonder about the validity of one of the basic assumptions in the description of DIET processes, namely the existence of a well-defined resonance state with a reasonable sharp maximum in the excitation probability at $V_i(R)$. Perhaps also the change of the peak energy and width of the relative ESD yield seen, e.g., in Fig. 5 is an indication of a breakdown of the concept of a well-defined resonance.

From a theoretical point of view quite large widths can be
expected if the coupling of the excited state to the s-p holes
is taken into account [7] (see Fig. 8). It will be difficult,
however, to determine the spatial variation of the width from
first principles. Unfortunately the results not only for W_O
but also for V_O depend quite sensitively on β.

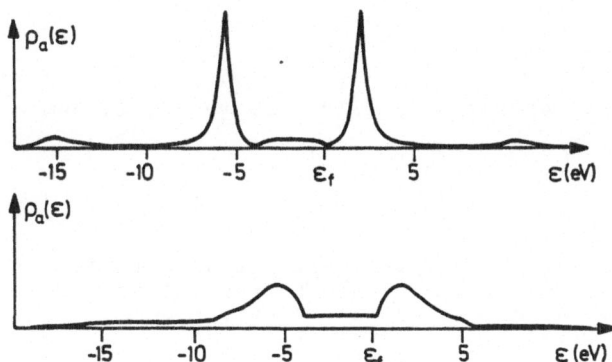

Fig. 8. Local density of states as in Fig. 4 but with the coupling to s-p states taken into account. The lower strong coupling
situation is supposed to hold for the equilibrium distance of
H/Ni. Taken from [7]

At the present stage one therefore has to admit that due to
unusually large quantum effects and uncertainties in the spatial variation of the reneutralization width $\gamma(R)$, it appears
to be impossible to map out the spectroscopic parameters of
the excited state from the experimental data in a precise and
unique way.

5. Desorption of Neutrals

The results of the foregoing sections can be obtained in an
approximation where the wave function factorizes into a product
of an ionic part Φ_{ik} and an electronic part ϕ_i, containing the
ionic coordinates only as parameters [8]

$$\Psi_{ik} = \Phi_{ik}(R) \phi_i(r;R) . \qquad (5.1)$$

Here ϕ_i is precisely the Schrödinger representation of (2.1),
and Φ_{ik} is a stationary solution of (4.1) with energy $E=k^2/2m$.

Eq.(5.1) may be called Born-Oppenheimer approximation although
the imaginary part of the potential in (4.1) amounts to taking
into account non-adiabatic corrections [9] .

If an approximation analogous to (5.1) is used for the desorption of neutrals via the reneutralization of the ionic

states one finds a probability zero for such processes. The only allowed decay mode in this approximation is the decay back into the ground state Φ_0 [8] .

If, however, corrections to the Born-Oppenheimer approximation are taken into account the results for ionic desorption remain essentially unchanged while for neutral desorption probability p_c one finds an improved formula [9]

$$p_c = \int_0^\infty <\Phi_0 (t|\gamma P_c|\Phi_0(t)>dt ,$$ (5.2)

where P_c is the projection operator onto the neutral continuum states

$$P_c = \sum_k |\Phi_{nk}><\Phi_{nk}| .$$ (5.3)

Similarly the decay probability p_b into neutral bound states is obtained by replacing P_c in (5.2) by the corresponding projection operator P_b onto the bound neutral states. Since $P_b+P_c=1$ one finds using (1.1)

$$p_b + p_c = \int_0^\infty \gamma(t)P(t)dt = 1 - P(\infty) .$$ (5.4)

So the total probability for ionic and neutral (bound plus continuum) decay is unity as it must be. A separate evaluation of p_b and p_c has not been done yet. Two limiting situations, however, can easily be discussed: One is the classical limit in which P_c just has the effect of restricting the energy of the neutrals to positive values, hence (see Fig. 1) $R-R_c$ to positive values and $t-t_c$ to positive values also. Thus one finds the well-known M-G-R results [1]

$$p_b = 1 - P(t_c) , \quad p_c = P(t_c) - P(\infty) .$$ (5.5)

This result can hopefully be applied to situations where one finds identical threshold energies for ions and neutrals [10]. The other limiting case is obtained when the decay is very fast, so that the motion of the initial wave packet can be neglected in zeroth order. Then the leading contributions to (5.2) come from $\Phi_0(0)$ which is orthogonal to the scattering states in P_c. In this limit one thus recovers the null result of the Born-Oppenheimer approximation [8]. If this result holds one expects neutrals to be produced,e.g.,via direct excitation of anti-bonding neutral states with threshold energies different from the ionic states [10,11] .

References

1 D. Menzel, R. Gomer: J.Chem.Phys. 41, 3311 (1964)
  P.A. Redhead : Canad.J.Phys. 42, 886 (1964)
2 M.L. Knotek, P.J. Feibelman, Phys.Rev.Lett. 40, 964 (1978)
3 W. Brenig: Proc. 7th Intern.Vac.Congr. & 3rd Intern.Conf.
  Solid Surfaces (Vienna 1977), Vol. I p. 719

4 C.F. Melius, R.H. Stulen, J.O. Noell: Phys.Rev.Lett.$\underline{48}$,1429 (1982)
5 P.Schuck, W. Brenig: Z.Phys. $\underline{B-46}$, 137 (1982)
6 M. Nishijima, F.M. Probst: Phys.Rev. $\underline{B2}$, 2368 (1970)
7 K. Schönhammer: Sol.State Comm. $\underline{22}$, 51 (1977)
8 W. Brenig: Z.Phys. $\underline{B23}$, 361 (1976)
9 W. Brenig: J.Phys.Soc.Japan $\underline{51}$, 1915 (1982)
10 P. Feulner, R. Treichler, D. Menzel: Phys.Rev. $\underline{B24}$, 7427 (1981)
11 D. Menzel: Ber.Bunsenges.Phys.Chem. $\underline{72}$,591 (1968)

Part 3

Desorption Spectroscopy

3.1 On the Nature of the ESD Active Species on Metal Surfaces

E. Bauer

Physikalisches Institut, Technische Universität Clausthal
D-3392 Clausthal-Zellerfeld und
Sonderforschungsbereich 126 Göttingen - Clausthal, Fed. Rep. of Germany

1.   Introduction

This paper discusses the question whether the ESD active species on metal surfaces are physisorbed molecules, chemisorbed atoms or compounds. This problem is intimately connected with that of the ESD mechanism and that of the relation between ESD and surface structure. Much can be learned about it in particular by studying this relation and this will be emphasized here. The flux of desorbed particles is usually a complex function of many variables:

$$I_i^\nu = f(E_o, I_o, t, A, E, \vartheta, \varphi, \Theta_i, \Theta_k, T, T_a, \ldots).$$

The first four variables characterize the incident electron beam (energy, current, irradiation time and area), the next three and the indices ν, i the desorbed species (energy, polar and azimuthal angle of emission, charge state: $\nu = +, o, -$ and i the chemical nature). The final parameters describe the surface: Θ_i coverage of the species of interest, Θ_k coverage of intentional and frequently unintentional (surface cleaning, residual gas pressure!) coadsorbates, T observation temperature and T_a adsorption or annealing temperature if higher than T.

The yield $Y_i^\nu = I_i^\nu / I_o$ may be large, e.g. from ionic compounds such as WO_3 or "polar" molecules such as α-CO, so that the desorbed species can be detected easily with a channel plate-fluorescent screen combination or even with a fluorescent screen alone. But frequently Y_i^ν is small, e.g. from many chemisorbed atoms, and single pulse counting is necessary as well as very low background pressures ($<5 \times 10^{-11}$ Torr) to reduce ESD from background gas adsorption. Ideally Y_i^ν (E, ϑ, φ) should be measured with sufficient energy and angular resolution but this has been up to now more the exception [1] than the rule and has been limited to high yield charged ESD species, i.e. positive ions.

The angular dependence of Y_i^ν will be discussed in MADEY'S paper [2]. The present paper will, therefore, be limited to the energy dependence at fixed angles or angular ranges, and be concerned with two main topics: i) the present frontiers of the field which in the authors opinion are neutrals, negative ions and co-adsorption; ii) the classical but still controversial systems oxygen on W(100), (110) and Mo(100) in which the relation between ESD and structure can be studied particularily well.

2.   The Frontiers

2.1.   Neutrals

Due to the difficulty of detecting neutrals all of the past
studies have been indirectly, measuring the depletion of the
adsorbate. Only recently work has started analyzing the desorbed
particles [3]. The results published up to now - for N_2 and CO
on Ru(001) [3] - show a) that the threshold for neutral desorption is lower than for positive ion desorption and b) that only
valence electrons are involved in the excitation process but no
core electrons. The desorption process proceeds via direct excitation to neutral repulsive states (N_2,CO) and in addition via
valence ionized states, e.g. CO^+, followed by reneutralization.
Not much more is known at present from experiment about ESD
active species and process.

2.2.   Negative Ions

ESD of negative ions has been observed during the last few years
for a number of adsorbate/substrate systems but only CO on W [4]
and oxygen on polycrystalline W [5], Mo [5,6] and on Mo(100) [7]
will be discussed here. The following nomenclature will be used
for O on W and Mo: β_2 for the low yield state generally ascribed
to the (unreconstructed or reconstructed) chemisorption layer,
β_1 for the "slow" high yield state attributed here to two-dimensional oxide and α for the "fast" high yield state which is
suggested here to be three-dimensional oxide. For saturation at
300K the yield ratios $O^+:O^-$ are about 120 on Mo(100) [7], 20 on
polycrystalline W [5] and 50 for CO on polycrystalline W [4].

The dependence of ion yield an exposure [5,7] as shown for
Mo(100) in Fig.1 suggests that in the case of oxygen the β_2
state is O^- ESD inactive. The presence of O^- at 1000K, a temperature at which α state formation is considerably reduced if
not eliminated, shows that the β_1 state is active as is the α
state formed in addition at 300K. The ESD cross sections are
approximately the same for O^+ and O^-, 2×10^{-18} cm^2 [5]. The
temperature dependence of the high yield O^+ state (β_1,α) and
of O^- emission differ but little approaching zero yield at about
700K [5]. These three features, exposure dependence, cross-
section and temperature dependence strongly indicate that both
ions, O^+ and O^-, originate from the same species. The energy
distribution of the O^- ions is, however, more asymmetric than
that of O^+ and peaks at a higher energy on Mo(100), 7.5eV vs.
5.8eV [7]. Furthermore, the threshold for O^- ESD is much lower
than that for O^+ ESD, $\approx$4eV vs. $\approx$20eV on W, and no evidence is
seen for core ionization as a source of O^- [5]. These two observations suggest a different mechanism for O^+ and O^- production.

The origin and mechanism favored by the authors [6] is mole-
cular oxygen and dissociative attachment according to $e^- + O_2 \rightarrow (O_2^-)$
$\rightarrow O + O^-$ with a threshold at 3.64eV. In addition to this process
a second one, polar dissociation according to $e^- + O_2 \rightarrow O^+ + O^- + e$
with the threshold at 17.18eV, is postulated to explain a broad
peak in the yield vs. energy curve. The similarity between O^-
and high yield O^+ emission and the assumption that the latter

105

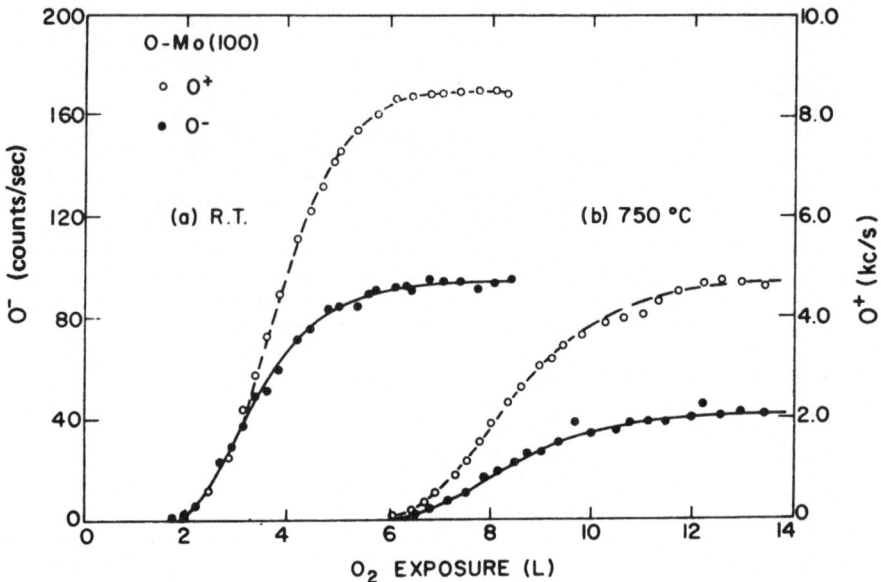

Fig.1. O^- and O^+ ESD from Mo(100) as a function of O_2 exposure at (a) 300K and (b) 1000K [7]

originates from 2d and 3d oxide suggests, however, another
mechanism: electronic interband transitions in WO_3 which are
connected with charge redistribution resulting in ejection of
O^-. In a crystal like WO_3 the local density of states of the
valence electrons is centered on the O ions, that of the conduction electrons on the W ions. A valence band - conduction
band transition in a purely ionic picture, therefore, weakens
the ionic O-W bond ($O^{2-}+W^{6+} \rightarrow O^- + W^{5+}$) so that the O ion at the
apex of the WO_6 octahedron may see a sufficiently repulsive
overall potential to cause its desorption. The threshold for
this process is expected to be at 2.6eV but valence band XPS
spectra [8] indicate an initially slow increase of the density
of states with increasing binding energy which is compatible
with the yield curve [6]. The additional features of the yield
curve can be qualitatively associated with other interband transitions seen in the electron energy loss spectrum of WO_3 [9]
whose analysis is still hampered by the limited knowledge of the
band structure of WO_3 [10]. Future studies with better signal to
noise ratio and on better defined surfaces including WO_3 surfaces will have to decide which interpretation is correct.

The observation of O^- ESD from CO on W [4] shows that negative ion ESD is certainly also possible from adsorbed molecules.
80% of the O^- ions were found to originate from the CO^+ yielding
α_1 state, the rest from the O^+ yielding α_2 state. The O^- ESD
cross-section lies between that for O^+ and CO^+ ESD as does the
temperature dependence. This indicates that the same species is
involved in all cases. However, the energy distribution of the

O$^-$ ions is more asymmetric than that of the O$^+$ ions and peaks
at a much higher energy, $\approx$ 16eV vs. $\approx$ 8eV, which again requires
a different but still unknown mechanism.

Summing up, negative ion ESD appears very interesting from
the point of view of the ESD mechanism but does not seem to give
new information on surface species at present.

2.3. Coadsorption

Although the large influence which coadsorption may have on ESD
has been recognized long ago (see the references in [11]) very
little systematic work has been done on this topic. Also the
examples discussed here are mostly more or less coincidental.
The best known case is probably the influence of pre- and coadsorbed oxygen, in particular in the form of H$_2$O, on H$^+$ ESD.
This influence is illustrated in Fig.2 which shows the variation
of the H$^+$ signal from a polycrystalline Nb surface with exposure
time to H$_2$ at various pressures [12]. Similar curves had been
reported before for H on Ni(100) and attributed to adsorption
of pure H$_2$ [13]. In the case of Fig.2 it could be shown convin-
cingly, however, that the large H$^+$ signal came from H in the
environment of O supplied to the surface by coadsorption of
H$_2$O which was produced in the ion pump. Adsorption of pure H$_2$
produced only a negligible H$^+$ signal, preadsorption of oxygen
enhanced the H$^+$ yield significantly but much less than H$_2$O coadsorption. A much higher H$^+$ ion yield from adsorbed H$_2$O than from
adsorbed hydrogen has also been reported for polycrystalline W
[14], as well as a significant enhancement of the H$^+$ yield by
preadsorption of oxygen [11]. In the case of H$_2$O coadsorption
on Nb the optimum O coverage for maximum H$^+$ yield is 1/2-2/3
monolayers [12], while on W the H$^+$ yield increases monotonically
with O$_2$ exposure [11].

For the understanding of the origin and mechanism of the H$^+$
yield enhancement the following observations are relevant.
i) There is strong off-normal H$^+$ emission from Nb indicating
that this species does not come from OH groups standing normal
to the surface; the observations rather suggest that the O atoms
are incorporated in the topmost layer and that the H atoms are
bonded both to O and Nb. ii) The H$^+$ ion energy distribution from
the H$_2$O-induced, high yield state on Nb peaks at a much higher
energy than that from the low yield H$^+$ state on W(100) after
exposure to clean H$_2$, 5.3eV vs. 1.6eV. This suggests a stronger
repulsion in the excited state. iii) The yield of the high yield
H$^+$ state on Ni(100) increases strongly at 23eV after an initially
slow increase from the threshold at 15eV. If the presence of
oxygen is invoked, the strong increase can be understood easily
in terms of primary ionization of a neighbouring O 2s level followed by H ionization, e.g. via an interatomic Auger process. If
such a two-electron ejection process occurs the H$^+$ ion sees a
rather repulsive potential which could account for the high energy
of the ion and this in turn for the high survival rate against
neutralization. More systematic studies are necessary to verify
this explanation.

Another interesting but still poorly understood coadsorbate
system is O + Cs. ESD in this system was studied on a W field

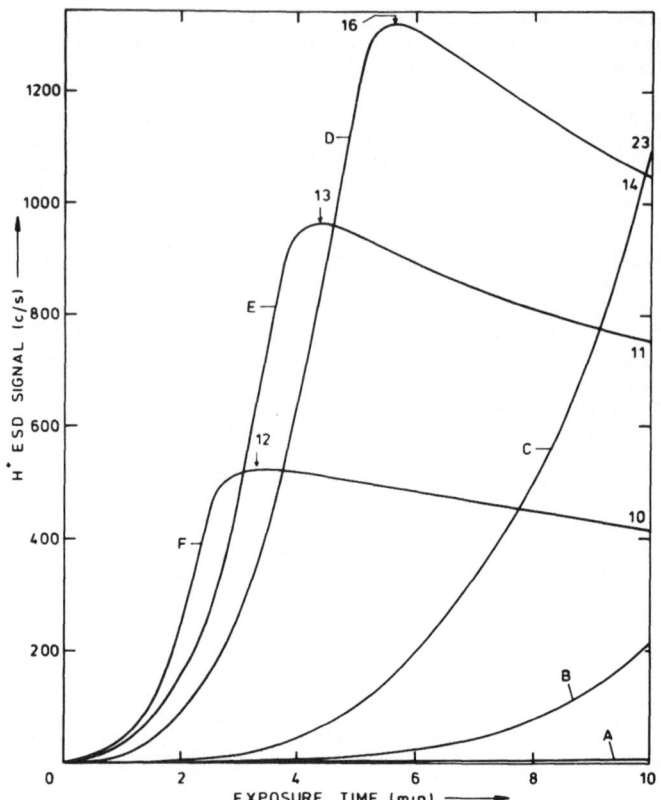

Fig. 2. H$^+$ ESD from polycrystalline, (110)-textured Nb as a function of exposure time to H$_2$ at pressures of (A) 0.5, (B) 0.75, (C) 1.0, (D) 1.25, (E) 1.5 and (F) 2.0x10^{-6} Torr. 700eV electrons; the beam current changes with pressure and time due to poisoning of the emitter by the H$_2$O impurity and is indicated at various times in nA [12]

emitter by monitoring the average work function [15] and on Mo(100) by measuring the O$^+$ and O$^-$ current [16]. In the first case it is concluded that Cs, whose ESD cross-section on the clean metal surface is immeasurably small like that of other metallic adsorbates, is desorbed by 250eV electrons with a cross-section as high as 8x10^{-20} cm^2 if oxygen is preadsorbed. The mechanism responsible for this phenomenon may be the same as the one suggested for H+O coadsorption, ejection of Cs$^+$ ions with high kinetic energy resulting from the repulsion in a two-hole environment created by an Auger process in O. Support for this hypothesis is that Ba$^+$ ESD ions from Ba+F and Ba+CO coad-sorbates have high kinetic energies (4eV and 10eV) [17] which probably is sufficient to avoid neutralization inspite of the high mass (low velocity!). It should be pointed out that the ion created initially may very well be doubly charged if it was singly charged (Cs^+, Ba^+) before the ionization process because the 5p electrons of Cs and Ba are bound sufficiently weakly to be able to fill the O 2s hole and still allow ejection of an O 2p electron. Thus one electron could be captured during ion emission and still leave a singly charged ion. The high energy of the ions allows to exclude another possible mechanism which was suggested to explain the high cross-section for neutral Xe ESD from W(110) precovered with oxygen: desorption of oxygen which in turn kicks off the weakly bound Xe on top of it [18]. The desorbed oxygen would have to have much higher energies than observed to transfer the energies observed for Ba^+.

Of course, not only Cs is desorbed but also oxygen as seen in the O^+ and O^- signals on Mo(100) [16]. Cs decreases O^+ emission at all Cs coverages and O^- emission at all but at very low coverages where a slight increase with Cs coverage can be seen. This can be accounted for by two effects: i) the increasing neutralization probability of O^+ and decreasing ionization probability of O^- with decreasing workfunction (increasing coverage) and ii) by the shielding of the O atoms by Cs adsorbed on top of them.

The observation that coadsorbed oxygen can cause ESD of electropositive adsorbates whose ESD cross-section is immeasurable on the clean surface suggests the question whether or not this phenomenon does also occur with electronegative adsorbates. For the only case studied up to now, Cl+O on W(100), the answer is no: the total Cl ESD cross-section for 2keV electrons is rather reduced from about 5×10^{-22} cm^2 to about 1.2×10^{-22} cm^2 by oxygen coadsorption [19]. On the other hand, Cl can be completely desorbed in the presence of K as demonstrated by thermal desorption and workfunction measurements of KCl adsorbed on W(110) and (111) after exposures to 100eV electrons [20]. While in pure Cl adsorption no Cl ESD could be detected with the experimental set-up used, high Cl cross-sections (8×10^{-17} cm^{-2} on (110) and 8×10^{-18} cm^{-2} on (111)) were obtained for the K+Cl adsorbate. No K desorption could be seen. It is tempting to invoke an interatomic Auger process again to explain this observation, but now with the roles of electropositive and electronegative partner exchanged: a hole is created in the 3p shell of K^+ which can be filled only by substrate or Cl^- 3p electrons. If the energy freed in this transition is used to eject another Cl electron the resulting Cl^+ may see a sufficient repulsive potential to be desorbed. If not, neutral Cl is desorbed.

It should be noted that not only primary hole creation in the coadsorbate may cause enhanced desorption but also primary hole creation in the substrate if the coadsorbate can transfer the substrate atom excitation to its partner. This may not only occur when the maximal valency condition is fulfilled, i.e. when the substrate atom is stripped of all its valence electrons by the coadsorbate such as W in a WO_3 configuration. An example is F on W(100) which has even at very low coverages a very high F^+ yield and is an excellent ESD-inducing coadsorbate for electropositive atoms as indicated earlier. Although an interatomic Auger process has been negated earlier for F on W(100) [21], a more detailed study under extremely clean conditions has given clear evidence for W core level thresholds in the F^+ ESD yield

[22]. The discrepancy has probably to be attributed to coadsorption in the earlier studies. This can wipe out easily the structure in the yield curve which is much weaker than in the case of oxygen on W at high coverages.

The examples presented here should be sufficient to show that coadsorption studies are extremely well suited to obtain a deeper understanding of the ESD mechanism if done systematically with the tools presently available. Once the mutual influence of coadsorbed atoms on ESD is understood ESD could become a powerful tool for the determination of local configurations. However, there are still many open questions in adsorption systems involving only a single species as shown in the following section.

3.   ESD of Positive Ions in "Simple" Systems

3.1.   O/W(110)

This system has been studied widely over a wide coverage range up to thick oxide layers and over a wide temperature range from 20K [23-25] to 1860K [26]. Here only the results between 300K and 1500K which throw some light on the O$^+$ ESD-active species and on the mechanism will be discussed. Two models of the species exist presently: small W oxide clusters or crystallites [27,28] and chemisorbed O atoms in singly coordinated sites at steps [29] or on top of atoms on the flat surface [30]. The evidence supporting the W oxide model is overwhelming: i) The

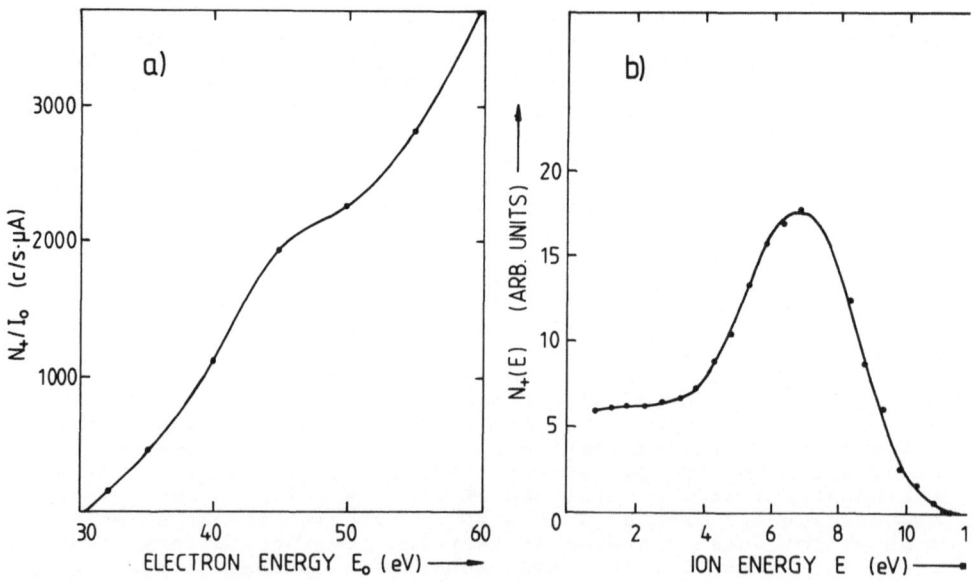

Fig.3.   O$^+$ ESD from polycrystalline WO$_3$ [31]. a) O$^+$ yield as a function of the energy E$_O$ of the primary electrons, b) O$^+$ energy distribution, E$_O$=70eV, T=250°C

O^+ signal vs. coverage is the same whether oxygen is supplied in form of O_2 or W oxide [27]. ii) Upon heating the O^+ signal disappears in the temperature region ($\approx 1000K$) in which WO_3 begins to dissociate or desorb [27] iii) The first strong threshold for O^+ emission from chemisorbed oxygen at 32eV [30] agrees well with that from WO_3 [31,32]. iv) The O^+ ESD cross-sections of chemisorbed oxygen and WO_3 are approximately the same, $\approx 1 \times 10^{-18}$ cm^2 [27,31]. v) The O^+ energy distributions are very similar in the two cases, except for the low energy part [28,31]. vi) The ESDIAD patterns are very similar in the two cases [29,33,34]. vii) There is no measurable O^+ ESD from the complex superstructure obtained at high oxygen coverage after annealing although they should contain a significant number of atoms in on top sites [35]. viii) The O coverage at which O^+ ESD sets in is proportional to the step density [29] while the inverse relationship would be expected in the singly coordinated site model [30]. For a W oxide cluster a minimum number of O atoms are necessary which have to be supplied from the terrace; the narrower the terrace the later the clusters are formed.

Since there are no data which contradict the W oxide cluster/crystallite model in the temperature range considered, this model is adopted here in view of the strong evidence. The O^+ ESD-active state on W(110) is therefore the α state, i.e. small compound particles which form preferentially at steps; these determine the azimuthal orientation of the particles and consequently, together with the shape of the particle, the ESDIAD pattern. Chemisorbed oxygen, i.e. the β_2 state, is not O^+ ESD-active on W(110) in contrast to W(100), but why? The primary excitation mechanism is undoubtedly Auger double ionization either intraatomically in the O 2s shell as suggested long ago [36,31], causing the weak threshold at about 22eV [27,32] or interatomically (KF mechanism) [32] causing the strong threshold at 32eV and additional thresholds at higher energies. These thresholds are also seen on W(100) at high coverage [37] in the fast high yield state (α state) which is also a minority state and attributed to oxide crystallites.

3.2. O/W(100)

At lower coverages this system does not show minority state ESD but majority state ESD so that the ESD results can be related to LEED, $\Delta\phi$, TDS and other data. In this way it becomes possible to determine the species which produce the various ESD states. The α state has already been attributed to 3d W oxide particles, the β_1 state can be associated with the 2d oxide believed to consist of a monolayer of WO_6 octahedra [28,38]. Most of the adsorbate structure - ESD correlation is based on the O^+ angular distribution (ESDIAD) [2,28,39] but there is of course also a correlation between O^+ energy distribution (ESDIED) and structure [28,39]. Here only two examples will be discussed: i) the 5.3eV state which is connected with the irreversible order-order transition between the unreconstructed and the reconstructed adsorbate in the coverage range above 1/2 monolayer [40,41] and ii) the 6.3eV high temperature state which is connected with the reversible order-disorder transition of the reconstructed p(2x1) structure [41-43].

When an adsorption layer with a coverage $\theta>1/2$ monolayer is annealed above $T_a \approx 500K$ a O^+ peak appears at 5.3eV in the ESDIED spectrum in normal emission and dominates the spectrum at $\theta \approx 0.6$. At $\theta>0.7$ it disappears again upon heating above approximately 900K but for $0.5<\theta<0.7$ is remains until T_a is high enough to cause thermal desorption of atoms in excess of 1/2 monolayer (Fig.4a). The threshold temperature of this peak decreases strongly with the amount of oxygen in excess of $\theta=1/2$ (Fig.4b) while its intensity increases up to $\theta \approx 2/3$. It can also be seen after annealing high coverage adsorbates at $T_a>1500K$ so that θ passes through the critical range with increasing T_a causing a well pronounced maximum in the 5.3eV O^+ yield as a function of T_a. The comparison with detailed LEED and $\Delta\phi$ measurements shows that this state is due to those oxygen atoms which cannot be incorporated into the reconstructed domains. The number of these atoms depends upon $\theta-1/2$ as well as upon annealing temperature T_a and time t_a. The smallest (Fig.4c), second smallest (Fig.4d) and infinitely large domain can accomodate 2/3, 4/7 and 1/2 monolayer of oxygen, respectively. The strong O^+ emission from the excess oxygen normal to the surface suggests that these atoms are adsorbed in on top sites.

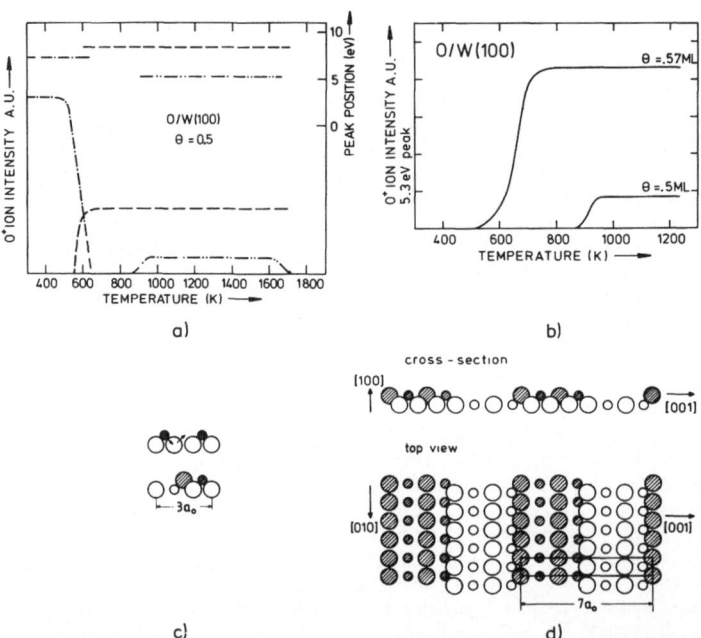

Fig.4. O^+ ESD from W(100) [41]. a) O^+ intensity (left-hand scale) and energy (right-hand scale) upon annealing a surface covered with $\theta \gtrsim 1/2$ monolayer of O at successively higher temperatures T_a. b) 5.3eV O^+ ESD peak hight for two coverages above $\theta=1/2$ as a function of T_a. c) Smallest reconstructed domain; the initial elementary step of the reconstruction is indicated on top. d) Second smallest domain

The reconstructed p(2x1) structure at $\theta \approx 1/2$ shows a reversible order-disorder transition starting at about 700K. This transition is associated with a significant work function increase and with the appearance of a 6.3eV peak in the O^+ ESD spectrum whose intensity rises rapidly with T. The size and location of the 8.3eV and 5.3eV peaks observed below the phase transition is unaffected by heating at least up to 1100K [41]. Only at much higher temperatures, e.g. at 1670K, is the 8.3eV peak completely replaced by the 5.3eV peak [43]. No continuous shift of the peak from 8.3eV to 5.3eV nor a broadening with increasing T is seen [43]. This excludes an interpretation in terms of increasing population of higher vibrational levels of the 8.3eV state and shows that the 5.3eV state is a new state presumably due to O atoms in saddle positions over which they have to pass during the disordering process. The activation energy for producing this state is 0.34eV [41,43]. A similar high temperature O^+ ESD state with an activation energy of 0.6eV and a peak energy shift of about 1eV to lower values between 700°C and 1300°C has been reported for O/Mo(100) [7]. This state is probably associated with the order-disorder transition of the c(4x4) structure [7,44].

3.3. O/Mo (100)

Although oxygen on Mo was the subject of the first very detailed study of ESD [45] only recently this system has been re-investigated combining ESD with LEED, $\Delta\phi$ and TDS measurements [7,44,46]. It turned out to be the most complex monatomic ESD active adsorbate/substrate system studied up to now. If only the high yield states ($\beta_1 + \alpha$) are considered then the situation is quite similar to oxygen on W(100). These states begin to form at 300K after an exposure of about 2L (Fig.1) [7,44] which produces approximately one monolayer [44]. The ion energies of these states ($\approx$ 6eV) [7,44,46] are similar to those of Mo oxide as are the ESD cross-sections ($\approx 1 \times 10^{-18}$ cm^2) [31,45]. Therefore it is very probable that these states are 2d and 3d oxides as in the case of W(100).

In the submonolayer range Mo(100) and W(100) are quite different, however. There is no measurable O^+ ESD below $\theta = 1/3$ and $\theta = 1/2$ at 300K upon adsorption at 300K and 1100K, respectively, and the ions observed after the onset of ESD have unusually low energies (1.5eV and 2.6eV) [44]. As in the case of W(100) the yields of these β_2 states are much smaller than that of the β_1 state. Emission is in normal and off-normal directions but the off-normal emission has been deduced indirectly and still needs confirmation by direct O^+ measurements [44]. The growth and disappearance of the 1.5eV and 2.6eV state is closely related to changes of the adsorbate structure with coverage, temperature and time. An example is shown in Fig.5. The O^+ ESD energy distributions in a coverage range in which structure and work function are changing significantly are measured sequentially and show a strong yield increase with time. The 1.5eV peak grows without electron irradiation while nearly no growth of the 2.6eV peak was detectable without irradiation. Although post-adsorption of a very high yield impurity, e.g. H in the presence of O or F (see Sect. 2.3.), can not be excluded completely, all evidence speaks against it. But even if such a coadsorbate would be responsible for the ESD signal in Fig.5 its change with O

113

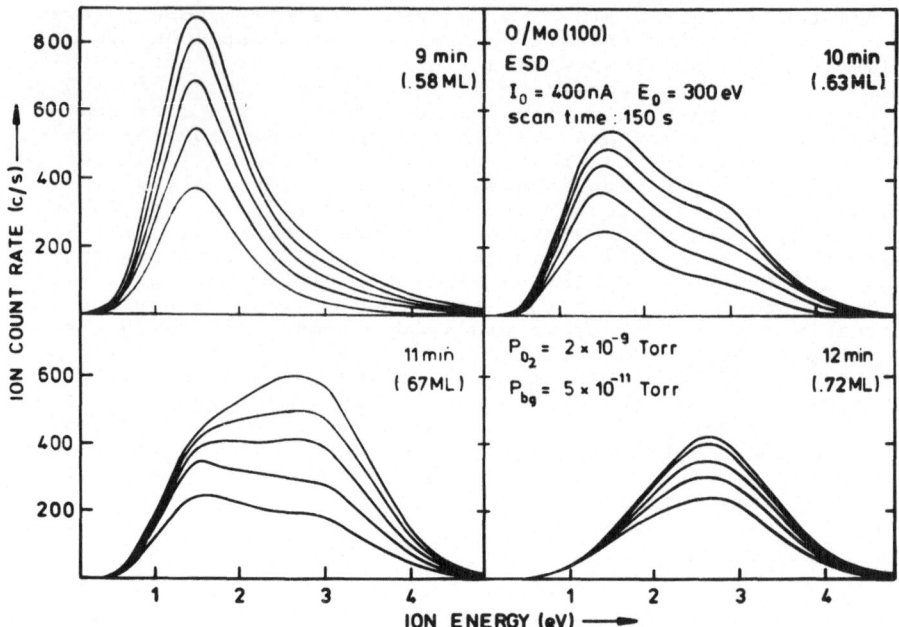

Fig.5. ESD from oxygen on Mo(100) between $\Theta=.58$ and $\Theta=.72$. The ion energy is scanned continuously, the lowest curves being the first ones. Background pressure 5×10^{-11} Torr. At this pressure the 1.5eV peak increases on the clean surface to 30 c/s and 60 c/s within 300 and 1200 sec, respectively [44]

coverage would be a sensitive indicator of the structural changes. Obviously more detailed studies of this complex system are still needed.

4.   Summary

In the frontiers of ESD the field of neutrals is just opening up and its yield is still uncertain. Although negative ion ESD at present does not seem to give additional information on the ESD active species, this phenomenon is interesting from the point of view of the ESD mechanism. The tremendous yield enhancements caused by coadsorption make this field exciting from the point of view of the ESD mechanism and of the local configuration of the adsorbate.

The best-studied systems, oxygen on refractory metals, turn out to be not as well understood as one would expect. For example, the nature of the ESD active species is still controversial - oxide or singly coordinated chemisorbed atoms - and the correlation between ESD ion angular and energy distribution on the one hand and adsorption site and bonding on the other is still highly speculative, in particular for the low yield states. Furthermore, some basic questions are still unanswered. For example, it is understandable that oxygen incorporated into the topmost substrate layer is ESD-inactive (Mo(100)) but why is it ESD-active on W(100) not only on the unreconstructed surface but also in the reconstructed state; and why is O adsorbed on top of the substrate (W(110)) ESD-inactive? Nevertheless, there is already ample reliable evidence that ESD is very sensitive to minority species temporarily or permanently formed during structural changes and, thus, is a very useful tool for studying such changes.

References

1   H. Niehus and B. Krahl-Urban, Rev. Sci. Instrum. <u>52</u>, 56 (1981)
2   T.E. Madey, this volume
3   P. Feulner, R. Treichler and D. Menzel, Phys. Rev. B<u>24</u>, 7427 (1981)
4   J.L. Hock, J.H. Craig, jr. and D. Lichtman, Surface Sci. <u>87</u>, 31 (1979)
5   J.L. Hock, J.J. Craig, jr. and D. Lichtman, Surface Sci. <u>85</u>, 101 (1979)
6   Liu Zhen Yiang and D. Lichtman, Surface Sci. <u>114</u>, 287 (1982)
7   M.L. Yu, Phys. Rev. B<u>19</u>, 5995 (1979)
8   G.K. Wertheim, M. Campagna, J.-N. Chazalviel and D.N.E. Buchanan,Appl. Phys. <u>13</u>, 225 (1977)
9   J.J. Ritsko, H. Witzke and S.K. Deb, Solid State Comm. <u>22</u>, 455 (1977)
10 L. Kopp, B.N. Harmon and S.H. Liu, Solid State Comm. <u>22</u>, 677 (1977)
11 W. Jelend and D. Menzel, Surface Sci. <u>42</u>, 485 (1974)
12 E. Bauer and H. Poppa, Surface Sci.<u>99</u>, 341 (1980)
13 D. Lichtman, F.N. Simon and T.R. Kirst, Surface Sci. <u>9</u>, 325 (1968)
14 M. Nishijima and F.M. Probst, Phys. Rev. B<u>2</u>, 2368 (1970)
15 C.J. Bennette and L.W. Swanson, J. Appl. Phys. <u>39</u>, 2749 (1968)
16 M.L. Yu, Surface Sci. <u>84</u>, L493 (1979)
17 G.R. Floyd and R.H. Prince, Surface Sci. <u>59</u>, 631 (1976)
18 Q.J. Zhang and R. Gomer, Surface Sci. <u>109</u>, 567 (1981)
19 H.M. Kramer and E. Bauer, Surface Sci. <u>107</u>, 1 (1981)
20 F. Bonczek, T. Engel and E. Bauer, Surface Sci. <u>94</u>, 57 (1980)
21 D.P. Woodruff, P.D. Johnson, M.M. Traum, H.H. Farrell, N.V. Smith, R.L. Benbow and Z. Hurych, Surface Sci. <u>104</u>, 282 (1981)
22 Ch. Park, M. Kramer and E. Bauer, Surface Sci. <u>109</u>, L533 (1981)
23 C. Leung and R. Gomer, Surface Sci. <u>59</u>, 638 (1976)
24 C. Leung, Ch. Steinbrüchel and R. Gomer, Appl. Phys. <u>14</u>, 79 (1977)
25 C. Wang and R. Gomer, Surface Sci. <u>74</u>, 389 (1978)
26 Ch. Steinbrüchel and R. Gomer, Appl. Phys. <u>15</u>, 141 (1978)
27 T. Engel, H. Niehus and E. Bauer, Surface Sci. <u>52</u>, 237 (1975)
28 E. Bauer, J. Electron Spectrosc. Rel. Phenom. <u>15</u>, 119 (1979)
29 T.E. Madey, Surface Sci. <u>94</u>, 483 (1980)

30 S.-L. Weng, Phys. Rev. B$\underline{23}$, 1699 (1981)

31 L. Ohse, M.S. thesis, Clausthal 1977

32 M.L. Knotek and P.J. Feibelman, Phys. Rev. Letters $\underline{40}$, 964 (1978)

33 H. Niehus, Ph.D. thesis, Clausthal, 1975

34 H. Niehus, Surface Sci. $\underline{78}$, 667 (1978)

35 E. Bauer and T. Engel, Surface Sci. 71, 695 (1978)

36 R.E. Howard, S. Vosko and R. Smoluchowski, Phys. Rev. $\underline{122}$, 1406 (1961)

37 J. Kirschner, D. Menzel and P. Staib, Surface Sci. $\underline{87}$, L267 (1979)

38 E. Bauer, H. Poppa and Y. Viswanath, Surface Sci. $\underline{58}$, 517 (1976)

39 E. Preuss, Surface Sci. $\underline{94}$, 249 (1980)

40 S. Prigge, H. Niehus and E. Bauer, Surface Sci. $\underline{75}$, 635 (1978)

41 H.-M. Kramer and E. Bauer, Surface Sci. $\underline{92}$, 53 (1980); 93, 407 (1980)

42 E.N. Kutsenko and N.D. Potekhina, Sov. Phys. Techn. Phys. $\underline{18}$, 250 (1973)

43 V.N. Ageev, S.T. Dzhalilov, N.I. Ionov and N.D. Potekhina, Sov. Phys. Techn. Phys. $\underline{21}$, 596 (1976)

44 E. Bauer and H. Poppa, Surface Sci. $\underline{88}$, 31 (1979)

45 P.A. Redhead, Can. J. Phys. $\underline{42}$, 886 (1964)

46 H. Niehus, private communication

3.2 Photodesorption and Negative Ion ESD

D. Lichtman

Physics Department and Surface Studies Laboratory, University of
Wisconsin-Milwaukee, Milwaukee, WI 53201, USA

I would like to take this opportunity to discuss briefly two topics which I believe
are relevant to this workshop. The first topic I would like to consider is that of
photodesorption. I believe this is a process that is sufficiently different from that
of photon-stimulated desorption that its characteristics should be understood and
the nomenclature describing various experiments kept clear. All photodesorption
measurements made to date show the following common characteristics:

1) The substrate or the substrate surface is a semiconductor.

2) The substrate surface contains impurity carbon atoms.

3) To observe photodesorption, the substrate must first be exposed to a
   considerable oxygen flux, e.g., $\sim 10^6$ Langmuir.

4) The oxygen-exposed-substrate must be irradiated with photons with
   energy equal to or greater than the band gap of the semiconductor
   material.

5) Mass analysis indicates that, under band-gap irradiation and at room
   temperature, neutral CO_2 is either the only or the dominant desorbing
   species.

6) The desorption signal is linearly proportional to the photon flux over
   many orders of magnitude (the power density in the beam is such that
   thermal effects are negligible).

7) Surface conductivity measurements, where taken, indicate a decrease
   during oxygen adsorption and an increase during CO_2 desorption.

These effects have been observed on a number of materials including ZnO [1],
CdS [2], TiO_2 [3], Nb_2O_5 [4], $SrTiO_3$ [5], V_2O_5 [6], Al_2O_3 [7], the chrome oxide
surface of stainless steel [8] and of black chrome [9], and elemental silicon [10].
A review paper [11] describes virtually all publications on photodesorption from
1952 to 1977. Based on the observations made, the following picture of
photodesorption has emerged. Oxygen adsorbs, most probably, on the impurity
carbon atoms, taking an electron from the conduction band and producing a
strongly bound negative ion complex, perhaps CO_2^-. Band-gap radiation produces
electron hole pairs. Some of the holes produced near the surface migrate to the
negative ion complex and recombine with the electron leaving a weakly bound,
probably physisorbed, species which then thermally desorbs as a neutral CO_2
molecule.

All the data obtained, including the kinetics of the reaction as a function of time, seem to be fully explainable by this model [12]. The process seems quite efficient, indicating a cross section in some systems in the order of 10^{-17} cm^2. The desorption is clearly a quantum process and, as such, should be included in an overall listing of desorption induced by electronic transitions. It is sufficiently different from those photon-induced processes which lead to the desorption of adsorbed species or substrate components as is seen, for example, in PSD. I propose that we restrict the word "photodesorption" to the specific process as described above and use "PSD" for other photon-induced processes that are essentially similar to ESD processes.

The second topic I would like to discuss briefly is the detection of negative ions using ESD or PSD. This particular experiment has now been done by several people using ESD [13-20]. Based on the limited number of experiments to date, it seems reasonably clear that the negative ions come from some of the same states which give rise to positive ions (i.e., O$^-$ and O$^+$, H$^-$ and H$^+$). Therefore, information from negative ion ESD experiments may not provide new information about adsorption states. However, the details of the negative ion signal, e.g., threshold energies, cross section data as a function of bombarding energy, ion energy distribution, and even ESDIAD of negative ions, etc., may provide significant clues as to the mechanisms involved in the ESD process. Since there is presently very great interest in the determination of basic mechanisms, I believe it would be very helpful if more experimenters presently obtaining positive ion data would also attempt to obtain negative ion data, not only in ESD measurements but also in PSD measurements. This information would be of considerable value to those attempting to develop models of the desorption process.

There are many references to ESD work in this volume, and I include here only a list of most of the earlier review papers on ESD [21-29] for reference convenience.

References

1. Y. Shapira, et al., Surface Sci. 54 (1976) 43.

2. S. Baidyaroy, W. R. Bottoms and P. Mark, Surface Sci. 28 (1971) 517.

3. N. Van Hieu and D. Lichtman, Surface Sci. 103 (1981) 535.

4. D. Lichtman and T. Liu, Proc. 7th Intern. Vac. Congr. & 3rd Intern. Conf. Solid Surfaces (Vienna, 1977) Vol. 11, p. 1277.

5. N. Van Hieu and D. Lichtman, J. Catal. 73 (1982) 329.

6. N. Van Hieu and D. Lichtman, J. Vac. Sci. Technol. 18 (1981) 49.

7. S. Brumbach and M. Kaminsky, J. of Applied Physics 47 (1976) 2844.

8. G. W. Fabel, et al., Surface Sci. 40 (1973) 571.

9. G. Zajac, A. Ignatiev, and G. B. Smith, J. Vac. Sci. Technol. 18 (1981) 379.

10. Submitted for publication.

11. D. Lichtman and Y. Shapira, CRC Chemistry and Physics of Solid Surfaces, Vol. II (1979) 397.

12. Y. Shapira, et al., Phys Rev. B $\underline{15}$ (1977) 2163.

13. G. E. Moore, J. Appl. Phys. $\underline{30}$ (1959) 1086.

14. A. Kh. Ayukhanov and E. Turnashev, Soviet Phys. Tech. Phys. $\underline{22}$ (1977) 1289.

15. J. L. Hock and D. Lichtman, Surface Sci. $\underline{77}$ (1978) L184.

16. M. L. Yu, J. Vac. Sci. Technol. $\underline{16}$ (1979) 518.

17. M. L. Yu, Phys. Rev. B $\underline{19}$ (1979) 5995.

18. J. L. Hock, et al., Surface Sci. $\underline{85}$ (1979) 101.

19. J. L. Hock, et al., Surface Sci. $\underline{85}$ (1979) L218.

20. Liu, Z. X. and D. Lichtman, Surface Sci. $\underline{114}$ (1982) 287.

21. D. Lichtman and R. B. McQuistan, Prog. Nuclear Energy (Series IX) Vol. 4, Pt. 2, (1965) 95.

22. P. A. Redhead, J. Vac. Sci. Technol. $\underline{7}$ (1970) 182.

23. T. E. Madey and J. T. Yates, Jr., J. Vac. Sci. Technol. $\underline{8}$ (1971) 525.

24. J. H. Leck and B. P. Stimpson, J. Vac. Sci. Technol. $\underline{9}$ (1972) 293.

25. V. N. Ageev and N. I. Ionov, Prog. Surf. Sci. 5, Pt. 1 (1974) 1.

26. R. Gomer, Solid State Physics $\underline{30}$ (1975) 93.

27. D. Menzel, Surface Sci. $\underline{47}$ (1975) 370.

28. M. J. Drinkwine and D. Lichtman, Prog. Surf. Sci. $\underline{8}$ (1977) 123.

29. T. E. Madey and J. T. Yates, Jr., Surface Sci. $\underline{63}$ (1977) 203.

3.3 The Determination of Molecular Structure at Surfaces Using Angle Resolved Electron- and Photon-Stimulated Desorption

T.E. Madey, F.P. Netzer[1], J.E. Houston[2], D.M. Hanson[3], and R. Stockbauer

Surface Science Division, National Bureau of Standards[4]
Washington, DC 20234, USA

Abstract

We review recent data and theoretical models related to the use of angle-resolved electron-and photon-stimulated desorption in determining the structures of molecules at surfaces. Examples include a variety of structural assignments based on ESDIAD (electron-stimulated desorption ion angular distributions), the observation of short-range local ordering effects induced in adsorbed molecules by surface impurities, the influence of electron-beam damage on surface structure, and a direct comparison of ESD and PSD ion yields for the same system.

1. Introduction

Some key questions which arise in studies of molecules on surfaces concern the geometrical structure of the adsorbed species: at which site is a molecule bonded to the surface, what are the orientations of the bonds between molecule and surface, and what are the bond directions of the ligands in surface molecular complexes? It is by now well established that the electron-stimulated desorption ion angular distribution (ESDIAD) method and the angle-resolved photon-stimulated desorption (PSD) method using synchrotron radiation are of great use in providing direct information about site location and geometrical structure of molecules on surfaces [1,2].

In these methods, electronic excitation of surface molecules by a focused electron beam or a photon beam can result in desorption of atomic and molecular ions from the surface. The ions desorb in discrete cones of emission in directions determined by the orientation of the bond which is ruptured by the excitation. For example, as illustrated in Fig. 1, ESD or PSD of adsorbed CO bound in a standing-up configuration on a metal surface will result in desorption of O^+ in the direction of the surface normal, while ESD/PSD of H^+ from H_2O adsorbed via the O atom will result in desorption of H^+ in an off-normal direction. Measurements of the patterns of ion desorption provide a direct display of the geometrical structure of surface molecules in the adsorbed layer [1-3].

Permanent address:
[1]Institute of Physical Chemistry, University of Innsbruck, A-   Innsbruck, Austria
[2]Sandia National Laboratory, Albuquerque, NM 87185, USA
[3]Department of Chemistry, SUNY, Stony Brook, NY 11794, USA

[4]Supported by the U.S. Department of Energy under contract number DE-AC04-76DP00789

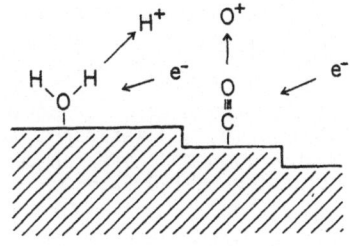

Fig. 1. Schematic bonding configurations for adsorbed CO and H$_2$O, showing relationship between surface bond angle and ion desorption angle in ESDIAD

In the present paper, we shall review recent experimental results detailing the utility of ESDIAD and angle resolved PSD in studies of molecules on surfaces. In particular, we shall emphasize that

(a) ESDIAD and angle resolved PSD provide <u>direct</u> information about the structure of surface species without complex mathematical analysis, and

(b) these methods provide information regarding the local bonding geometry of surface species, even in the <u>absence</u> of long-range order.

Several reviews of ESDIAD have appeared recently [1-3], and the fundamentals of the method and experimental procedures are described extensively therein. The physical principles of ESD and PSD are treated comprehensively elsewhere in the present volume.

The plan of this review is as follows: Section 2 contains a brief discussion of experimental procedures, and the application of ESDIAD and PSD to specific molecular systems is described in Section 3. Examples include a variety of structural assignments based on ESDIAD, the observation of short-range local ordering effects induced in adsorbed molecules by surface impurities, the influence of electron-beam damage on molecular structure, and a direct comparison of ESD and PSD ion yields for the same system. The emphasis will be on the structure of adsorbed molecules rather than atoms. ESDIAD studies of adsorbed atoms, including the role of steps and defects in ESD, have been discussed recently [1,2]. Theoretical concepts are treated in Section 4, which also includes a discussion of factors which perturb the trajectories of desorbing ions. Future directions are outlined in Section 5.

2.   Experimental Procedures

An extensive discussion of experimental methods of ESD and PSD is given in [4], and these will not be considered in detail here. Briefly, two types of detection schemes have been employed for most of the published ESDIAD studies: an area detector with visual display and a scanning ion detector, as shown in Fig. 2.

In the display-type apparatus developed at NBS [1,3,4], a focused electron beam bombards a single crystal sample. The ion beams which desorb via ESD pass through a hemispherical grid and impinge on the front surface of a double microchannel plate (MCP) assembly. The output electron signal from the MCP assembly is accelerated to a fluorescent screen where it is displayed visually (the ESDIAD pattern) and photographed. By changing potentials, the elastic low-energy electron diffraction (LEED) pattern from the

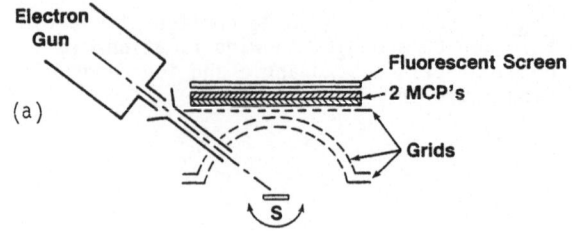

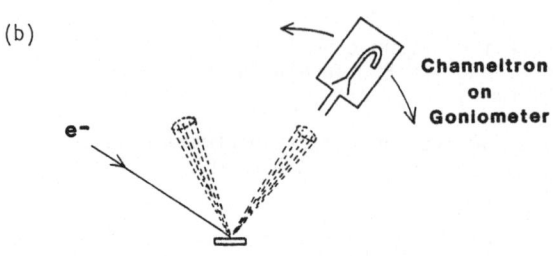

Fig. 2. (a) Schematic of NBS ESDIAD apparatus. ESDIAD patterns are displayed using the grid-microchannel plate-fluorescent screen detector array [1,3,4]. (b) Schematic of computer-driven moveable ion detector for ESDIAD [2,5]

sample can be studied. Mass identification of ESD ions is accomplished using a quadrupole mass spectrometer, which is also used as a detector in thermal desorption studies from the adsorbed layer.

NIEHUS [2,5] uses a channeltron multiplier mounted on a computer-driven goniometer as a moveable ion detector for ESDIAD (Fig. 2b). His data are in the form of computer-generated plots of ion intensity vs. desorption angle; ion mass is determined using a time-of-flight method.

In PSD of ions, the low signal levels and high background signals from scattered light cause special problems for angle-resolved measurements. An ellipsoidal mirror analyzer, a display-type instrument designed by EASTMAN and colleagues [6], has been used successfully for several angle resolved PSD studies [7,8]; however most PSD measurements reported to date do not include angle resolved measurements. At NBS, a double pass cylindrical mirror analyzer (CMA) is used for mass and energy analysis of PSD ions [4,9] similar to the apparatus of TRAUM and WOODRUFF [10].

3. Experimental Results

A. Calibration of the ESDIAD Method

Shortly after the first report of ESDIAD patterns for adsorbed species, it was postulated that the direction of ion desorption was determined by the orientation of the surface bond ruptured by the electronic excitation [11]. Theoretical model calculations [12,13] yielded results which were consistent with this postulate, but a general theory of ESD angular distributions applicable to a wide class of surface systems was not (and is not) available (cf. Section 4 below for a discussion of theory, as well as of factors which perturb the ion trajectories). In order to test the postulate, the NBS group adopted an empirical approach, viz., to apply the ESDIAD method to adsorbate-substrate systems whose surface geometries have already been predicted or determined based on other structure-sensitive surface probes.

Table 1 summarizes and lists molecular adsorbates which have been studied in various laboratories using ESDIAD. The top part of the Table, (a), indicates those molecules which have been used to "calibrate" the ESDIAD method, i.e., molecules for which a structural identification has been made using methods such as high resolution electron energy loss spectroscopy (EELS), reflection-absorption infrared spectroscopy (IR), ultraviolet photoemission spectroscopy (UPS) and angle resolved UPS (ARUPS), or low energy electron diffraction (LEED).

In each case listed in Table 1(a), the ESDIAD results are consistent with the other techniques. E.g., for standing-up CO on Ru(001), Ni(111) and W(110), the molecule is bound via the carbon atom to the metal surface with its molecular axis perpendicular to the surface; ESD O^+ (and CO^+) ions desorb in the direction perpendicular to the surface. For the bridge-bonded CO structure identified on Pd(210) using IR, the molecular axis is inclined from the perpendicular to the macroscopic surface, and ESD O^+ emission occurs in the predicted off-normal direction. A "halo" of H^+ ion emission is seen for ESD of NH3 on Ni(111), consistent with the ARUPS identification of ammonia bonded to Ni via the N atom with the H atoms oriented away from the surface. For NO on Ni(111), evidence for both standing-up and inclined NO was found [28], and the structures varied with coverage and temperature as seen in an EELS study [29] of the same system. ESDIAD was used to predict that C_6H_{12} was adsorbed on Ru(001) in the "chair" form, with the plane of the molecule parallel to the surface [33]; this structure is consistent with the later EELS work of HOFFMANN et al. [34]. The complex coverage and temperature dependence of H_2O on Ru(001) seen in an early ESDIAD study [30] has been examined in detail recently by DOERING and MADEY [31]; the structural predictions are completely consistent with the EELS data of THIEL et al. [32]. Finally, NIEHUS has used ESDIAD to determine the structure of the surface of a $WO_3(111)$ crystal grown as an overlayer on a W(110) surface; he concluded that the structure was consistent with bulk WO_3 [35].

Part (b) of Table 1 contains a list of other adsorbed molecules for which structural assignments have been made using ESDIAD. In many cases, the structures were not previously suspected (e.g., inclined CO on Ni(111), Mo(100) or W(100); oxygen-stabilized ordering of NH3 and H_2O on Ni(111) - see below, Section 3 B) but none are inconsistent with experimental results.

It appears from the evidence in Table 1(a) that for all molecular systems where we have a priori knowledge of the surface structure, the ion desorption angle observed in ESDIAD is related to the surface bond angle. In particular, it is found that the expected azimuthal angle of the surface bond is preserved in ESDIAD, but the polar angle is increased for the ion trajectory, due largely to image charge effects (cf. Section 4 below). To date, no exceptions have been found.

Whereas the desorption direction of an ESD ion beam contains information about bond orientation, it appears that the profile of the beam (beam width and shape of cross section) carries information about surface vibrational dynamics, i.e., bending vibrational motions perpendicular to the desorption direction. In a simple physical picture, desorption lifetimes ($\sim 10^{-14}$s) are much shorter than the period of low-frequency bending vibrations ($\sim 10^{-12}$s), so that the electron beam used for excitation samples a statistical array of surface oscillators "frozen" into a distribution of orientations. According to this model, the beam widths should increase with temperature due to the increased amplitude of bending vibrations; such behavior has been observed experimentally for many systems, and characterized quantitatively for CO/Ru(001) [14]. In addition, asymmetric ESDIAD beams have been seen for O^+ from CO/Pd

Table 1. Structural assignments of adsorbed molecules based on ESDIAD measurements

---

(a) Structures of surface molecules which have been identified using various sensitive methods as well as ESDIAD

System	ESDIAD Ref.	Structural Assignment	Other Methods and Ref.
*CO/Ru(001)	[14]	standing-up CO	EELS [15], IR[16] ARUPS [17]
CO/Ni(111)	[18]	standing-up CO	EELS [19], ARUPS [20]
virgin-CO/ W(110)	[1,21]	standing-up CO	UPS [22]
CO/Pd(210)	[23]	inclined CO	IR [24]
NH_3/Ni(111)	[25,26]	bonded via N atom, H atoms oriented away from surface	ARUPS [27]
NO/Ni(111)	[28]	standing-up and inclined NO	EELS [29]
H_2O/Ru(001)	[30,31]	coverage and temperature dependent	EELS [32]
C_6H_{12}/Ru(001)	[33]	"chair" form of C_6H_{12} parallel to surface	EELS [34]
WO_3(111)	[36]	inclined W-O	LEED; crystallography

(b) Other surface structures identified using ESDIAD

System	Ref.	Structures
CO/Ni(111)	[18]	standing-up CO and "inclined" CO both observed under different conditions of temperature and coverage.
virgin-CO/stepped W(110)	[1,21]	
α-CO on W(110)	[36]	
CO/W(100),(111)	[37-39]	
CO/Mo(001)	[40]	
NH_3/Ru(001)	[41]	similar to NH_3/Ni(111)
NH_3 and H_2O on oxygen-dosed Ni(111),Al(111)	[26,42,45,46]	azimuthal ordering

---

*Also verified using Angle Resolved PSD [43].

(210) and associated with differing vibrational amplitudes in-plane and out-of-plane of the bridging CO ($>$C=O) [23]. CLINTON [44] has pointed out that, depending on the model assumed for the final state interaction in ESD, the beam profile can image the initial state vibrational wave function of the desorbing species in either coordinate space (as assumed above) or momentum space. It is clear that much more needs to be done to understand the factors influencing beam profiles.

We note another example of structural information available from ESD studies of CO on transition metal surfaces. Terminally bonded CO, singly-coordinated to metal surface atoms via the C atom, yields both O^+ and CO^+ as ion desorption products; this has been seen for CO on Ru(001) and Ni(111) [14,18]. In contrast, multiply-coordinated, bridge-bonded CO yields only O^+, as seen for CO on Pd(210), and at low coverages on Ni(111) [18,23]. Increased coordination appears to correlate with a higher neutralization rate for CO^+, so that its yield is suppressed from multiply-coordinated CO.

B. Short-range Local Ordering in Absence of Long-range Order

In a particularly interesting application of the ESDIAD method, it has been found that traces of preadsorbed oxygen on a metal surface will induce a high degree of azimuthal order in adsorbed molecules which are disordered azimuthally on the clean surface [26,42,45,46]. This azimuthal-ordering effect has been studied for NH_3 on Ni(111) and Al(111) and H_2O on Ni(111).

Figure 3 is taken from [26], and contains a sequence of LEED and ESDIAD patterns for the adsorption of NH_3 and H_2O on clean and oxygen-dosed Ni(111). Fig. 3(a) is a LEED pattern from the clean surface, and Fig. 3(b) is an H^+ ESDIAD pattern for NH_3 adsorbed on Ni(111). As discussed in Section 3 A and in [25], the "halo" pattern of Fig. 3(b) is characteristic of NH_3 adsorption at T $\geq$ 140 K for fractional monolayer coverages and is consistent with an array of NH_3 molecules bonded via the N atoms, with no preferred azimuthal orientation of the H ligands which point away from the surface. If the Ni(111) surface is predosed with a small coverage of oxygen (heated to 600 K to anneal, then cooled to 80 K) and then dosed with ammonia to yield fractional monolayer coverages, an ESDIAD pattern exhibiting three-fold symmetry is observed. Heating to ~140 K causes desorption of all NH_3 except the chemisorbed layer in contact with the substrate, and a well-ordered H^+ ESDIAD pattern illustrated in Fig. 3(c) results. The ion emission appears as a "broken" halo, with intense off-normal emission along the [$\bar{1}\bar{1}2$] azimuths. Even at the lowest oxygen coverages used in this study [26] ($\theta_0 \leq 0.03$) evidence for this oxygen-induced azimuthal ordering of adsorbed NH_3 was observed. Upon heating to T $\geq$ 300 K, the ESD ion emission disappears as the NH_3 desorbs. No improvement in the long-range ordering of NH_3 molecules was induced by coadsorbed oxygen as evidenced by LEED observations.

Similar results are seen for H_2O adsorbed on Ni(111) [26,42]. Fig. 3(d) is an ESDIAD pattern characteristic of adsorbed H_2O on Ni(111) at 80 K, for coverages less than saturation of the chemisorbed H_2O layer bonded directly to the Ni(111). The center of the pattern is dim, with most of the H^+ ion emission directed away from the normal. This pattern is consistent with bonding of H_2O to Ni(111) via the O atom, with H atoms oriented away from the surface. The lack of well-defined halo [as seen in Fig. 3(b)] suggests that the array of adsorbed hydrogen-bonded H_2O molecules contains a distribution of "tilt" angles measured with respect to the surface normal, as well as a random distribution of azimuthal angles.

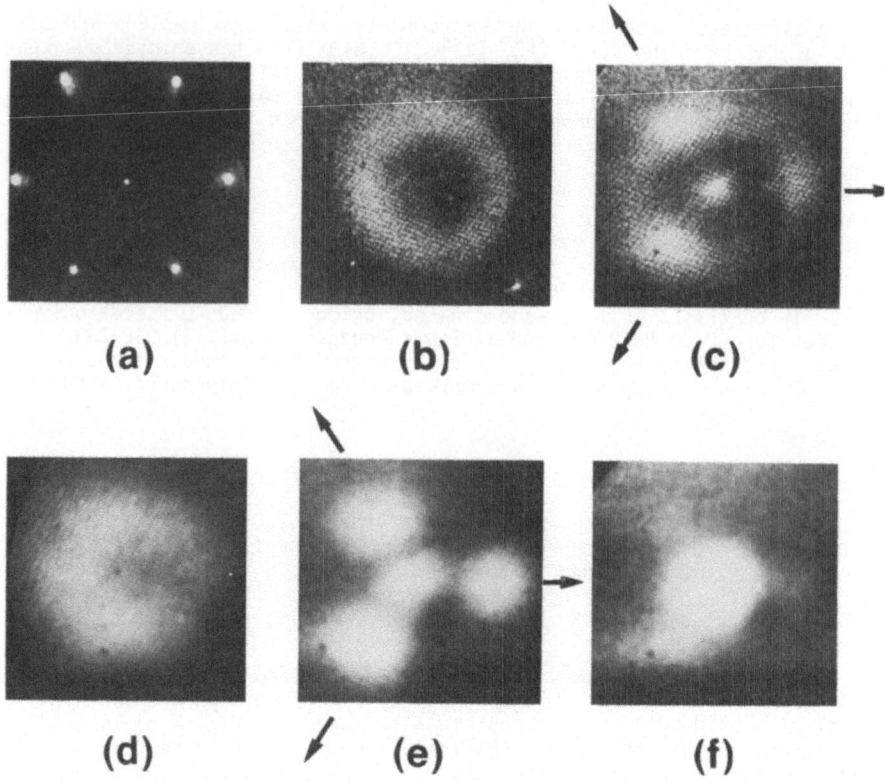

Fig. 3. LEED and ESDIAD patterns for NH_3 and H_2O on Ni(111): (a) Clean LEED, electron energy V_e = 120 eV. (b) H^+ ESDIAD pattern, NH_3 on clean Ni(111), V_e = 300 eV. (c) ESDIAD pattern, NH_3 on O/Ni(111), $\theta_O \lesssim$ 0.05 monolayer, V_e = 300 eV. (d) ESDIAD pattern, H_2O on clean Ni(111) at 80 K, V_e = 300 eV. (e) ESDIAD pattern, H_2O on O/Ni(111) at 80 K, $\theta_O \lesssim$ 0.05 monolayer, V_e = 300 eV. (f) ESDIAD pattern, H_2O on O/Ni(111) after heating to 120 K, V_e = 300 eV. The arrows on the figure are oriented along $[\bar{1}\bar{1}2]$ azimuths

In contrast, the adsorption of low coverages of H_2O at 80 K onto the oxygen-predosed and annealed Ni(111) surface produces a highly ordered three-fold-symmetric ESDIAD pattern [Fig. 3(e)] with intense ion emission along the same azimuths as seen for NH_3 [Fig. 3(c)]. The threefold pattern seen for the C_{2v} molecule H_2O is apparently a consequence of different degenerate ordered configurations on the threefold substrate. Unlike the behavior for NH_3, heating does not simply result in gradual disappearance of the symmetric pattern. Upon heating to 120 K, a temperature well below the onset of desorption, the three outer lobes in the ESDIAD pattern [Fig. 3(e)] disappear, and only intense normal H^+ emission remains [Fig. 3(f)]; the normal emission disappears at $T \geq$ 300 K.

Both NH_3 and H_2O yield ESDIAD patterns that have the same threefold symmetry and the same azimuthal orientation. Both NH_3 and H_2O are adsorbed at 80 K mainly without dissociation, and if the bonding sites for preadsorbed

O atoms are not influenced by the adsorbed molecules, the respective three-fold (NH_3) and twofold (H_2O) symmetry axes of the two molecules demand that NH_3 and H_2O are bonded in different sites on the Ni(111) substrate.

In Fig. 4, a structural model is presented. From LEED, it is known that oxygen on Ni(111) is adsorbed in threefold hollow sites [47]. If the O atoms are located in threefold hollows above second-layer atoms, H_2O can be bonded in atop sites and NH_3 in other threefold hollows. The surface O-O distance for $O-H_2O$ is 2.87Å, consistent with the 2.75-2.96Å range seen for the O-O distance in hydrogen-bonded ice [48]. The O-N distance for $O-NH_3$ is also 2.87Å, consistent with typical H-bonding distances. If the O atoms are above the second-layer vacancies, the NH_3 and H_2O sites are shifted according-ly with no change in N-O or O-O separation. In this model, the non-hydrogen-bonded ligands are the ones which are seen in ESDIAD. Either the hydrogen-bonded ligands are more effectively neutralized following electron excitation, and are less likely to desorb as ions, or the H^+ ions from the hydrogen-bonded ligands follow shallow trajectories and are recaptured by the surface [see discussion in Section 4 below).

(a) Model for Orientation of NH_3　　　　　　**(b) Model for Orientation of H_2O**

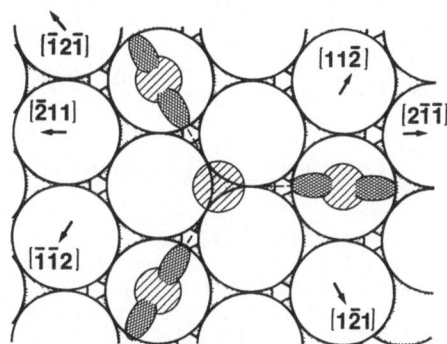

Fig. 4. Schematic models for bonding of H_2O and NH_3 to Ni(111) in the presence of oxygen. The dashed lines indicate hydrogen-bonding interactions

In Fig. 4, we show a single O atom influencing the azimuthal orientation of more than one molecule, consistent with the observations of BOWKER, BARTEAU, and MADIX [49], and KRETZSCHMAR et al. [50]. Whether this is true or whether O atoms simply "nucleate" ordered azimuthal layers by lateral interactions between molecules is not clear. It is apparent, however, that mere traces of O can order substantial coverages of H_2O or NH_3. Less than 0.05 monolayer of O will cause ordering of the saturation NH_3 layer at 140 K ($\theta \sim 0.25$ to 0.40).

We note that there was a previously reported contradiction [25] between ESDIAD and angle-resolved UPS regarding the presence or absence of azimuthal ordering in NH_3 on clean Ni(111) which has now been resolved: it is very likely that there were traces of O remaining on the crystal used for the UPS measurements [27,51]. The azimuthal orientations observed in the present work for NH_3 are the same as those seen in the UPS measurements [51].

For Ni(111), there is a close correspondence between the ideal hydrogen-bonding distance and the distance between high symmetry adsorption sites. In order to ascertain the influence of substrate geometry on the azimuthal ordering effect, recent experiments have been performed on Al(111) [46] (lattice constant ~ 15% greater than Ni) and Ru(001) [31] (lattice constant ~ 8 % greater than Ni). Interestingly, the azimuthal ordering is seen for NH_3 on O-dosed Al(111), despite the lack of ideal geometry [46]. It appears that hydrogen-bonding interactions cause NH_3 to be "pulled off" the high symmetry atop positions located along [$\bar{1}\bar{1}2$] azimuths from O atoms adsorbed in three-fold hollows. The corresponding ESDIAD patterns are less well-defined than shown in Fig. 3 but the azimuthal ordering is clear. For H_2O on O-dosed Al(111), no improvement in azimuthal ordering is seen, presumably because a majority of the H_2O dissociates upon adsorption [52].

In very recent work, DOERING and MADEY [31] have verified that H_2O on Ru(001) is azimuthally ordered on the clean surface, due to a near-epitaxial relationship between the two-dimensional hydrogen-bonded ice layer and the hexagonal Ru substrate. In this case, the adsorption of oxygen induces disorder into the H_2O layer, because the H_2O - O interaction results in a less favorable adsorption geometry for H_2O.

We note that these observations of steric effects in coadsorption may have implications for our understanding of the mechanisms by which catalyst promoters and poisons function. It may be that a promoter or poison can alter a reaction pathway by inducing the formation of a surface structure which has a low probability of formation in the absence of the promoter or poison.

C. Beam Damage in ESD: Evidence for Adsorbate Rearrangement and Short-range Diffusion

In all of the ESDIAD studies discussed in the previous sections of this paper, the total electron dose was sufficiently small that beam damage was a minor perturbation. In many cases, however, electron beam damage occurs in the form of extensive decomposition and desorption of adsorbed species or of the substrate; the effects of beam damage in Auger spectroscopy and other surface analysis methods have been reviewed by PANTANO and MADEY [53]. (In general, it is advisable to keep the electron dose below ~ 10^{-3} Coul/cm^2 to minimize damage in surface analysis.)

Although electron beam-induced decomposition and desorption are well known, there are few systematic studies of beam-induced surface diffusion. NAUMOVETS, FEDORUS and coworkers [54] have reported electron beam-induced disordering of hydrogen and Li on W(110) under conditions where desorption apparently did not occur; they interpreted these results as being due to beam-induced short range diffusion. Other examples of beam-induced changes in LEED patterns are well known, but decomposition and desorption generally accompany the disordering.

We have recently found evidence for the formation of new surface structures due to beam-induced surface rearrangement and short-range diffusion in an ESDIAD study of CO on stepped W(110) surfaces [1,21]. The chemistry of CO on tungsten is very complex, and has been discussed in detail [22,55,56]. For adsorption at ~ 77 K, however, the adsorption behavior is relatively simple: CO is adsorbed almost exclusively in molecular form, in the so-called "virgin-CO" state. We have studied the adsorption of virgin-CO on a planar W(110) surface and on stepped surfaces containing W(110) terraces [1,21]. The experiments were performed on a 7 mm diam. tungsten crystal cut in the form of a truncated pyramid to expose 5 separate facets - a W(110) facet and 4 stepped surfaces of different step densities (6° and 10° off the W(110) plane) with step orientations parallel to [100] and [110] directions [57].

The ESDIAD data for molecular virgin-CO are consistent with a simple model (Fig. 5) in which CO is bonded in a standing-up mode with its molecular axis perpendicular to the W(110) surface, not only on the (110) flat, but on the (110) terraces on the stepped surfaces also [21]. Moreover, on the stepped facets, a fraction of the CO is tilted away from the normal to the terraces by ~ 40°. There is an activation barrier for adsorption of the inclined CO at step edges, so that the population of CO at step-edge sites increases with temperature.

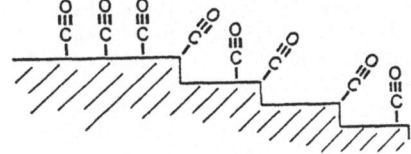

Fig. 5. Bonding model for standing-up CO on W(110) terraces and inclined CO at step edges

Fig. 6 illustrates the effect of beam damage on the ESDIAD patterns for virgin-CO (the coverage was ~ 0.8 of the saturation value) adsorbed on 3 facets of the W crystal described above and indicated schematically in the lower part of the figure: C refers to the central W(110) plane, while T and R refer to stepped surfaces. For each of the stepped surfaces, one effect of prolonged electron bombardment is the appearance of ESDIAD beams in directions away from the surface normal, in "downstairs" directions. The

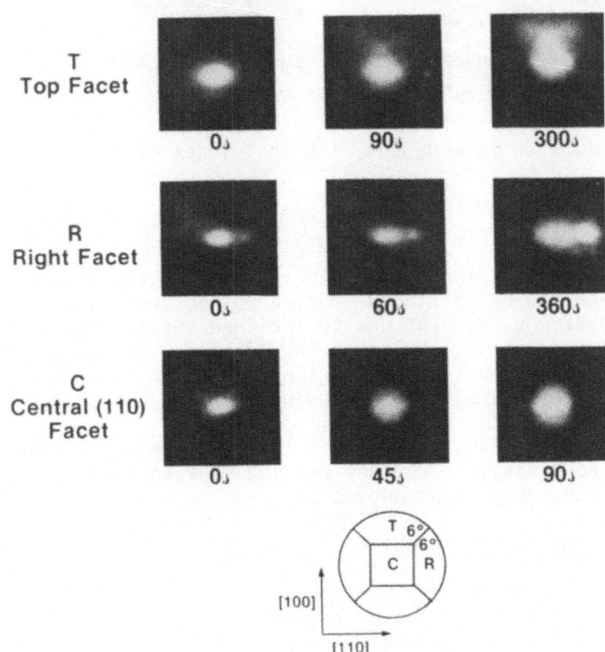

Fig. 6. Electron beam damage in ESDIAD of virgin-CO on multifaceted W(110) crystal; T = 80 K, $J_e \approx 5 \times 10^{-5}$ A/cm^2, V_e = 360 V

129

resultant ESDIAD patterns for facets T and R are qualitatively similar to but different in detail from the thermally-induced patterns [21]; more structure is seen in the damaged patterns. On the central facet, the normal beam becomes more broad under electron bombardment (in addition, the ratio of O^+/CO^+ changes with time [55]).

The data of Fig. 6 clearly indicate that there are electron-induced rearrangments of a fraction of the adsorbed CO. Since virgin-CO is known to dissociate under electron bombardment [55,56] and convert to other bonding forms, it is possible that some of the extra ESDIAD beams in Fig. 6 are due to dissociation products (although the patterns are different from oxygen on stepped W surfaces [57]). The cross sections for these processes are $\geq 3 \times 10^{-17}$ cm^2 on all facets, suggesting that valence electron excitations (and not the low cross section C 1s excitation) are responsible. We suggest that short-range diffusion and molecular rearrangements can occur on the surface via recapture of an electronically and vibrationally excited neutral species which can "hop" a few Angstroms before its energy is dissipated in the solid. The highly excited CO can also assume configurations at step edges which are not thermally populated. Furthermore, the rate at which damage occurs, as well as the appearance of the patterns, is a function of CO coverage: the rate is ~5 times faster than that given above for CO coverages ≤ 0.3 monolayers [21].

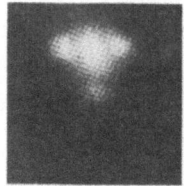 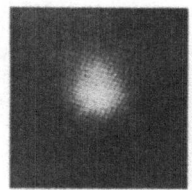 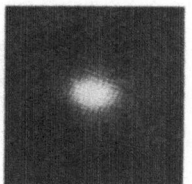

$\theta \cong 0.3$
(Damaged)

Readsorb
to $\theta \cong 0.8$

Undamaged
$\theta \cong 0.8$

Fig. 7. Effect of readsorption on beam damage: CO on T facet of multi-faceted W(110), 80 K

Finally, Fig. 7 demonstrates that the beam damage induced on the T facet for a low initial CO coverage can be partially "healed" by adsorption of additional CO. Some of the damage features due to inclined species disappear with further adsorption. The additional CO can either displace molecules from inclined configurations, or simply block the ion desorption by intermolecular neutralization [18] or geometrical shadowing at high coverages.

A major conclusion to draw from Figs. 6 and 7 is that beam-induced desorption and conversion of adsorbed species from one bonding mode to another can be accompanied by profound structural rearrangements.

D. Structural Information from Photon-Stimulated Desorption

There are only three examples of angle resolved PSD which have appeared in the literature to date: PSD of O^+ from cleaned V_2O_5 [7], O^+ from oxygen adsorbed on W(111) [8] and O^+ from CO adsorbed on Ru(001) [43]. In the latter cases, the ion desorption patterns were in excellent agreement with previous ESDIAD data. Moreover, the measurements of ion yield vs. photon energy clearly demonstrated the role of metal core hole excitations in PSD of O^+ from O/W and V_2O_5, whereas intramolecular excitations are responsible for PSD of O^+ from chemisorbed CO [43,58].

In a recent study in our laboratory [59], a comparison between ESD and PSD for H_2O adsorbed on Ti illustrated the power of PSD in its sensitivity to a specific bonding ligand (OH) in different configurations, viz., OH adsorbed on Ti(001) as Ti-OH, in comparison to condensed water layers (HOH). The ESD/PSD measurements were not angle resolved (in the sense that complete ion angular distributions were not measured); however, ion desorption in the direction of the surface normal was measured using the double pass cylindrical mirror analyzer mentioned in Section 2.

In a variable-wavelength UPS study of H_2O on Ti(001) using synchrotron radiation, it was established that H_2O adsorbs dissociatively to form adsorbed OH, H and O for H_2O exposures $\leq$ 1L (1L = 10^{-6} Torr sec) at 90 K [59]. For higher exposures, the adsorption of molecular H_2O is observed, and multilayers of ice can be grown. Figure 8 shows the PSD H^+ ion intensity as a function of H_2O exposure for a photon energy of 47.5 eV. The H^+ ion yield rises initially upon exposure of the surface to H_2O, passes through a sharp maximum at $\sim$ 0.75 L, and falls as the dose increases. The high H^+ yield at low exposures appears to be related to the formation of OH (ads.) and the decrease at high doses is due to the formation of an ice multilayer. The ESD H^+ yield as a function of H_2O exposure is also shown in Fig. 8. As in the PSD case, the H^+ yield rises to a maximum for low H_2O exposure where OH is formed; however, the ESD yield rises significantly at higher exposures as the molecular H_2O multilayers are formed, in marked contrast to the PSD ion yield.

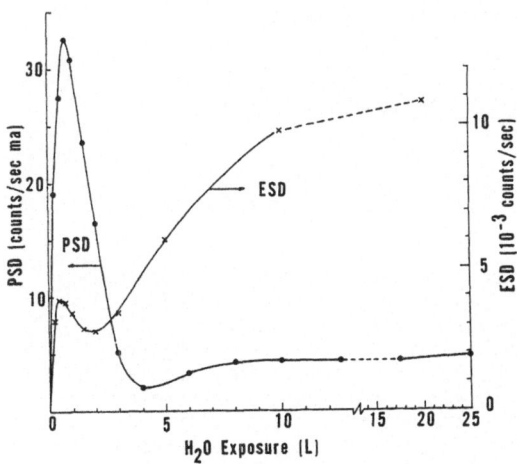

Fig. 8. Comparison of PSD and ESD H^+ yield as a function of H_2O exposure for adsorption on Ti(001). For exposures $\leq$ 1 L, the H^+ signal is due to OH (ads). For higher exposures, the H^+ signal is due to condensed H_2O. (1 L = 1 x 10^{-6} Torr s) [59]

It is clear that bombardment of an ice multilayer by 500 eV electrons results in electronic excitations caused by both primary and secondary electrons which lead to appreciable H^+ emission, and that 47.5 eV photons do <u>not</u> cause such excitations. This can be seen in Fig. 9 which compares

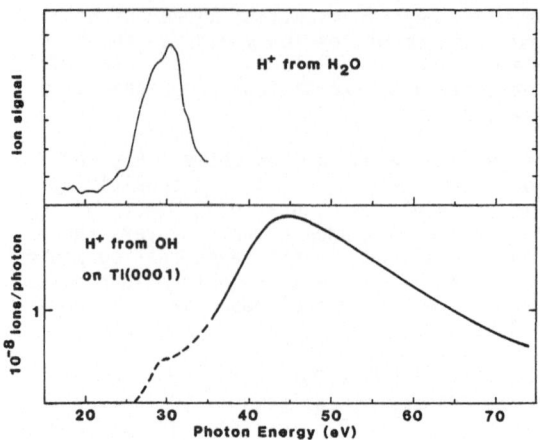

Fig. 9. Comparison of H⁺ ion yield as a function of photon energy for condensed H_2O (top, from [60]) and for OH on Ti(001) (bottom, from [59])

the data of ROSENBERG et al. [60] for PSD of H⁺ from amorphous ice with the data of STOCKBAUER et al. [59] for PSD of H⁺ from OH on Ti. For ice, the maximum ion yield occurs at $h\nu \sim 30$ eV; for OH on Ti, the maximum ion yield is ~ 45 eV.

The differences in PSD of Ti-OH and H - OH are the subject of a recent study by D. E. RAMAKER [61], who suggests that multielectron excitations lead to different two-hole final states [excitation of $3a_1^{-1}$ $1b_2^{-1}$ to produce the 30 eV peak in H_2O, and ionization of the $2a_1$ orbital + shakeup to produce the 45 eV peak in OH]. Whether this interpretation is correct, or whether the suggestion that excitation of the Ti 3p level is responsible for PSD of H⁺ from Ti - OH [59] is currently under study.

4.   Theory of the Origin of Angular Effects in ESD and PSD

There have been few advances in the theory of angle-resolved ESD and PSD since these topics were last reviewed [1,2], but a brief guide to the appropriate literature is in order.

GERSTEN, JANOW and TZOAR [12,62] were the first to perform model calculations using dynamical arguments in which they related the ion angular distributions to the positions of adatoms in different surface binding sites. They reported angular distributions for O⁺ ions desorbing from O atoms adsorbed at different sites on W(100) and W(111). They modeled the O⁺ - W repulsive potential using atomic wave functions and an unrelaxed lattice; they calculated asymptotic ion trajectories classically and displayed them graphically. Details of the resultant angular patterns of ion desorption were found to vary sensitively with binding site: ion beams desorbing perpendicular the surface were due to adsorption at sites atop substrate atoms whereas off-normal ion beams were due to adsorption at non-high-symmetry sites. PREUSS [63] performed improved calculations using screened atomic Coulomb potentials for the repulsive potentials between ion and surface, and generated model ion desorption patterns which agreed closely with experimental results [2].

132

All of the above calculations are based on a final-state picture of the desorption process, in which ion trajectories are determined solely by the three-dimensional repulsive potential contours.

CLINTON [13] has presented a quantum scattering theory of ESD in which he assumes that the repulsive final state potential acting on a desorbing ion is a sum of central potentials. He concludes that desorption will occur in a direction where the vibrational wave function of the substrate-atom bond has its maximum value, i.e., in the direction of the chemical bond "broken" by the excitation. He indicates that ESDIAD processes should be dominated by initial state (ground state) structures of atoms and molecules on surfaces. In more recent work, CLINTON [64] has used a frozen-orbital approximation to show that the instantaneous force on an ESD ion will be repulsive, in the direction of the original bond: he concludes that ESDIAD images the initial state vibrational wave function. These calculations do not include reneutralization processes nor the influence of the image potential on ion trajectories.

Even in the initial state picture in which the desorption impulse is along the bond direction, final state effects will influence the trajectories and the resultant ion desorption patterns [1]. Possible final state effects include anisotropy in the ion reneutralization rate, "focusing" effects due to curvature in the final state potential, and deflection of the escaping ions by the electrostatic image potential. Since we have no detailed knowledge of final state potentials, we have no basis (other than the previously-mentioned model calculations [12,62,63]) for considering their influence. There have been several studies recently in which evidence for anisotropies in ion neutralization rates have been reported [65,66]. In both cases, high energy ion beams scattered from single crystal surfaces were preferentially neutralized along scattering azimuths where the scattered ion interacts most strongly with substrate atoms. It appears that such effects are most important near grazing scattering angles, and should not play a major role in producing contrast in ESDIAD patterns.

Several authors have discussed the influence of the image potential on ion trajectories [1,67,68]. Despite the different approaches, the results are identical: the image potential causes an increase in the polar angle for ion desorption. CLINTON's [67] calculations indicate that an ion desorbing with an initial angle α_i with respect to the surface normal will arrive at the detector with an apparent desorption angle α_0 given by

$$\cos \alpha_0 = \cos \alpha_i \left[\frac{1 + V_I/[(E_K-V_I) \cos^2 \alpha_i]}{1 + V_I/(E_K-V_I)} \right]^{1/2} . \tag{1}$$

In this expression, V_I is the screened image potential [69] at the initial ion-surface separation Z_0^I, and E_K is the final, measured kinetic energy of the ion. We note that V_I^0 is always a negative quantity, so that

$$|V_I/(E_K-V_I)| \leq 1 \text{ and } \alpha_0 \geq \alpha_i .$$

A discussion of the location of the image plane and the assumptions leading to Eq. (1) are given in [67,70].

In order for equation (1) to have a real physical solution, the following relationship must be satisfied:

$$\left| \frac{V_I}{(E_K - V_I) \cos^2 \alpha_i} \right| \leq 1 \; . \tag{2}$$

This means that there will be a cut-off angle for ion desorption defined by the value of α_i for which $\alpha_0 = 90°$. For values of α_i greater than the cut-off angle, the ions will follow low trajectories and be recaptured by the surface. The possibility of beam-damage induced by bombardment of surface species by low-energy ESD ions is very likely.

As discussed previously [1], it appears that the major final-state perturbation of ion trajectories is in the polar direction. In general, for a perfectly planar surface or for desorption along an azimuth of symmetry, the perturbation of the azimuthal angle should be minimal.

Finally, we note that photo-dissociation studies of non-linear gas phase triatomic molecules (NO_2, NOCl) [71] demonstrate that dissociation occurs rapidly, before statistical equilibrium is attained. The angular and energy distributions of fragments indicate that dissociation occurs from "bent" final state structures. A modified impulsive model for NOCl dissociation, in which fragment recoil lies approximately along the direction of the breaking NO--Cl bond, is consistent with the data.

5. Summary

We summarize the structural information obtainable using ESDIAD and PSD as follows:

1) ESDIAD provides direct evidence for the structures of surface molecules and molecular complexes, and in certain cases (e.g., oxides) can provide structural information about the substrate surface.

2) ESDIAD is particularly sensitive to the orientation of hydrogen ligands in adsorbed molecules. In general, electron scattering from surface H is sufficiently weak that LEED is not very useful.

3) ESDIAD is sensitive to the local bonding geometry of molecules on surfaces even in the absence of long-range translational order.

4) The identification using ESDIAD of impurity-stabilized surface structures (H_2O and NH_3 on O-dosed surfaces) is of relevance to surface reaction mechanisms, as well as to catalyst promoters and poisons.

5) ESDIAD provides a direct view of short-range surface diffusion and molecular rearrangement induced by electron beam damage.

6) Photon Stimulated Desorption, although not yet as extensively applied as ESDIAD in structural studies, has all of the capabilities of ESD. A particular advantage of PSD is the use of tuneable photon sources to excite specific surface bonds selectively.

In the future, we anticipate that ESDIAD will find wide use as a technique to complement and extend other surface-sensitive structural probes, particularly in studies of local structure in the absence of long-range order. The correlation of PSD ion angular distributions with specific valence and core hole excitations in surface species should be particularly fruitful, especially for complex systems (mixed oxides, coadsorption systems, catalysis). The combination of angle-resolved PSD with SEXAFS (surface x-ray absorption fine structure) will provide an opportunity to measure both bond orientation and bond length. Finally, the angle resolved desorption of neutrals (including metastables) and negative ions will provide new insights into both structure and mechanisms of excitation.

6. Acknowledgements

FPN is grateful to the Max Kade Foundation, NY, for fellowship support during his term at NBS. DMH wishes to acknowledge support provided by the SURF Fellowship Program through the University of Maryland. This work was supported in part by the Department of Energy, Office of Basic Energy Sciences and in part by the Office of Naval Research.

References

1) T. E. Madey in Inelastic Particle-Surface Collisions, Springer Series in Chemical Physics 17, eds, E. Taglauer and W. Heiland (Spring-Verlag, Berlin, 1981) p. 80.

2) H. Niehus, Appl. Surface Sci. (in press).

3) T. E. Madey and J. T. Yates, Jr., Surface Sci. 63, 203 (1977).

4) T. E. Madey and R. Stockbauer, in Methods of Experimental Surface Science, eds. R. L. Park and M. G. Lagally, in press.

5) H. Niehus and B. Krahl-Urban, Rev. Sci. Instr. 52, 68 (1981).

6) D. E. Eastman, J. J. Donelon, N. C. Hien and F. J. Himpsel, J. Nucl. Instr. Methods 172, 327 (1980).

7) J. F. van der Veen, F. J. Himpsel, D. E. Eastman, and P. Heimann, Solid State Comm. 36, 99 (1980).

8) T. E. Madey, R. Stockbauer, J. F. van der Veen, and D. E. Eastman, Phys. Rev. Lett. 45, 187 (1980).

9) D. M. Hanson, R. Stockbauer and T. E. Madey, Phys. Rev. B 24, 5513 (1981).

10) M. M. Traum and D. P. Woodruff, J. Vac. Sci. Technol. 17, 1202 (1980).

11) T. E. Madey, J. J. Czyzewski, and J. T. Yates, Jr., Surface Sci. 49, 465 (1975).

12) J. I. Gersten, R. Janow and N. Tzoar, Phys. Rev. Lett. 36, 610 (1976).

13) W. L. Clinton, Phys. Rev. Lett. 39, 965 (1977).

14) T. E. Madey, Surface Sci. <u>79</u>, 575 (1979).

15) G. E. Thomas and W. H. Weinberg, J. Chem. Phys. <u>70</u>, 1437 (1979).

16) H. Pfnur, D. Menzel, F. M. Hoffmann, A. Ortega, and A. M. Bradshaw, Surface Sci. <u>93</u>, 431 (1980).

17) J. C. Fuggle, M. Steinkilberg, and D. Menzel, Chem. Physics <u>11</u>, 307 (1975).

18) F. P. Netzer and T. E. Madey, J. Chem. Phys. <u>76</u>, 710 (1982).

19) W. Erley, H. Wagner, and H. Ibach, Surface Sci. <u>80</u>, 612 (1979).

20) G. Apai, P. S. Wehner, R. S. Williams, J. Stöhr, and D. A. Shirley, Phys. Rev. Lett. <u>37</u>, 1497 (1976).

21) T. E. Madey, J. E. Houston and S. C. Dahlberg, in Proc. Fourth Intern. Conf. Solid Surfaces and Third European Conference on Surface Science, eds. D. A. Degras and M. Costa (Supplement a la Revue "Le Vide, les Couches Minces" No. <u>201</u>, 205 (1980).

22) E. W. Plummer, B. J. Waclawski, T. V. Vorburger and C. E. Kuyatt, Prog. in Surface Sci. <u>7</u>, 149 (1976).

23) T. E. Madey, J. T. Yates, Jr., A. M. Bradshaw and F. M. Hoffmann, Surface Sci. <u>89</u>, 370 (1979).

24) A. M. Bradshaw and F. M. Hoffmann, Surface Sci. <u>72</u>, 513 (1978).

25) T. E. Madey, J. E. Houston, C. W. Seabury and T. N. Rhodin, J. Vac. Sci. Technol. <u>18</u>, 476 (1981).

26) F. P. Netzer and T. E. Madey, Phys. Rev. Lett. <u>47</u>, 928 (1981).

27) C. W. Seabury, T. N. Rhodin, R. J. Purtell and R. P. Merrill, Surface Sci. <u>93</u>, 117 (1980).

28) F. P. Netzer and T. E. Madey, Surface Sci. <u>110</u>, 251 (1981).

29) S. Lehwald, J. T. Yates, Jr. and H. Ibach, in Proc. IVC-8, ECOSS-3, Cannes, 1980, eds. D. A. Degras and M. Costa (Le Vide, les Couches Minces, Suppl. <u>201</u>, 221 (1980)).

30) T. E. Madey and J. T. Yates, Jr., Chem. Phys. Lett. <u>51</u>, 77 (1977).

31) D. L. Doering and T. E. Madey, Surface Sci. (in press).

32) P. A. Thiel, F. M. Hoffman, and W. H. Weinberg, J. Chem. Phys. <u>75</u>, 5556 (1981).

33) T. E. Madey and J. T. Yates, Jr., Surface Sci. <u>76</u>, 397 (1978).

34) F. M. Hoffmann, T. E. Felter, P. A. Thiel and W. H. Weinberg, J. Vac. Sci. Technol. <u>18</u>, 651 (1981).

35) H. Niehus, Surface Sci. <u>78</u>, 667 (1978).

36) T. E. Madey and J. E. Houston, unpublished.

37) H. Niehus, Surface Sci. 80, 245 (1979).

38) T. E. Madey, J. J. Czyzewski and J. T. Yates, Jr., Surface Sci. 57, 580 (1976).

39) R. Jaeger and D. Menzel, Surface Sci. 93, 71 (1980).

40) H. Niehus, Surface Sci. 92, 88 (1980).

41) T. E. Madey and J. T. Yates, Jr., Proc. 7th Intern. Vac. Congress and 3rd Intern. Conf. Solid Surfaces, Vienna (R. Dobrozemsky et al., Vienna, 1977) p. 1183.

42) T. E. Madey and F. P. Netzer, Surface Sci., 117,   (1982).

43) T. E. Madey, R. Stockbauer, S. A. Flodström, J. F. van der Veen, F. J. Himpsel and D. E. Eastman, Phys. Rev. B 23, 6847 (1981).

44) W. L. Clinton, to be published.

45) F. P. Netzer and T. E. Madey, Surface Sci. 119,   (1982).

46) F. P. Netzer and T. E. Madey, Chem. Phys. Letters, 88, 315 (1982).

47) L. D. Roelofs, A. R. Kortan, T. L. Einstein and R. L. Park, J. Vac. Sci. Technol. 18, 492 (1981).

48) B. Kamb, in Water and Aqueous Solutions, ed. by R. A. Horne (Wiley-Interscience, New York, 1972), p. 9.

49) M. Bowker, M. A. Barteau and R. J. Madix, Surface Sci. 92, 528 (1980).

50) K. Kretzschmar, J. K. Sass, A. M. Bradshaw and S. Holloway, Surface Sci. 115, 183 (1982).

51) W. M. Kang, C. H. Li, S. Y. Tong, C. W. Seabury, K. Jacobi, T. N. Rhodin, R. J. Purtell, and R. P. Merrill, Phys. Rev. Lett. 47, 931 (1981).

52) F. P. Netzer and T. E. Madey, Surface Science Letters, to be published.

53) C. G. Pantano and T. E. Madey, Applic. Surface Sci. 7, 115 (1981).

54) A. G. Naumovets and A. G. Fedorus, Zh. Eksp. Teor. Fiz. 68, 1183 (1975). [Soviet Physics-JETP 41, 587 (1976)]; V. V. Gonchar, O. V. Kanash, A. G. Naumovets and A. G. Fedorus, Soviet Physics-JETP Letters 28, 330 (1978).

55) R. Gomer, Japan J. Appl. Phys., Suppl. 2, Pt. 2, 213 (1974).

56) J. R. Chen and R. Gomer, Surface Sci. 81, 589 (1979).

57) T. E. Madey, Surface Sci. 94, 483 (1980).

58) D. E. Ramaker, in press.

59) R. Stockbauer, D. M. Hanson, S. A. Flodström and T. E. Madey, Phys. Rev. B , in press; also, J. Vac. Sci. Technol. 20, 562 (1982).

60) R. A. Rosenberg, V. Rehn, V. O. Jones, A. K. Green, C. C. Parks, G. Loubriel, and R. H. Stuhlen, Chem. Phys. Lett. 80, 488 (1981).

61) D. E. Ramaker, to be published.

62) R. Janow and N. Tzoar, Surface Sci. 69, 253 (1977).

63) E. Preuss, Surface Sci. 94, 249 (1980).

64) W. L. Clinton, to be published.

65) D. J. Godfrey and D. P. Woodruff, Surface Sci. 105, 438 (1981).

66) I. S. T. Tsong, N. H. Tolk, T. M. Buck, J. S. Krauss, T. R. Pian, and R. Kelly, Nucl. Instr. and Methods 194, 655 (1982).

67) W. L. Clinton, Surface Sci. 112, L791 (1981).

68) R. A. Gibbs, S. P. Holland, K. E. Foley, B. J. Garrison, and M. Winograd, Phys. Rev. B24, 6178 (1981).

69) J. W. Gadzuk, Surface Sci. 67, 77 (1977).

70) J. W. Gadzuk, Phys. Rev. B 14, 2267 (1976).

71) G. E. Busch and K. R. Wilson, J. Chem. Phys. 56, 3626 (1972); 56, 3638 (1972); 56, 3655 (1972).

3.4 Stimulated Desorption Spectroscopy*

M.L. Knotek

Sandia National Laboratories, Albuquerque, NM 87185, USA

As the field of stimulated desorption has progressed, a variety of different techniques with which to extract information on the surface chemical bond have evolved. One of the most useful has been the use of stimulated desorption in a spectroscopic mode where the ion yield vs. energy of the exciting particle is measured [1-20]. Spectral thresholds and shapes provide information on both the basic excitations initiating the desorption process and, conversely, the nature of the chemical bond from knowledge of the desorption process. Thus we perform two types of desorption spectroscopy experiments from surfaces. The first involves well-characterized systems that have been studied by other techniques, to gain information about the nature of the desorption process. In the second, we use relatively well-defined desorption processes to gain information on the nature of the surface bond. In both cases, it is necessary to pay close heed to the details of the physics of the excitation process since they are crucial to our ability to interpret spectra. The object of this paper is to bring together some of the elementary phenomena encountered in electron- and photon-stimulated desorption (ESD and PSD) in order to contrast the two excitation methods and outline the potential usefulness of desorption spectra.

One of the underlying physical facts which gives us hope for the use of stimulated desorption as an analytical probe of surfaces is the very highly localized energy deposition and desorption phenomena which have been observed in many classes of materials [21-29]. The most important point is that excitation of a given molecular species or site-adsorbate pair results in the bond breakage on that particular excited species. This means that spectral information observed in our desorption yield curves is specific to the species being desorbed as opposed to a more general excitation of the surface or bulk. At present it has not been rigorously established that nearest neighbor bond breakage is being observed to the exclusion of more delocalized processes. However, the evidence to date and the emerging theories go together to suggest that in many cases this is so. The prospect, then, is that we can examine surfaces that have

---

*This work performed at Sandia National Laboratories supported by the U.S. Dept. of Energy under contract # DE-AC04-76DP00789.

a multiplicity of bonding sites and adsorbates with the objective of determining two things: first of all, from threshold energies we can determine the bonding sites for desorbing species [21]. This is the first step in unravelling the chemistry of complex surfaces. Next, the analysis of the spectral dependence of the yield of a given desorbing species can provide the electronic and structural information on its bonding site, by exciting either the desorbing species or an atom in its near environment, be it an intramolecular excitation in an adsorbed complex or a substrate atom in the desorbing species bonding site.

We start with the general assumption that PSD and ESD are initiated by the same elementary excitation of the surface and that differences between the PSD and ESD spectral yield curves reflect differences in the physics of the excitation process. The equivalence of the ESD and PSD excitation process has been established by the comparison of: a) thresholds for desorption; b) ion energy distributions; c) ion angular distributions, and; d) the nature of the chemical states leading to desorption. In comparing the thresholds for PSD and ESD, it has been found that in all measured examples the ESD and PSD thresholds are essentially equal to within the uncertainty of determining thresholds in ESD,[10][16][29]. In addition Woodruff et al. have shown that derivative ESD data show similar spectral features to the equivalent PSD data for O^+ from O/W(100) [10]. PSD and ESD ion energy distributions for O^+/W(100) show a like correspondence suggesting that the same final state is accessed in both cases [10]. Ion angular distributions for O^+ from O/W measured using both an imaging analyzer [11] and an angle resolved cylindrical mirror analyzer [10] again suggest that ESD and PSD are accessing equivalent final states of the desorbing chemical configuration. Finally, measurements of yield vs coverage for O/W [11], CO/Ru and CO/Ni [4][30][31] and the chemical state of the TiO_2 surface [21][29] show that the same chemical state exhibits high yield in both PSD and ESD. These all suggest that: a) the excitations leading to desorption map one-to-one for ESD and PSD with neither exhibiting any additional processes over the other, and; b) both show the same detailed evolution to the final desorptive state. As we show below, however, this might be somewhat surprising given the difference in detail of the two excitation processes.

Excitation Physics

We now briefly examine electron and photon excitation cross sections from an atomic viewpoint. We show that there are differences in the spectral dependence of the excitation cross section, in the final states accessed, in the way we define the energy deposited in the excitation process, and in the magnitude and energy dependence of the maximum cross section we can expect for electron- and photon-stimulated desorption processes. Finally, the density of final state information conveyed by the two techniques show distinct differences as well.

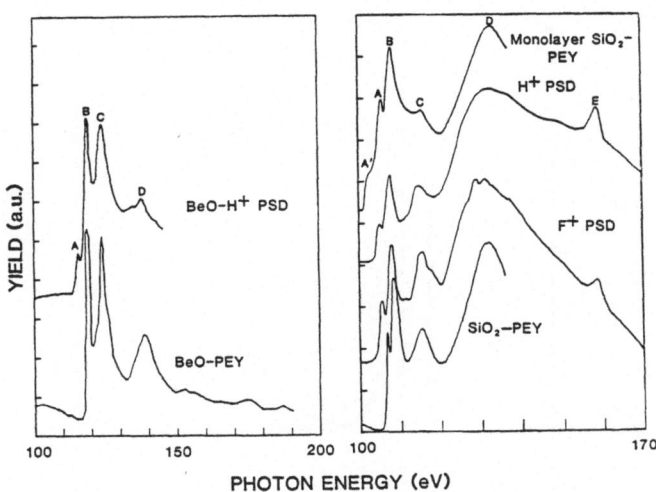

PHOTON ENERGY (eV)

Fig.1 PSD of H^+ and F^+ from BeO (1120) and fused SiO_2 compared to equivalent photoelectron yield data at the Be(K) and
$Si(L_{23})$ excitation thresholds, respectively. The envelope of
the Be(K) edge shows the typical <u>abrupt</u> K-edge threshold while
the $Si(L_{23})$ shows a delayed onset typical of p→d excitations.
Details of the fine structure are discussed in the text (from
ref. [32])

Consider Figures 1 and 2, which show PSD spectra from BeO and
SiO_2 [32], and ESD spectra from clean and oxidized Si(111) [33],
respectively. We notice that PSD spectra have very sharp thresholds, followed by peaks in desorption and a relatively abrupt
decay of the signal above threshold. By contrast, ESD spectra
have weak thresholds and rise smoothly above threshold. At
higher energies ESD spectra peak out at something on the order
of several times the threshold energy and then saturate or fall
off smoothly with a weak dependence on energy [34]. In electron stimulated desorption we seldom observe peaks or other
strong structure conveying geometric or electronic structural
information. As already mentioned, however, derivative ESD
spectra do have some equivalence to PSD spectra [10].

We next examine the origin of the spectral shapes in the
two cases, by looking at Figure 3, which shows the electron and
photon excitation cross sections for helium and neon atoms
[35][36]. There are four distinct spectral shapes seen in
these electron and photon excitation functions. The simplest
excitation, the photoexcitation of the He(1s), which can be
understood on the basis of a simple Coulomb interaction between
the electron and the core or nucleus of the excited atom
[35][36], consists of an extremely abrupt threshold followed
by a sharp decay above threshold with an energy dependence of
the order of $1/E^n$ where $2 < n < 3$. In the excitation of a state
of higher angular momentum, the potential between the nucleus
and the region away from the core takes on a more steep func-

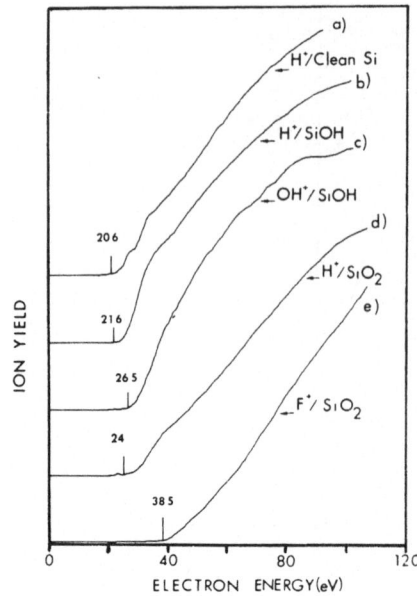

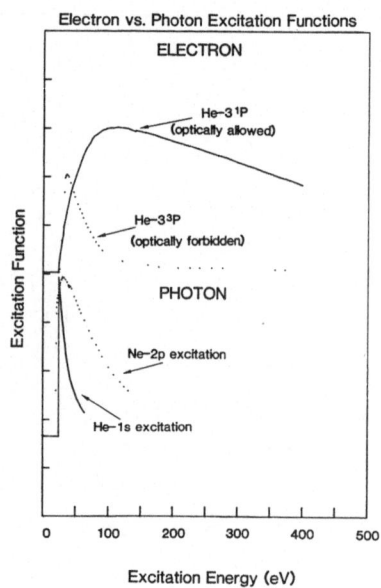

Fig. 3

Fig. 2. A series of ESD spectra from Si(111) and oxidized Si(111).
All of the spectra show relatively weak onsets and rise to a peak
at higher energies. The threshold is often followed by an inflection point and then a more gradual rise, indicative of nondipole
and dipole contributions, respectively (from ref.[33])

Fig. 3. Electron and photon excitation functions of the He(1s)
and Ne(2p) levels displaying atomic effects in absorption spectra
(from refs. [35,36]

tional dependence. It is of the order of Ze^2/r very close to
the nucleus and more like e^2/r well away from the nucleus, when
the electron is outside the electronic shells. In the region
in between the functional dependence behaves more like $\propto e^2/r^3$.
In addition, there is another contribution to the effective potential; the so-called centrifugal <u>force</u> which has the functional
form

$$F = \ell(\ell+1)h^2/mr^3 \ . \tag{1}$$

The sum of these two terms results in a peak in the effective
potential within 1 or 2 atomic units of the nucleus. This can
result in the so called delayed onset for photon excitation;
the peak in the photo excitation cross section then occurs at
an energy well above threshold with a subsequent $1/E^n$ decay,
like the simple hydrogenic case, as illustrated in the neon 2p
excitation cross section.

142

For electron excitation dipole selection rules are not in effect when the Born-Oppenheimer approximation does not apply. However, there are differences between dipole-allowed and forbidden excitation functions [37]. The He [1]s to [1]p excitation, which is dipole allowed, has an excitation spectrum which rises weakly from threshold, peaks at 3-4 times the threshold energy and decays like $\ln E/E$ at higher energies. By contrast, the dipole-forbidden [1]s to [3]p excitation peaks more rapidly at ~ 1.5 times the threshold energy and then falls off like $1/E$. While this is but the simplest case, this general behavior is widely observed.

We now reexamine Figures 1 and 2. In Fig.1 the BeO PSD and PEY show an envelope typical of K-edge absorption; an abrupt edge followed by a sharp falloff. The SiO_2 L_{23} edge shows a delayed onset typical of $p \rightarrow d$ excitations, with a peak at ~ 30 eV above edge. The Si L_1 edge is weak but not as weak as equivalent PEY spectra. This is because Koster-Kronig deexcitation, which lifetime broadens the L_1 edge, leads to an Auger cascade, $(L_1L_{23}V + L_{23}VV)$, generating an extra valence-hole and enhancing desorption. The origin of the sharp peaks inside the envelopes is not atomic and will be discussed below.

The ESD spectra of Fig.2 are typical of ESD spectra observed from TiO_2 [13], $SrTiO_3$ [14], condensed alkanes [18], and a variety of oxidized metals [10]. We see a relatively abrupt threshold followed in some cases by an inflection point at ~ 1.5-2 times the threshold energy followed by a more gradual increase in yield. In some cases there is discernable near-edge structure within ~ 20 eV of the edge (due to solid state of molecular interactions), but the overall envelope is generally maintained with varying relative strengths to the two components. In desorption of H^+ from metals [34] and metal-like organic molecules [38] the spectrum is superlinear throughout the whole energy range with no apparent structure. The simplest interpretation of these general features is that the initial rise can have contributions from dipole forbidden transitions and the second is due to dipole allowed. This suggests that ESD may provide complementary information to PSD in the near threshold region.

The most elementary excitations are discussed here to give the reader some insight into the nature of the problem of interpreting desorption spectra. In the absorption by deeper core-levels with higher angular momentum initial and final states, with higher degree of localization, strong departures from these simple forms are encountered [36]. However, due to these more complex edge effects, excitations from higher angular momentum states are of limited utility in an analytical sense because these strong atomic effects mask features due to the atoms environment.

The next distinct difference between electron and photon stimulated processes is quantifying the energy adsorbed by the electronic structure of the solid. Most electron-stimulated desorption experiments have been carried out using a thermionic emitter. A simple potential diagram of this experimental arrangement is shown in Figure 4. The electron has an energy relative to the Fermi level of the sample given by the sum of the applied potential, eVapp, plus the work function of the cathode, $e\phi_c$, plus an energy, kT, where T is the temperature of the cathode. When the electron strikes the sample, it can give up an energy equal to this entire energy if the electron falls to the Fermi level of the substrate upon which the desorbed species is bound. Then at threshold we equate the energy of the excitation, E_{exc}, to the sum which we will call the available electron energy, E_e, given by

$$E_e = eVapp + e\phi_c + kT = E_{exc}. \qquad (2)$$

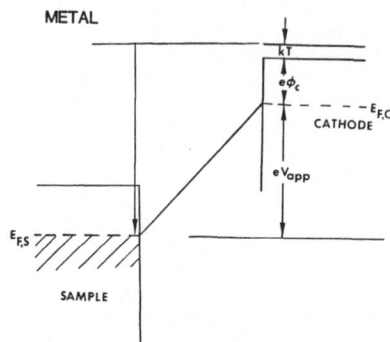

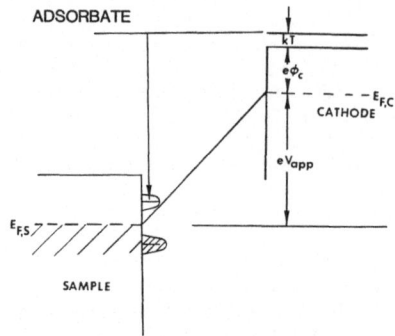

Fig.4. Energy diagram appropriate for ESD when the primary electrons final state is at the Fermi level of the sample. An energy E_e = eVapp + e ϕ_c + kT is released

Fig.5. Energy diagram appropriate for ESD when the primary electron has a final state above the samples Fermi level in an adsorbate. An energy E = eVapp + e ϕ_c + kT − Δ is released

This simple treatment of the electron-stimulated desorption data has given reasonable, and understandable, agreement in thresholds between ESD and PSD data [10][13][14][29][39]. Some deviations from this have been noticed, for example in SrTiO$_3$ where it is known that the bands bend near their surface it has been shown that chemical induced shifts in the conduction band position relative to the Fermi level result in energy shifts between the ESD and PSD thresholds [14]. The important difference between ESD and PSD is that at any photon energy hv, all of the energy is absorbed so that at threshold, the excitation energy is equal to hv,

$$E_{exc.} = (hv)_{th}. \qquad (3)$$

An adsorbate on the surface may have an electronic structure such that incoming electrons do not end up at the Fermi level, but in an empty final state on the excited species (Fig.5). The energy deposited in the excitation process is equal to E_e

144

minus some energy Δ which is the energy difference between the Fermi level and the final state. Thus in an ESD spectrum we will find that the threshold for an excitation with an energy $E_{exc.}$ is shifted upward by Δ, so that

$$(E_e)_{th} = E_{exc.} + \Delta. \tag{4}$$

Hence, differences in ESD and PSD thresholds can tell us something about the location of the empty final state relative to the Fermi level of the solid, assuming they both involve the same electronic excitation. In some cases this may not be true especially when there is a strong non-dipole component to the excitation process near threshold. We also must not neglect the interaction between the primary and excited electrons, which may cause shifts in ESD thresholds.

A further complication in ESD or any appearance potential spectroscopy is that the primary electron must have an empty final state to occupy in the solid [40]. If there is some distribution of final states to be accessed both by the primary and the excited electrons, above threshold both electrons will have access to a range of final states dictated by energy balance, with the incoming electron giving up varying amounts of its total energy depending on the nature of its final state and its position in energy. The probability for excitation to some energy E_O then has the functional form

$$P_e(E_O) \; \alpha \; \int_0^{E_O} N(E)N(E_O-E)dE \; . \tag{5}$$

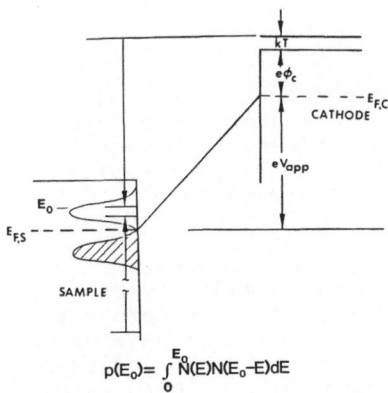

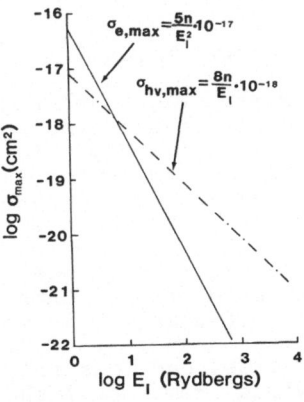

Fig.6. An energy diagram showing that both primary and excited electrons can have a range of final states dictated by energy balance. The excitation probability contains a convoluted density of states function

Fig.7. The maximum excitation cross section as a function of the excitation energy E_I for electrons and photons. At low energies e-cross sections dominate, at high energies hν-cross sections are larger

In PSD we find that the simple one electron nature of the excitation results in the excitation spectrum reflecting the final state density of the excited species;

$$P_{h\nu}(E) \propto N(E).$$ (6)

Hence, ESD spectra suffer from two problems in comparison to PSD; first, the the thresholds are very weak and spread out; second, the incoming electron must have a final state in the solid, which convolutes our final state density information. The upshot is that the PSD spectra are in general much richer sources of surface electronic structure information and in the near term at least we will find that when looking for information in desorption spectra, PSD will be the technique of choice.

Aside from the different functional forms, the electron and photon ionization cross sections exhibit differences in magnitude for a given excitation. An approximate form for the maximum electron-excited cross section for a level with an ionization energy E_I containing n electrons is [41][42]

$$\sigma_{e,max} = \frac{5n}{E_I^2} \cdot 10^{-17} \text{ cm}^2 \text{(Rydberg)}^2$$ (7)

while the equivalent maximum cross section for photons is

$$\sigma_{h\nu, max} = \frac{8n}{E_I} \cdot 10^{-18} \text{ cm}^2 \text{ Rydberg}.$$ (8)

These are shown in Fig.7.

By examining Fig.7 and Fig. 3 we can understand several points. First, since even the best photon source has a "monochromatic" flux of only $\sim 10^{10}$-10^{11} photons/sec. (i.e., equivalent to ~ 1 nA of electrons) as opposed to $\sim 1 \mu A$ in a typical electron beam experiment, PSD yield can be exceedingly small by comparison to ESD in the region below 10 Rydbergs. Hence, the time lag between the first observations of ESD and PSD. Second, one can readily see why beam damage by an electron beam is so much greater than by a photon beam [43]. By Fig.3 an electron of energy E is very effective in exciting processes with excitation energies $<<E$. Photons, by contrast, have appreciable cross sections only down to $\cong E/3$, say. In addition, it is obvious from Fig. 7 that the greatest desorption/damage will occur by processes with an excitation energy ≤ 10 Rydbergs where the electron cross sections are greater than the equivalent photon cross sections. Finally, we can get some feel for the relative role of desorption by secondary electrons vs the primary electron or photon. While the number of true secondaries from any primary event are relatively small and heavily weighted to energies < 2 Rydbergs, the cross sections for lower lying levels which they can excite can be quite large in comparison to the primary event. To determine relative yields

we weigh the primary excitation cross section times its ion
yield probability against the excitation cross section for
lower energy processes, which give the same ion, integrated
across that part of the secondary electron distribution which
has energies in excess of the ESD threshold for desorption.
As reported by Jaeger et al. at this meeting, secondary electrons can have a significant ESD contribution, even in PSD experiments where the secondary yield is lower than in ESD, and
must be considered when using stimulated desorption as a spectroscopy.

Examples

The simplest spectra from which to extract information on the
excited atom are those for which the desorption yield is directly proportional to the excitation cross section, e.g., in
a system where creation of a core-hole leads to desorption by
an Auger decay. In that case, the desorption probability $P(E)$
at an excitation energy E is given by

$$P(E) \; \alpha \; A(1-f) \, \theta(E) , \qquad (9)$$

where A is the probability that Auger decay results in a desorptive final state, f is the reneutralization probability,
and $\theta(E)$ is the core-hole ionization cross section at the
energy E. Since to a first approximation only θ is a function
of E, the desorption spectrum will directly reflect the excited
atoms x-ray absorption spectrum. At present we have only a
few examples of different desorption spectral phenomena which
are more "proof of concept" than finished analytical studies.
Indeed, the message of this proceedings is that little has
been done and opportunities abound. We will examine some of
these examples looking first at the region near the ionization
threshold and then proceeding to the energy regions progres-
sively higher above threshold. In the energy region < 50 eV
and > 50 eV above threshold the acronyms XANES and EXAFS for
x-ray absorption near edge structure and extended x-ray absorption fine structure, respectively, are commonly used. When an
electron is excited from a core level, there are a variety of
bound or semi-bound and continuum final states to be assessed.

Autoionization Resonances

When the atomic final state is highly localized, strong resonances can be observed in the near-edge region. While these
are atomic in nature, they can be affected by the environment
as typified in Fig.8 which shows H^+ from oxidized cerium near
the excitation threshold for the Ce(4d) ionization compared to
optical absorption in Ce metal and CeO_2 [17]. We observe a
series of sharp atomic-like resonances which arise from the
strong coupling between the Ce 4d and 4f shells [44][45]. Each
feature in these spectra is due to a dipole transition of the
form

$$4d^{10}4f^n \rightarrow 4d^9 4f^{n+1} . \qquad (10)$$

147

The exchange interaction between the f electron and the remaining d electrons splits the $4d^9 4f^{n+1}$ configuration giving rise to a series of multiplet states. These intermediate bound excited states then autoionize to a final state from which ion desorption occurs. The multiplet structure is critically dependent on the valency of the metal atom. Curve (a) in Figure 8 shows the H^+ PSD ion-yield spectrum after 750-L O_2 exposure at 300 K, while curve b is the H^+ desorption after heating to 475 K. Curve (c) and (d) are the optical absorption spectra for CeO_2 (Ce^{4+}) and Ce metal (Ce^{3+}), respectively. The primary difference between (a) and (b) is crosshatched in (a) and shows that annealing removes Ce^{4+} valency from the surface. Analysis of this experiment and accompanying measurements showed that the Ce oxidizes to a CeO_2 stoichiometry which reduces to Ce_2O_3 on the surface due to the high oxygen conductivity of CeO_2.

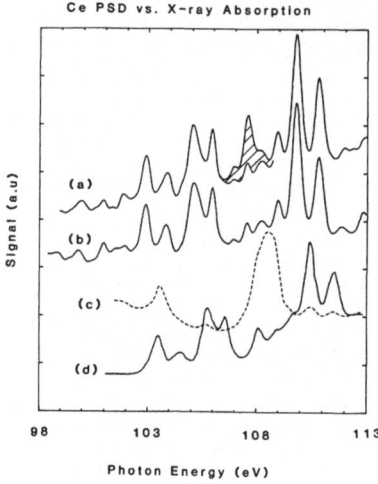

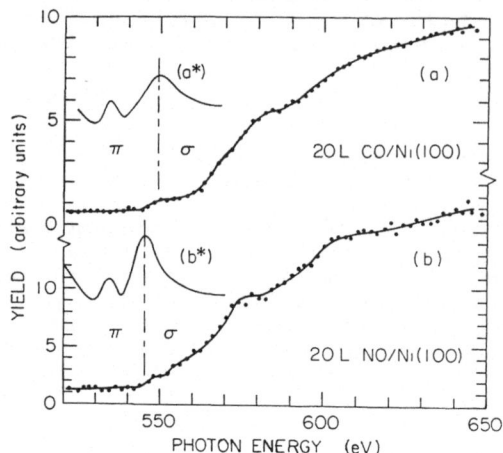

Fig. 8   Fig. 9

Fig. 8. A comparison of H^+ PSD from (a) freshly oxidized Ce and (b) oxidized Ce annealed at 475 K compared to x-ray absorption of (c) CeO_2 and (d) Ce metal. The shaded peak in (a) is due to desorption from Ce^{4+} site which are absent after annealing (from ref. [17])

Fig. 9. Comparison of partial PEY (a*) and (b*) and O^+ PSD (a) and (b) from CO and NO on Ni(100), respectively. The PSD does not exhibit the π resonance and has a weak σ resonance (from ref. [16])

Core Excitons

In many insulating materials or wide band semiconductors, the first excitation out of a core level is to the so-called core excitonic state. This electronic state is equivalent to a Z+1 impurity orbital whose binding energy and radius are determined by the dielectric properties of the surrounding solid

and the effective mass of the particular conduction band to
which the electron was excited. This is not unlike the Rydberg
levels observed in molecules. Chemically-induced changes in
the excitons energy position are effected by shifts in the
core-position, in the conduction band edge, and in the binding
energy of the exciton relative to the conduction band edge
[46][47]. If the exciton's position is dominated by the core's
position then information equivalent to that from XPS is present. An example of a shift in this core exciton as a function
of different adsorbates at a bonding site are shown in Fig.1b
showing PSD spectra for H^+ and F^+ from a fused SiO_2 surface
in the vicinity of the silicon L_{23} edge. The core exciton
for the H^+ PSD occurs at 105.5 eV while the core exciton for
the F^+ PSD occurs at 105.9 eV. By contrast, the core exciton
in photo yield from SiO_2 has an energy at the peak of 106.5
eV, while a monolayer of SiO_2 on Si has a core exciton in the
vicinity of 105.4 eV [46][47]. From this standpoint H^+ is
bonded to a silicon site very similar to the top most layer of
oxide on an oxidized surface. F adsorbed at the site shifts
it more towards bulk values. Such core excitons are also
observed in BeO, Fig.1a, where peak A is not observed
in the PEY, in Al_2O_3 [1][32] and alkalide halides such as NaF
[48]. In all of these cases the core exciton, due to its extreme sharpness, is a better gauge of the desorption threshold
than simply trying to determine an onset.

Inner-Well and Shape Resonances

In the energy region from the continuum threshold to an energy
up to ~ 50 eV there are a variety of very complex structures
which are due to the interaction of the electron with the neighboring atoms. These structures are due to density of states
structure coupled with scattering from the lattice. An example
of such interactions for covalent adsorbates is the data of
JAEGER et al. [16] in Fig.9 showing O^+ PSD and partial photoelectron yield (PEY) for CO and NO on Ni(111) in the vicinity
of the O(K) edge. The PEY's in the near edge (a* and b*) are
dominated by the π and σ structures. The π structure is due
to a 1σ–π bound state excitation while the σ is a 1σ–σ shape
resonance, both intramolecular on the CO molecule. The striking
feature of the O^+ PSD data is that it does not follow the PEY
(i.e., the x-ray absorption). Instead the PSD shows no structure at the π resonance, only a weak σ structure and a main
onset which can be assigned to a 1σ+3σ shakeup excitation.
This implies that in such adsorbates an excitation more complex
than core ionization is necessary to initiate desorption. Recent Auger data by KOEL et al. [49] show that for CO adsorbed
on metals the hole-hole repulsive energy of the Auger final
state is reduced to ~ 0 eV from the gas phase value of 15 eV.
Thus while a simple Auger decay of a O(K) core hole results in
efficient gas phase dissociation of CO [50],such a simple
process is not expected to give desorption on a surface.
Hence, a more complicated excitation seems to be necessary.

In the more ionic materials such as Al_2O_3, BeO and SiO_2, the
interaction with the anion shell around the excited metal atom
leads to peaks called inner-well resonances or Tossel structures

[46] which can yield information on geometric details of the
environment of the excited atom. These structures are considerably more complex to analyze than are EXAFS structures due to
multiple scattering processes [51]. In general, to analyze
these spectra model site geometries are treated and the derived
structures are compared to the experimental data. In principle
these structures are more informative than EXAFS in that multiple atom position correlations can be drawn together with
detailed information on the bond angles and site symmetries.
One promising empirical way to analyze these PSD structures is
to make direct comparisons to the x-ray absorption spectra for
model compounds when the dominant features in the structure
are determined by the first neighbor shell. In Al_2O_3, SiO_2,
BeO and other ionic materials where the interaction of the
emitted electrons is predominantly with the negative anion
shell around the excited atom, the inner-well resonance structures are determined mainly by the immediate environment of
the atom [46][47]. In covalent materials, or in metals, the
scattering from several far nearest-neighbor atomic shells are
important in determining the structure, so the use of model
compounds to mimic the site environment cannot be so readily
employed [52].

Figures 1 and 10 show examples of the use of model spectra
to analyze the inner well resonance peaks in PSD spectra.

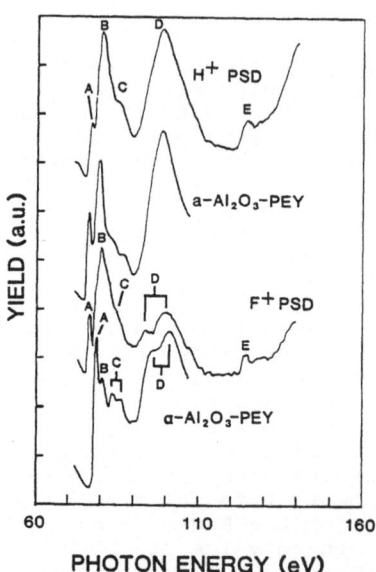

PHOTON ENERGY (eV)

Fig. 10. A comparison of H^+ and
F^+ PSD from α-Al_2O_3 (0001) and
PEY from a-Al_2O_3 (tetrahedral)
and α-Al_2O_3 (octahedral). The H
bonding site is purely tetrahedral while the F shows a mixed
signature (from ref. [32])

Fig.1a shows the H^+ PSD and PEY for BeO and demonstrates that
the surface Be site to which the H is bonded (either directly
to the Be or to the oxygen in BeO) has the same electronic and
geometric environment as the bulk. Likewise in Fig.1b the Si
bonding site is tetrahedrally coordinated in all cases, but

150

the positions of the peaks are shifted relative to the bulk
site for the surface oxide PEY and the H^+ and F^+ PSD. Thus
all of the "surface" structures have a similar (lower) valence
relative to the "bulk" site. Fig.10 on the other hand, which
shows H^+ and F^+ PSD from an $\alpha-Al_2O_3(0001)$ surface, and PEY
from amorphous Al_2O_3 ($a-Al_2O_3$) and $\alpha-Al_2O_3$, indicate H^+ and F^+
bond at quite different sites [32]. The $a-Al_2O_3$ PEY has the
distinctive tetrahedral signature, similar to SiO_2. The H^+
PSD from $a-Al_2O_3$ also shows a tetrahedral signature. In
$\alpha-Al_2O_3$ the bulk site is octahedral and the PEY shows a split
peak at D. The F^+ PSD shows a similar split peak at D indicating that the molecular field at the aluminum atom to which
the F is bonded is similar to the octahedral site. Note, however, that the peaks A, B and C of the F^+ PSD are not unlike
those for the H^+ PSD indicating that the F site has some similarities to the tetrahedral site as well. One important fact
pointed out by this data is that further modeling of such systems is needed and would be most fruitful.

An example of PSD from an "oxidized" surface was presented
by JAEGER et al. [5] who analyzed the desorption of O^+ from
adsorbed O_2 on Mo(100) at the molydenum 3D and 3P edges. Delayed onsets were observed in both edges with additional sharp
near edge structures due to $Mo(3p_{1/2})$, $(3p_{3/2}) \rightarrow 4d$ excitation.
In the ion yield, thesholds were observed at the Mo(3d), (3p)
and (3s) and O (1s) edges. By contrast, the total photo electron yield (PEY) showed no measureable photo electron yield
from the oxygen 1s edge, indicating a lack of surface sensitivity for total PEY. The O^+ PSD threshold was shifted to higher
energy relative to the PEY from the metal underlayer, indicating
that the O^+ yielding sites were in an "oxidic" state.

Similar recent work on oxidized Ti(0001) show that the
oxidized surface contained hydrogen as well as oxygen [53].
The H was present in both hydride and hydroxide forms. Comparison of ion yields at the Ti(L) and O(K) edges to equivalent
soft x-ray absorption (SXA) data for bulk TiO_2 show that the
O^+ yield is from a TiO_2-like site while the OH^+ come from
a site of distinctly different valence. H^+ appears to be
bonded to both terminal O in a hydroxide state as well as to
lattice oxygen [2][54]. These studies demonstrate the utility
of making detailed spectral comparisons both between different
desorbed species on a given surface and to bulk "model" spectra.

EXAFS

In the energy region further above threshold, the so-called
extended x-ray absorption fine structure (or EXAFS) region, the
excited electrons interaction with the environment is simpler
to analyze [15]. When the photoemitted electron leaves the
excited atom it is scattered by the cores of the atoms in its
environment such that the wavefunction at the excited core
contains a factor

$$1 + \sum_R A_{k,R} \sin[2kR + \phi(k)] , \qquad (11)$$

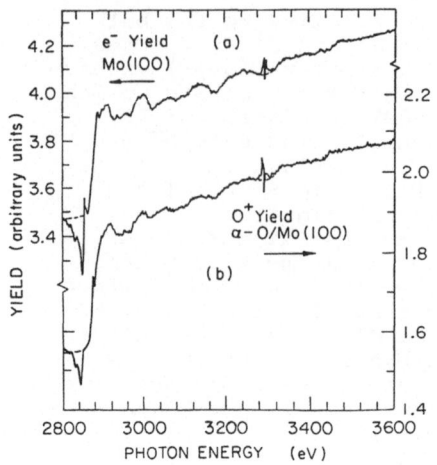

Fig. 11. A comparison of e⁻ yield from Mo(100) and O⁺ PSD yield from α-O/Mo(100) showing that O-bonded Mo has same Mo-Mo separation as bulk Mo but 1/2 the bulk coordiantion (from ref. [15])

where k is the wavevector of the excited electron, R is the radial distance of a specific coordination shell, and $\phi(k)$ is the scattering phase shift. For a given coordination shell

$$\sum_{shell} A_{k,R} = \frac{N}{kR^2} \ f(k,\pi) \ e^{-2\sigma^2 k^2 - 2R/\lambda(k)} \ , \tag{12}$$

where the first term arises from the spherical wave nature of the outgoing electron, f is a scattering amplitude, the σ^2 term in the exponential is the Debye-Waller factor and λ is the inelastic mean free path. From these equations we see that the oscillator structures in the EXAFS region of the spectrum can be related to the Fourier transform of the radial distribution function of the excited atom. Figure 11 shows the first example of the use of PSD to obtain EXAFS data which can be analyzed to obtain an RDF [15]. This data shows the O⁺ PSD yield and the e-yield for an oxygen exposed Mo(100) surface. Analysis of this data showed a) the Mo-Mo separation for the excited atom is essentially identical to the bulk Mo-Mo separation; b) using the general form

$$N = \text{effective coordination number} = \sum_{\substack{atoms \\ in \ shell}} (\frac{1}{3} + \vec{\varepsilon} \cdot \vec{r}_\ell) \ ,$$

where $\vec{\varepsilon}$ is the polarization vector and $\vec{r}_\ell$ is the unit vector of the ℓth atom, the data show that the coordination of the excited Mo atom is 1/2 that of the bulk; c) from the position of the threshold of this and the other core edges just discussed [5] it was determined that the excited Mo was in an "oxidic" state. These results go together to suggest that a Mo atom in a 4-fold site on the surface can be coordinated by up to 3 oxygens and change in valency up to ~ 6+ without a measurable change in the Mo-Mo separation; quite remarkable.

Conclusion

This paper breafly acquaints the reader not so much with the details involved in the analysis of ESD and PSD spectra, but with some of the ideas we use to understand their general features. While our level of accomplishment in this area is limited due to the limited data base, there is a great deal of potential physical and chemical insight to be gained from further work. This, together with the use of other desorption techniques such as ion angular and energy distributions promise to make stimulated desorption studies a major contributor to our understanding of surfaces.

References

1. M. L. Knotek, V. O Jones, and V. Rehn, Surf. Sci. 102, 566 (1981).

2. D. M. Hanson, R. Stockbauer, and T. E. Madey, Phys. Rev. B10, 5513 (1981).

3. P. Feulner, R. Treichler and D. Menzel, Phys. Rev. B 24, 7427 (1981).

4. T. E. Madey, R. Stockbauer, S. A. Flodström, J. F. van der Veen, F. J. Himpsel, and D. E. Eastman, Phys. Rev. B 23, 6847 (1981).

5. R. Jaeger, J. Stöhr, J. Feldhaus, S. Brennan and D. Menzel, Phys. Rev. B, 23, 2102 (1981).

6. H. Niehus and W. Losch, Surf. Sci. 111, 344 (1981).

7. R. A. Rosenberg, V. Rehn, V. O. Jones, A. K. Green, C. C. Parks, G. Loubriel, and R. H. Stulen, Chem. Phys. Lett. 80, 488 (1981).

8. S. L. Weng, Phys. Rev. B 23, 3788 (1981).

9. S. L. Weng, Phys. Rev. B 23 1699 (1981).

10. D. P. Woodruff, M. M. Traum, H. H. Farrell, and N. V. Smith, Phys. Rev. B 21, 5642 (1980).

11. T. E. Madey, R. L. Stockbauer, J. F. van der Veen, and D. E. Eastman, Phys. Rev. Lett. 45, 187 (1980).

12. E. Bauer and H. Poppa, Surf. Sci. 99, 341 (1980).

13. M. L. Knotek, Surf. Sci. 91, L17 (1980).

14. M. L. Knotek, Surf. Sci. 101, 334 (1980).

15. R. Jaeger, J. Feldhaus, J. Haase, J. Stöhr, Z. Hussain, D. Menzel, and D. Norman, Phys. Rev. Lett. 45, 1870 (1980).

16. R. Jaeger, R. Treichler and J. Stǒhr, Surf. Sci. <u>117</u>, 533 (1982).

17. B. E. Koel, G. M. Loubriel, M. L. Knotek, R. H. Stulen, R. A. Rosenberg, C. C. Parks, Phys. Rev. B, <u>25</u>, 5551 (1982).

18. J. A. Kelber and M. L. Knotek, "Electron Stimulated Desorption of Condesed, Branched Alkanes." Sur. Sci., (In Press)

19. R. H. Stulen, T. E. Felter, R. A. Rosenberg, M. L. Knotek, G. Loubriel and C. C. Parks, Phys. Rev. B, <u>25</u>, 6530 (1982).

20. R. Franchy and D. Menzel, Phys. Rev. Lett. <u>43</u>, 865 (1979).

21. M. L. Knotek and P. J. Feibelman, Phys. Rev. Lett. <u>40</u>, 964 (1979).

22. P. J. Feibelman and M. L. Knotek, Phys. Rev. B <u>18</u>, 6531 (1978).

23. M. L. Knotek and P. J. Feibelman, Surf. Sci. <u>90</u>, 78 (1979).

24. P. J. Feibelman, Surf. Sci. <u>102</u>, L51 (1981).

25. M. L. Knotek, Nature Vol. <u>291</u>, 452 (1981).

26. D. E. Ramaker, C. T. White and J. S. Murday, J. Vac. Sci. Technol., 18(3) (1981) 748.

27. D. R. Jennison, J. Vac. Sci. Technol. <u>20</u>, 548 (1982).

28. C. F. Melius, R. H. Stulen, J. O. Poell, Phys. Rev. Lett. <u>48</u>, 1429 (1982).

29. M. L. Knotek, V. O. Jones, and V. Rehn, Phys. Rev. Letters <u>43</u>, 300 (1979).

30. T. E. Madey, Surf. Sci. <u>79</u>, 575 (1979).

31. R. Jaeger, J. Stǒhr, R. Treichler, and K. Baberschke, Phys. Rev. Lett. <u>47</u>, 1300 (1981).

32. M. L. Knotek, R. H. Stulen, G. M. Loubriel, V. Rehn, R. A. Rosenberg, C. C. Parks, submitted to Surface Science.

33. M. L. Knotek and J. E. Houston, J. Vac. Sci. Technol. <u>20</u>, 544 (1982).

34. M. Nishijima and F. M. Propst, Phys. Rev. B<u>2</u>, 2368 (1970).

35. H. S. W. Massey and E. H. S Burhop (Clarendon Press, Oxford, 1969). "Electronic Ionic Impact Phenomina, Vol. I, Collisions of Electrons with Atoms."

36. U. Fano and J. W. Cooper, Reviews of Modern Physics <u>40</u>, 141 (1968).

37. V. E. Henrich, G. Dreaselhaus, and H. J. Zeiger, Phys. Rev. B22, 4764 (1980).

38. J. A. Kelber and M. L. Knotek, to be published.

39. M. M. Traum and D. P. Woodruff, J. Vac. Sci. Technol. 17, 1202 (1980).

40. R. L. Park and J. E. Houston, J. Vac. Sci. Technol. 11, 1 (1974).

41. E. J. McGuire, Phys. Rev. A 16, 73 (1977).

42. E. J. McGuire, The Photoiionization Cross-section of the Elements Vol. III - Magnesium to Potassium, Report SC-TM-67-2955, Sandia National Laboratories, Albuquerque, NM (1967).

43. C. G. Pantano and T. E. Madey, Surf. Sci. 7, 115 (1981).

44. J. L. Dehmer, A. F. Starace, U. Fano, J. Sugar and J. W. Cooper, Phys. Rev. Lett. 21, 1521 (1971).

45. J. Sugar, Phys. Rev. B 5, 1785 (1972).

46. A. Bianconi, Surf. Sci. 89, 41 (1979).

47. A. Bianconi, Appl. of Surf. Sci. 6, 392 (1080).

48. C. C. Parks, D. A. Shirley, Z. Hussain, M. L. Knotek, G. M. Loubriel and R. A. Rosenberg, to be submitted to Phys. Rev. B, "Photon Stimulated Desorption from a Sodium Fluoride (100) Surface".

49. B. E. Koel, J. M. White, G. M. Loubriel, to be published.

50. R. B. Kay, Ph.E. Van der Leeuw and M. J. Van der Weil, J. Phys. B10, 2521 (1977).

51. P. J. Durham, J. B. Pendry, C. H. Hodges, Sol. State Comm. 38, 159 (1981).

52. G. N. Greaves, P. J. Durham, G. Diakun, P. Quinn, submitted to Nature.

53. C. C. Parks, D. A. Shirley, M. L. Knotek, G. M. Loubriel, B. E. Koel, R. A. Rosenberg and R. H. Stulen, to be submitted to Surface Science, "Photon-Stimulated Desorption from an Oxidized Ti(00o1) Surface".

54. "PSD and Ultraviolet Photoemission Spectroscopic Study of the Interaction of H_2O with a Ti(001) Surface," R. Stockbauer, D. M. Hanson, S. Anders Flodström and T. E. Madey, to be published in Phys. Rev.

3.5 The Electronic Desorption of Excited Alkali Atoms from Alkali Halide Surfaces

N.H. Tolk, W.E. Collins, J.S. Kraus, R.J. Morris, T.R. Pian, and M.M. Traum

Bell Laboratories, Murray Hill, NJ 07974, USA

N.G. Stoffel and G. Margaritondo

Department of Physics, University of Wisconsin, Madison, WI 53706, USA

1. Introduction

Alkali halides have long been known to erode under electron and photon bombardment. The detailed nature of the electronic mechanisms responsible for the erosion remains, however, a matter of vigorous discussion. In order to elucidate some of these mechanisms we present recent measurements of the incident beam energy dependence of excited alkali neutral desorption arising from electron and photon bombardment of alkali halide surfaces. In all cases, the yields for excited alkali neutrals are measured to be much greater than those obtained for positive ion ejection due to electron or photon stimulated desorption. Substrate core hole formation is seen to play an important role in the initial energy transfer process. In each case only the lowest resonance excited states are observed. The energy dependent yield of excited neutrals arising from electron bombardment is found to differ markedly from the corresponding photon bombardment data. Present theories dealing with (a) the production and migration of defects, and (b) the ejection of ions, are not adequate to explain the desorption of neutral alkalis in excited states. In both the incident electron and photon cases, our observations suggest microscopic models which take into account the mechanisms of initial electronic excitation of the surface, subsequent ejection of a surface particle and possible electron exchange as the particle leaves the surface.

2. Electron Stimulated Desorption of Excited Neutral Alkalis

As part of an effort to identify and study the final states of electronically desorbed particles, measurements were taken of the dependence on incident electron energy of excited alkali neutral desorption from alkali halides due to electron stimulated desorption (ESD). Specifically, the first resonance line photon yields of Na (589.6 nm), Li (670.7 nm) and K (766.4 nm), due to the ESD of NaCℓ, LiF and KCℓ, have been measured as functions of incident electron energy (0-500 eV). The photon emission is characteristic of the radiative decay of free excited neutrals; thus the resonance photon intensities are proportional to the yield of desorbed excited neutral alkalis. Our experimental apparatus has previously been described [1,2,3]. To clean and anneal the surface and to control the surface charging we heat our samples to 200—300°C during data collection. All data presented here are normalized to incident electron current.

In Fig. 1 is shown the electron energy dependence of excited alkali neutral emission due to ESD from NaCℓ, KCℓ and LiF. For the Na resonance yield from NaCℓ, we see an initial onset near 31 eV, definite enhancement about 63 eV, plus marked structure near 200 eV. A further feature emerges from the K excited neutral yield from KCℓ. Here the gradual rise in the photon yield is consistent with an enhancement near 200 eV. For the yield of Li excited neutrals desorbed from LiF, we see an initial onset around 63 eV, followed by a steady increase.

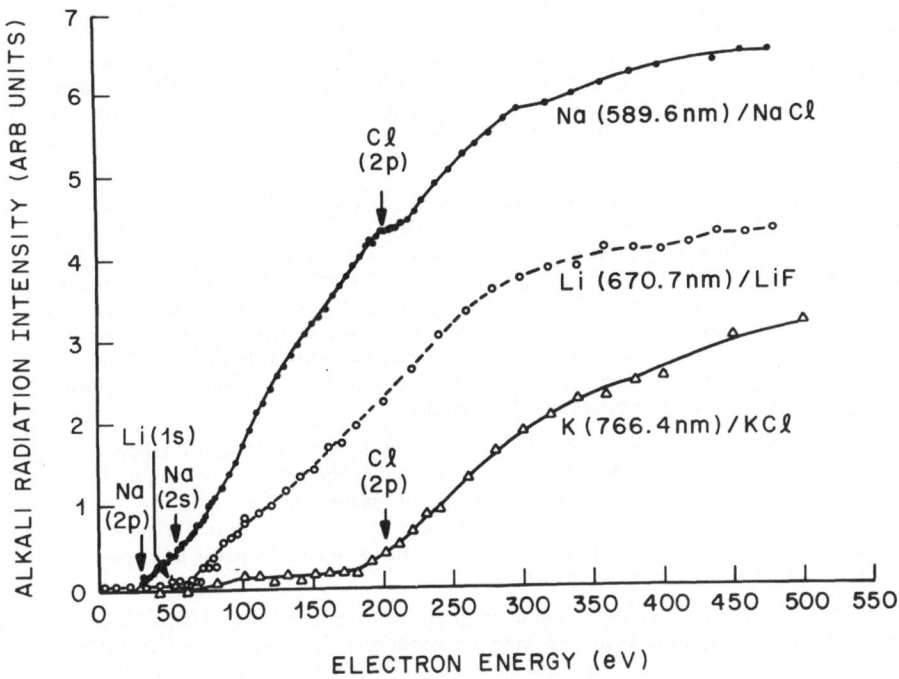

Fig.1. Electron Energy Dependence of Alkali Resonance Radiation from NaCℓ, LiF and KCℓ (Sample temperatures vary between 250°C and 300°C. Arrows depict positions of substrate core levels).

The data suggest that desorption of excited alkali neutrals may consist of a three stage sequence of events: 1) Energy absorption by the solid due to electronic transitions. For electron and photon bombardment we can eliminate momentum transfer as a significant factor leading to desorption. 2) Excitation of the particle-surface system into a non-binding state by either primary or secondary electronic processes, leading to alkali ejection from the crystal. 3) Possible electronic interaction with the surface which may alter the nature of the excited states of the desorbing particles.

We now discuss energy transfer to the solid by electron impact. For Na excited neutral desorption from NaCℓ we may discuss our results in relation to previous studies of electronic transitions in alkali halides due to electron and photon impact. The initial onset at 31 eV for Na (589.6 nm) corresponds to structure measured in optical absorption [4] and electron energy loss [5] studies. Such structure has been attributed to core exciton creation associated with the Na $L_{2,3}$ (2p) edge. We suggest that excited neutral Na desorption begins with the formation of this Na (2p) core exciton. The enhancement at 63 eV in NaCℓ may be due to the formation of a Na (2s) core exciton. Finally, we see structure near 200 eV, which corresponds to an apparent *reduction* in the excited neutral yield. This structure may be correlated to a CℓL$_{2,3}$ (2p) transition as discussed in optical absorption studies [6,7]. The excited neutral yield appears to rise smoothly above each onset. The results for K excited neutral emission from KCℓ are consistent with a strong *enhancement* of the photon yield with the production of a Cℓ (2p) core exciton, a transition measured by optical absorption [6,7]. In the absence of any structure related to

K core levels, this indicates that emission of excited K neutrals may depend strongly on energy transfer to the chlorine ion. For LiF we suggest that the desorption of excited neutral Li by electrons begins with the production of a Li (1s) core exciton. As shown in optical absorption [8], electron energy loss [9], and soft X-ray emission [10,11] studies, the exciton formation requires around 62 eV of energy. Again, there is a smooth rise above the onset indicating the ability of electrons above the threshold energy to create the core hole.

Bulk defects in alkali halides have been widely studied [12,13,14,16] in terms of characterizing point defect centers and as an explanation for electron "sputtering". The transport of these defects to the surface by focused collision sequences or diffusion may lead to particle ejection. An alternative view involves surface conditioning by the electron beam [1,14,16], where an alkali-rich layer may be formed due to preferential halogen ejection by electron bombardment, with subsequent thermal evaporation of the excess alkalis. Finally, one can view the phenomenon of excited neutral desorption along the lines of current ESD theories involving core [17] or valence [18] electron transitions which are generally invoked to explain ion desorption.

Our energy dependence results suggest that core hole formation in alkali halides is an important mechanism for transferring the electron energy to the solid. This would explain our observation of definite onsets for desorption of excited alkali neutrals. Valence electron excitations do not appear to play a major role in this energy transfer stage. We now discuss possible electronic mechanisms which lead to the subsequent desorption of alkalis from the surface in neutral excited states.

A possible electronic transition sequence may be hypothesized. Under electron bombardment a core exciton structure may be produced. Exciton formation implies localization and lifetime broadening (to 10^{-12} sec) of the excitation in an ionic crystal. By contrast in metals electron states are delocalized and core holes decay very quickly on the order of 10^{-16} sec. Eventually the long lived core hole in the alkali halide may undergo Auger decay with the upper electrons coming from the valence band principally associated with halogen (np) states. For a sufficient core level binding energy one valence electron may fill the core hole while another may be ejected into the continuum. The result is two holes in the valence band leaving a halogen with a positive charge. Recently jenmson et al. [19] have concluded that the mutual repulsion of the two holes increases their lifetime to the order of 10^{12} sec. This interaction implies localization and lifetime broadening for a period long enough to allow desorption if the hole states move into a forbidden gap region. The electron associated with the original core exciton may then become even more tightly bound and localized in the presence of two holes.

We stress that these experiments indicate rather high yields for the ESD of excited alkali neutrals. For each low energy (less than 500 eV) incident electron we measured more than 10^{-5} ejected excited alkali neutrals. We compare this figure to our recently measured Na^+ yield due to ESD of NaCl [20] which is $10^{-7} Na^+$/electron. The preference for neutral particle ejection is puzzling at first glance since alkali halides are believed to be the prototype of ionic crystals where the alkali cations and halogen anions exhibit a strongly ionic character. Neutral particle ejection cannot be explained by an Auger-decay electrostatic repulsion mechanism [17]. Such a model predicts positive anion emission due to Coulomb repulsion, leaving the cation ejection unspecified. We believe that the existence of strong Madelung attraction at the surface could inhibit the ejection of positive ions, which may help to explain the large disparity between neutral and ion yields. Our observations of excited neutral ejection suggest that electronic transitions (perhaps involving Auger process) to anti-bonding or energetic states may play a role in the overall mechanism for desorption.

3. Photon Stimulated Desorption of Excited Neutral Alkalis

We discuss here measurements of photon stimulated desorption (PSD) of excited neutral alkali atoms. This effect has been observed when photons with energy of 40-200 eV are incident on sodium chloride and lithium fluoride surfaces [21]. As in the electron bombardment case spectral analysis revealed two general classes of emitted radiation: (a) discrete line radiation from desorbed surface alkali species, and (b) broad band optical radiation arising from the solid which is known to occur when photons in this energy range are incident on some surfaces. For this work, our emphasis was on measurements of discrete line radiation. In addition to spectral studies, the intensity of optical radiation was measured as a function of incident photon energy and as a function of sample temperature (room temperature to 300°C). PSD measurements of positive ions were also performed on the same alkali halide samples.

In these experiments, sodium chloride and lithium fluoride targets maintained at ultrahigh vacuum were bombarded with photons (40 eV to 200 eV) using the Grasshopper monochromator at the Tantalus storage ring of the University of Wisconsin Synchrotron Radiation Center. The synchrotron radiation beam was incident at 57° to the surface normal. Optical photons were collected at about 90° to the surface normal. A cylindrical mirror analyzer was used for measurements of desorbed positive ions. The NaCl and LiF target (100) surfaces were polished prior to introduction into the ultrahigh vacuum system and then cleaned by heating to 500°C in vacuum (base pressure, 5×10^{-11} Torr) [22]. Optical radiation was observed with the use of a 0.3-m spectrometer with a resolution of 4.8 nm and spectral range 200-800 nm.

The field of view of the optical system included both the bulk and up to about 2mm in front of the surface. Consequently, radiation from desorbed free atoms as well as bulk luminescence was observed. In addition to the continuum, only the first resonance lines of sodium, the prominent NaD doublet at 589.0-589.6 nm, and lithium first resonance line at 670.7 nm were observed. This is similar to the electron bombardment case discussed above, but is in marked contrast to other means of excitation including photon- or electron-gas excitation and ion bombardment of alkali halide surfaces which result in detection of many atomic lines. The NaD photon signal was found to rise linearly with synchrotron photon flux, indicating that the creation of a free excited atom is an isolated single event. The rate of production of photon-bombardment-induced lithium resonance radiation at 61.5 eV incident photon energy was estimated from our measurements to be 7.2×10^{-3} emitted lithium photons per incident synchrotron photon. This rate is unusually high compared both with previously reported PSD measurements [23] which are typically less than 10^{-7} emitted ions per incident photon, and with the present PSD measurement on lithium fluoride which was found to be approximately 5×10^{-7} emitted positive ions per incident photon. No attempt was made to determine the mass of the emitted ions. The normalized photon and ion signals were found to be quite stable although a gradual dose dependent decay was observed. All measurements of photon intensity as a function of energy were taken with the target sample at room temperature. All PSD measurements of emitted ions however were obtained with a target sample temperature of approximately 250°C which was found to reduce surface charging.

Figure 2 shows an example of the data for the LiF case. Pronounced energy dependent structure appears in Fig.2a where the Li resonance radiation intensity is plotted as a function of incident photon energy. Also plotted are our PSD results for positive ions (Fig.2b) and soft x-ray absorption coefficient measurements (Fig.2c) obtained by F. C. BROWN et al. [24]. The most striking feature common to each measurement in Fig.2 is a narrow peak at approximately 61.5 eV ($<$ 2 eV FWHM) which is hypothesized to correspond to a Li 1s core exciton [24,25]. Although they have some features in common, the resonance radiation intensity measurements differ somewhat from the PSD results as well as absorption data.

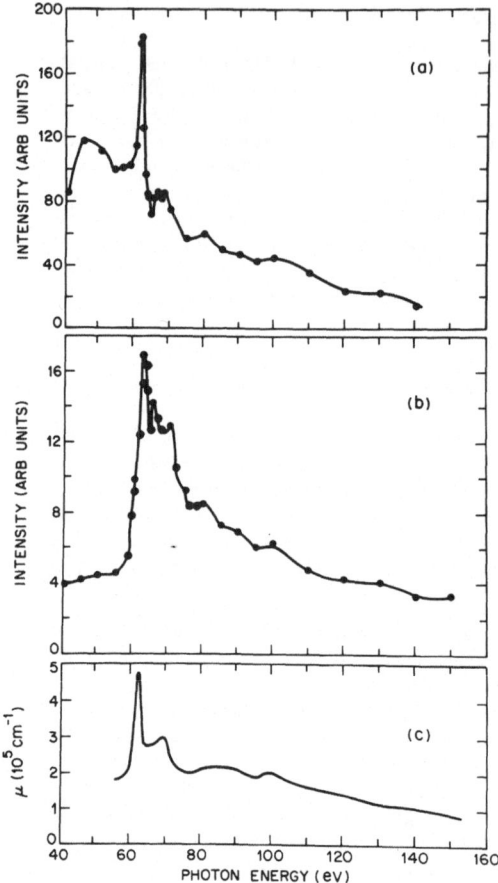

Fig.2. (a) Li* (670.7 nm) Optical Emission Dependence on Soft X-Ray Energy Using a (100) Single Crystal LiF Sample at Room Temperature
(b) Positive Ion yield as Function of Incident Photon Energy Using the Same Sample at about 250°C
(c) Photon Absorption Coefficient Measurements [24]

To attempt to understand this newly observed phenomena, both bulk and surface electronic processes must be considered. As discussed previously, electronic sputtering phenomena in alkali halides have been accounted for in the past by invoking the electron or photon impact induced production of defects (H or V_k centers) which may diffuse to the surface resulting in the ejection of a halogen [15,26], leaving excess metal to evaporate thermally. This is not adequate to explain the photon stimulated desorption of *excited* neutral alkalis but may be an important factor in determining the state of the surface.

Other desorption models currently employed to explain ion ejection include the MENZEL-GOMER-REDHEAD anti-bonding valence excitation [18] and the KNOTEK-FEIBELMAN inner shell excitation models [17], both of which involve specific excitation mechanisms. As shown in Fig.2, the desorption of excited neutrals is measured to be roughly proportional to the number of photons absorbed by the target material over a wide range of photon energy. This indicates that the process is non-specific and does not require a *particular* initial excitation.

4. Discussion

Clearly, the pronounced differences in energy dependence between the electron and photon bombardment cases as shown in Figures 1 and 2 are due to differences in the way that alkali halide solids initially absorb electron and photon energy. In spite of this difference there are marked similarities.

The fact that only the first alkali resonance transitions are detected in both our studies provides a clue to the interaction between the solid and the desorbing alkali. We consider the role of radiationless de-excitation processes in order to understand why photon emission from the lowest alkali resonance state is favored. Initially we may suppose that higher excited states are populated in the excitation process. One possible scheme is that the higher states can interact with the empty conduction band and thus may de-excite by tunneling processes. The first resonance excited state cannot decay in this way due to the forbidden gap in the crystal. It thus may survive allowing the excited neutral to leave the solid and decay radiatively. However, if this resonance ionization scheme were correct, we would expect a *higher* yield of ions than is measured [20]. It is more likely that in the early stages as the alkali departs the higher excited states which are more de-localized preferentially depopulate. Thus, the departing neutral alkali is left preferentially in its first excited state, and as a result only the first resonance level is observed.

5. Conclusion

In our analysis of the desorption of excited alkali neutrals arising from electron and photon bombardment of alkali halides we have characterized the first step: the energy transfer to the surface by production of substrate core holes. We have also discussed the possible long-lived and spatially localized anti-bonding states which may result in alkali desorption. Finally, we have considered electronic mechanisms associated with delocalization which may account for the preferential population of the first excited resonance state. These considerations provide a framework for understanding the processes of excitation, charge transfer, and excited neutral desorption occurring at an alkali halide surface under electron or photon bombardment.

We acknowledge discussions with R. Kelly and J. C. Tully. We also acknowledge the able assistance of E. M. Rowe, Dom. An., and the Staff of the University of Wisconsin Synchrotron Radiation Laboratory.

REFERENCES

1. T. R. Pian, N. H. Tolk, J. Kraus, M. M. Traum, J. C. Tully, and W. Eugene Collins, J. Vac. Sci. Tech., *20*, 555 (1982).

2. N. H. Tolk, L. C. Feldman, J. S. Kraus, R. J. Morris, T. R. Pian, M. M. Traum, and J. C. Tully, in W. Heiland and E. Taglauer, eds., *Inelastic Particle Surface Collisions*, Springer, NY (1981).

3. N. H. Tolk, L. C. Feldman, J. S. Kraus, R. J. Morris, M. M. Traum and J. C. Tully, Phys. Rev. Lett. *46*, 134 (1981).

4. R. Haensel, C. Kunz, T. Sasaki, and B. Sonntag, Phys. Rev. Lett. *20*, 1436 (1968).

5. M. Creuzberg, Z. Physik *196*, 433 (1966).

6. Y. Iguchi, T. Sagawa, S. Sato, M. Watarabe, and T. Oshio, Sol. St. Comm. *6*, 575 (1968).

7. T. Sagawa, Y. Iguchi, M. Sasanuma, and T. Oshio, J. Phys. Soc. Japan *21*, 2587 (1966).

8. C. G. Olson and D. W. Lynch, Sol. St. Comm. *31*, 51 (1979).

9. J. R. Fields, P. C. Gibbons and S. E. Schnatterly, Phys. Rev. Lett. *38*, 430 (1977).

10. O. Aita, K. Tsutsumi, K. Ichikawa, M. Kamada, M. Okusawa, H. Nakamura, and T. Watanabe, Phys. Rev. *B23*, 5676 (1981).

11. E. T. Arakawa and M. W. Williams, Phys. Rev. Lett. *36*, 333 (1976).

12. Y. al Jammal, D. Pooley and P. D. Townsend, J. Phys. C. Solid St. Phys. *6*, 247 (1973).

13. D. Pooley, Proc. Phys. Soc. *87*, 245 (1966).

14. L. A. Larson, T. Oda, P. Braunlich, and J. T. Dickinson, Sol. St. Comm. *32*, 347 (1979).

15. H. Overeijnder, M. Szymonski, A. Haring, and A. E. deVries, Radiat. Eff. *38*, 21 (1978).

16. P. W. Palmberg and T. N. Rhodin, J. Phys. Chem. Solids *29*, 1917 (1968).

17. P. J. Feibelman and M. L. Knotek, Phys. Rev. *B18*, 6531 (1978).

18. D. Menzel and R. Gomer, J. Chem. Phys. *41*, 3311 (1964); P. A. Redhead, Can. J. Phys. *42*, 886 (1964).

19. D. R. Jennison, J. A. Kelber and R. R. Rye, J. Vac. Sci. Technol. *18*, 466 (1981).

20. T. R. Pian, N. H. Tolk, M. M. Traum, N. Stoffel and G. Margaritondo, to be published.

21. N. H. Tolk, M. M. Traum, J. S. Kraus, T. R. Pian, W. E. Collins, N. G. Stoffel and G. Margaritondo, Phys. Rev. Lett. *49*, 812 (1982).

22. J. Estel, H. Hoinkes, H. Kaarman, H. Nahr and H. Wilsch, Surf. Sci. *54*, 393 (1976).

23. See, for example, D. Menzel, Surf. Sci. *47*, 370 (1975); T. E. Madey, and J. T. Yates, Jr., Surf. Sci. *63*, 203 (1977); P. D. Townsend, Surf. Sci. *90*, 256 (1979); P. J. Feibelman and M. L. Knotek, Phys. Rev. B*18*, 6531 (1978); D. P. Woodruff, M. M. Traum, H. H. Farrell, N. V. Smith, P. D. Johnson, D. A. King, R. L. Benbow, and Z. Hurych, Phys. Rev. B*21*, 5642 (1980).

24. F. C. Brown, C. Gähwiller, A. B. Kunz, and N. O. Lipari, Phys. Rev. Lett. *25*, 927 (1970).

25. J. R. Fields, P. C. Gibbons, and S. E. Schnatterly, Phys. Rev. Lett. *38*, 430 (1977).

26. P. D. Townsend, R. Browning, D. J. Garlant, J. C. Kelley, A. Mahjoobi, A. J. Michael and M. Saidoh, Radiat. Eff. *30*, 55 (1976).

Part 4

Molecular Dissociation

4.1 Dissociation in Small Molecules*

P.M. Dehmer

Argonne National Laboratory, Argonne, IL 60439, USA

The study of molecular dissociation processes is one of the most interesting areas of modern spectroscopy owing to the challenges presented by even the simplest of diatomic molecules. This paper reviews the commonly used descriptions of molecular dissociation processes for diatomic molecules, the selection rules for predissociation, and a few of the principles to be remembered when one is forced to speculate about dissociation mechanisms in a new molecule. Some of these points will be illustrated by the example of dissociative ionization in O_2.

Mechanisms for molecular dissociation resulting either from excitation out of the ground state or from the decay of a valence or Rydberg state can be classified as follows [1]:

(1) Direct dissociation. The most important case of dissociation continua in absorption is that in which a transition occurs from a stable lower state to a repulsive upper state or to the continuous portion of a stable upper state. The energy dependence of the continuum intensity is governed (to a first approximation) by the Franck-Condon principle, i.e., the most probable transition in absorption is that going vertically upward from the minimum of the lower potential energy curve.

(2) Radiative decay. Molecular dissociation results when a molecule in a stable excited state radiatively decays to a lower repulsive state. This is a common phenomena and the resulting continuous radiation is often used for laboratory light sources.

(3) Nonradiative decay. When a discrete level overlaps an energetically available dissociation continuum, the possibility exists that the state will decay without the emission of radiation. Such a state normally has a very short lifetime (much shorter than typical radiative lifetimes), and, as a result, the discrete level may appear broadened in absorption. Furthermore, there may be an absence or weakening of molecular emission, since only those molecules that do not decompose may radiate. This process, which is termed predissociation, occurs in diatomic molecules by the conversion of electronic or rotational energy and in polyatomic molecules by these mechanisms as well as the conversion of vibrational energy. Another nonradiative decay process which is analogous to predissociation and which is important in the dissociative ionization process in O_2 and other molecules is preionization. In this process, a discrete level overlaps an energetically available ionization continuum and decays via the ejection of an electron. The resulting parent ion may then predissociate to form a fragment ion.

164

Predissociation and preionization are special cases of perturbations, and therefore the same selection rules hold as for perturbations. For diatomic molecules in any coupling scheme [1], $\Delta J = 0$, $+ \not\leftrightarrow -$, and $s \not\leftrightarrow a$; in Hund's case (a) and (b), $\Delta S = 0$ and $\Delta\Lambda = 0, \pm 1$. If both states belong to case (a), $\Delta\Sigma = 0$; if both states belong to case (b), $\Delta K = 0$; and if both states belong to case (c), $\Delta\Omega = 0, \pm 1$ (rather than $\Delta S = 0$ and $\Delta\Lambda = 0, \pm 1$). Similar considerations apply to preionization; however the quantum numbers and symmetry properties in the continuous range of energy levels now refer to the system of molecular ion plus electron.

The selection rules for predissociation and preionization of triatomic and larger polyatomic molecules may be considerably more complex as the result of the interaction of electronic and vibrational angular momenta. GELBART [2] has discussed the photodissociation of polyatomic molecules in detail in a recent review article as has OKABE [3] in a recent book.

In addition to the energy, symmetry, and angular momentum restrictions imposed on predissociation and preionization processes, the Franck-Condon overlap of the potential energy curves of the interacting states imposes a further constraint that determines which of the allowed predissociations and preionizations will actually occur. It must be remembered that the Franck-Condon principle governs radiationless as well as radiative transitions. In general, a transition is possible if the relevant potential energy curves intersect or at least approach one another closely.

Some of the preceeding principles are illustrated using the example of photon-induced dissociative ionization in O_2 [4]. The threshold for O^+ production from O_2 occurs at 17.272 eV and corresponds to the ion pair formation process $O_2 + h\nu \rightarrow O^+ + O^-$. The cross section for this process is highly structured and the ratio of O^+ to O_2^+ is $\approx 1\%$ on the maxima of the peaks. Fig.1 gives a potential energy diagram for states of O_2 and O_2^+ [5], and it can be seen that molecular Rydberg states converging to the $b^4\Sigma_g^-$ state of O_2^+ will lie in the region of ion pair formation and are probably responsible for the structure in the ion pair cross section. The correlation of structure in the cross sections for ion pair formation and for parent ion formation is poor and the exact mechanism for this process is not known.

The threshold for O^+ production from O_2 via the dissociative ionization process $O_2 + h\nu \rightarrow O^+ + O + e$ occurs at 18.734 eV. The cross section for this process is also highly structured near threshold; however, the structure now appears on a more intense direct ionization continuum which rises to a maximum at approximately 640 Å and then declines and reaches an almost constant value in the wavelength region of the vibrational convergence limits of the O_2^+ $B^2\Sigma_g^-$ state. The relative photoionization cross sections for O^+ and O_2^+ formation are shown in the region near threshold in Fig.2. The ratio of O^+ to O_2^+ at 640 Å is $\approx 3\%$.

Appreciable direct dissociation may occur by excitation from the ground state of O_2 to the repulsive wall of O_2^+ $A^2\Pi_u$. Both theoretical [6] and experimental [7] Franck-Condon factors show that the O_2 $X^3\Sigma_g^-$, $v'' = 0 \rightarrow O_2^+$ $A^2\Pi_u$, v' transition has a long vibrational progression and that transitions to higher vibrational states ($v' \geqslant 15$) occur with approximately 20% of the intensity of the most probable transition. It is probable that direct transitions to the dissociation continuum also occur with significant intensity and account for the rising continuum near threshold. FREUND [8] has pointed out that this is the most likely mechanism for the production of near-zero kinetic energy O^+ ions observed in experiments measuring the

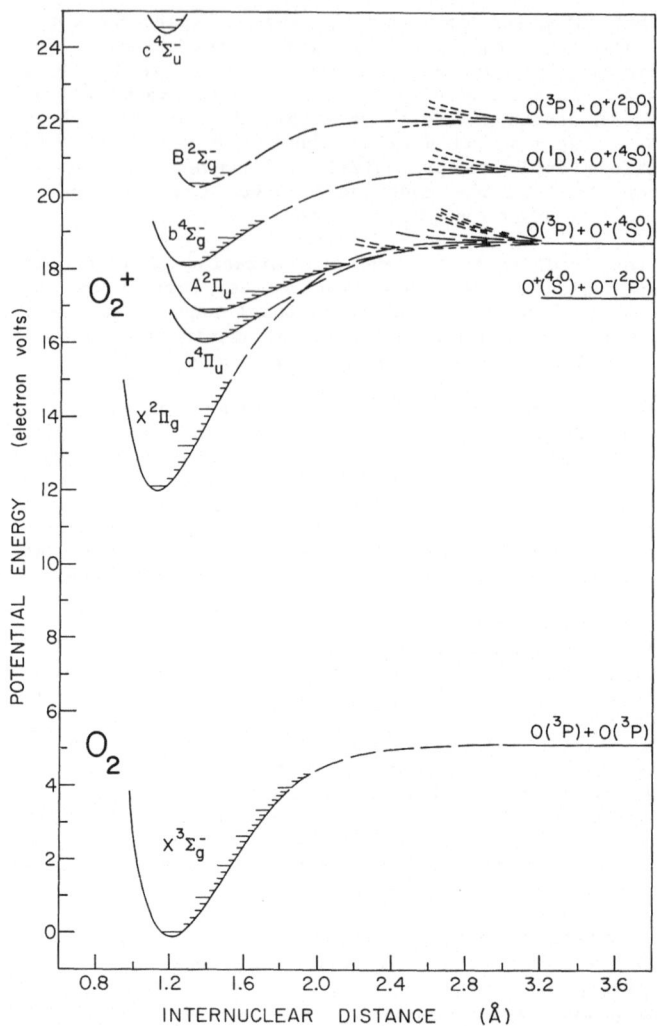

Fig.1. Potential energy curves for the ground state of O_2 and the lower-lying states of O_2^+, taken from the compilation of GILMORE [5]

kinetic energy distribution of O^+ from dissociative ionization in O_2. The amount of dissociation due to direct ionization will decrease (as does the total ionization cross section) with increasing energy above threshold and together with decreasing collection efficiency accounts for the decline in cross section above 640 Å. The increase in continuum intensity below about 610 Å probably is due to predissociation of O_2^+ formed in the $B^2\Sigma_g^-$ state by direct ionization.

The peaks in the O^+ photoionization cross section correspond to Rydberg states in O_2 converging to the $B^2\Sigma_g^-$ state of O_2^+. This structure is not

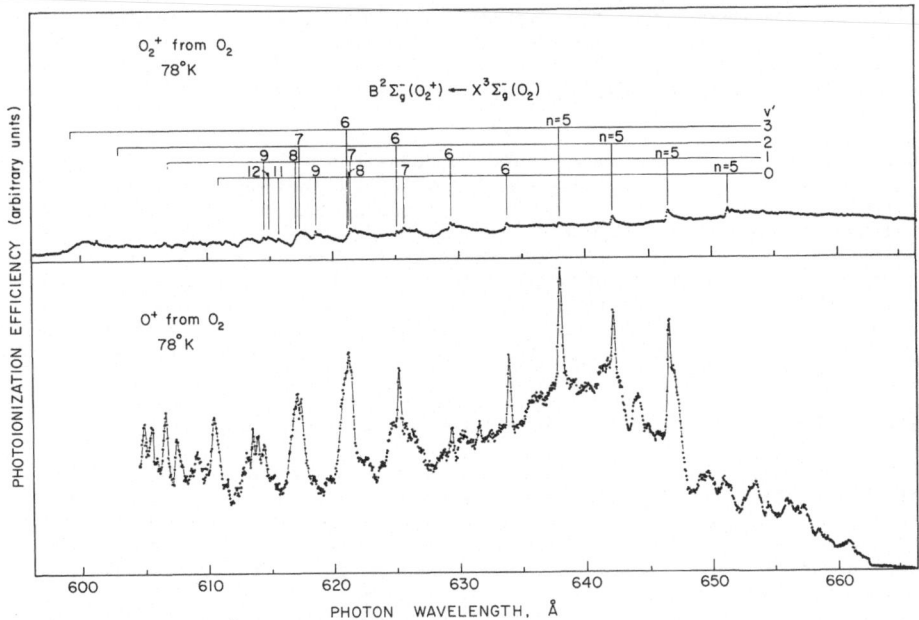

O_2^+ from O_2
78°K

$B^2\Sigma_g^-(O_2^+) \leftarrow X^3\Sigma_g^-(O_2)$

Fig.2. Relative photoionization cross sections for O^+ and O_2^+ from O_2 in the region of the threshold for dissociative ionization [4]

reflected in the O^- photoionization cross section and therefore cannot be due to predissociation of the molecular Rydberg states by ion pair states. Rydberg states converging to O_2^+ $B^2\Sigma_g^-$, v = 0 appear very weak in the O^+ spectrum. The only exception is the n = 6, v = 0 peak; however, this peak occurs at the expected energy of the n = 5, v = 4 state and therefore may not be due to a v = 0 transition. A plausible mechanism for O^+ production that is suggested by these observations is preionization of the Rydberg states to O_2^+ $b^4\Sigma_g^-$, v ⩾ 4 followed by predissociation to $O(^3P)$ + $O^+(^4S^0)$. The Rydberg states in question have approximately the same potential energy curve as the O_2^+ $B^2\Sigma_g^-$ state to which they converge. The Franck-Condon overlap between O_2^+ $b^4\Sigma_g^-$, v ⩾ 4 and the various vibrational states of O_2^+ $B^2\Sigma_g^-$ is poor for v = 0 but much improved for v = 1-4 [4]. This would explain the low intensity of the Rydberg states converging to O_2^+ $B^2\Sigma_g^-$, v = 0 in the O^+ photoionization cross section. Selection rules prohibit predissociation of the $b^4\Sigma_g^-$ state of O_2^+ by the $A^2\Pi_u$ or $a^4\Pi_u$ states. Predissociation by the ground $X^2\Pi_g$ state of O_2^+ is allowed but the crossing seems to be very poor; however, only a very small fraction of the states actually predissociate as is evidenced by the ratio of the O^+ to the O_2^+ ion intensities and by the observation of emission from the $b^4\Sigma_g^-$ state of O_2^+ [9].

At higher energies (E > 20 eV), quantitative information on the dissociative photoionization process in O_2 can be obtained from the determination of fast electron differential scattering cross sections at negligibly small momentum transfers [10]. Relative optical oscillator strengths are obtained by kinematic correction of the electron scattering intensities according to the Bethe-Born theory. The photoabsorption cross

167

section is obtained from the forward scattered electron energy loss spectrum; the photoelectron spectrum and the photoionization mass spectrum are simulated by the detection of fast scattered electrons in coincidence with ejected electrons (e,2e) and with ejected ions (e,e + ion), respectively.

BRION et al. [10] have used these "pseudo-photon" techniques to study the photoabsorption cross section of O_2 to 300 eV incident energy and the photoionization cross sections to 75 eV incident energy. They found that the cross section for O^+ formation increases abruptly at approximately 20 eV and shows a more gradual increase thereafter. The ratio of O^+ to O_2^+ is 0.08 at 20 eV, 0.60 at 50 eV, and 0.91 at 75 eV. The reason for this increase in O^+ formation is that the $B^2\Sigma_g^-$ state of O_2, which lies 20.3 eV above the ground state, is completely predissociated as are most of the higher lying ionic states [10]. Therefore, it is not surprising that the increase in total optical oscillator strength above 20 eV (i.e., the oscillator strength due to transitions to the $B^2\Sigma_g^-$ and higher lying excited ionic states) closely parallels the total O^+ production.

The preceeding example of dissociation in O_2 is among the simplest imaginable. The final (decomposing) states are well defined, the molecule undergoes no collisions with other molecules or surfaces, information on the decay process is available from many types of experiments, and, in many cases, potential energy curves are well known. In contrast, dissociation processes in larger molecules may appear much more complex. However, the basic principles remain the same. The dissociation process will be governed by energetics, angular momentum and symmetry selection rules, the Franck-Condon principle, and competition from other decay channels.

* Work performed under the auspices of the U.S. Department of Energy and the Office of Naval Research.

References

1. G. Herzberg: *Spectra of Diatomic Molecules* (Van Nostrand, Princeton 1950)
2. W. M. Gelbart: Ann. Rev. Phys. Chem. 28, 323 (1977)
3. H. Okabe: *Photochemistry of Small Molecules* (Wiley, New York 1978)
4. P. M. Dehmer, W. A. Chupka: J. Chem. Phys. 62, 4525 (1975)
5. F. R. Gilmore: J. Quant. Spectrosc. Radiat. Transfer 5, 369 (1965)
6. R. W. Nicholls: J. Phys. B 1, 1192 (1968)
7. O. Edqvist, E. Lindholm, L. E. Sellin, L. Asbrink: Phys. Scr. 1, 25 (1970)
8. R. S. Freund: J. Chem. Phys. 54, 3125 (1971) and references therein
9. P. H. Krupenie: J. Phys. Chem. Ref. Data 1, 423 (1972)
10. See C. E. Brion, K. H. Tan, M. J. van der Wiel, Ph. E. van der Leeuw: J. Electron Spectrosc. Relat. Phenom. 17, 101 (1975) and references therein

4.2 The Coulomb Explosion and Recent Methods for Studying Molecular Decomposition

T.A. Carlson

Oak Ridge National Laboratory, Oak Ridge, TN 37830, USA

Abstract

An account is given of the nature of a molecular Coulomb explosion, based on experimental studies of fragment ion spectra and their recoil energies resulting from photoionization of core shell electrons in gaseous molecules. The extent of multiple ionization in a vacancy cascade is discussed, and a simple model is used to predict the amount of kinetic energy possessed by the fragment ions. Data on both diatomic molecules containing light atoms and more complex molecules containing heavy atoms are reviewed.

In the second part of the paper a broad outline is presented of studies that can be carried out on molecular decomposition through the use of laser and synchrotron radiation. Some indications are given of how information on the decomposition of free molecules can be applied to the desorption of molecules adsorbed on surfaces.

1. Introduction

This paper is divided into two parts. The first deals with the subject of one of the more violent aspects of molecular decomposition: the Coulomb explosion. The second part attempts to give a more general view of molecular decomposition by discussing the various experimental techniques available for its study, in particular the recent use of laser and synchrotron radiation.

The Coulomb explosion of a molecule occurs when the molecule undergoes the loss of two or more electrons, because in this circumstance the presence of one or more fragment ions in close proximity causes the ions to recoil from one another with considerable Coulomb repulsion. An effective means of reaching multiple ionization is through the production of an inner shell vacancy followed by one or more Auger processes. If the initial vacancy occurs in a deep core shell of a heavy atom, a large number of Auger processes can take place, which is known as a vacancy cascade.

An early interest in the molecular effect of the Coulomb explosion came out of the study of the hot atom chemistry of atoms undergoing inner shell vacancies by means of internal conversion or electron capture [1]. Subsequent studies on gaseous molecules containing radioactive atoms gave information on the qualitative effects of the Coulomb explosion [2]. With the use of x rays for producing the inner shell vacancies and a spectrometer capable of handling the large recoil energies, it was, however, possible to study quantitatively the extent of the vacancy cascade in atoms and its consequences

169

in molecules that lead to Coulomb explosions into two or more fragment ions
[3,4,5]. It will be the main purpose of this paper to survey these latter
investigations.

2. Vacancy Cascades

Creation of an inner shell vacancy in a given atom is a highly efficient
means for producing a multiply-charged ion. A core hole will be filled by
a number of Auger processes, each Auger process giving rise to the ejection
of an electron. If a photon (x ray) is used to create the initial vacancy,
the process is highly selective, since the cross section will be highest for
one particular core shell (usually favoring the deeper core shell electrons,
whose binding energies are still lower than the x-ray energy). A description
of the vacancy cascade will be made with the help of Fig.1 [6]. In this case
we have chosen to illustrate the effects of producing a hole in the K shell of
xenon. If xenon is irradiated with x rays above the binding energy of the K
shell (34.6 KeV) the photoionization cross section is largest for that shell.
The vacancy cascade illustrated in Fig.1 is only one of a large number of
ways by which the K hole can relax, but it is one of the more probable ones.
The first transition is a radiative transition, merely transferring the K
hole to an L_{III} hole. The fluorescent yield for filling a K vacancy in
xenon is 89%. However, henceforth, the relative probability for Auger processes dominate over radiative processes, and for each hole filled, two are
created. It is to be also noted that Auger processes preferably occur between neighboring shells because of the greater overlap of the wave functions
for these orbitals, and thus the number of Auger processes is maximized. The
vacancy cascade comes to a halt when the vacancies have moved to the outer
shell and no more autoionization is energetically possible. (The final total

A VACANCY CASCADE IN Xe

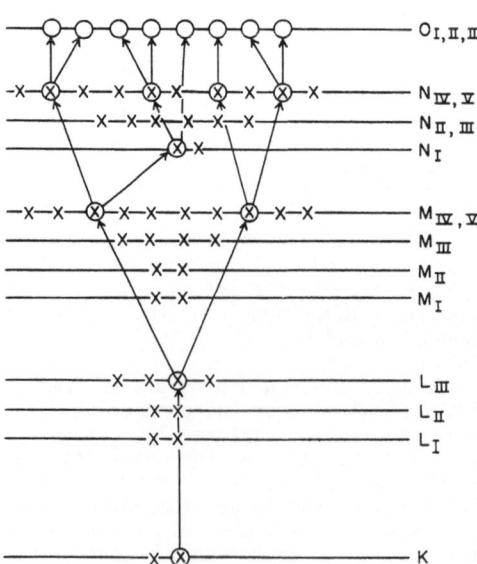

Fig.1. Schematic of vacancy cascade
arising from initial hole in K shell
of Xe. The X's represent electrons,
○'s vacancies and ⊗ the cases when
a vacancy has been made and then
filled by an electron. The arrow
indicates the direction of the vacancy cascade. In the case of Auger
processes, two vacancies are created
for each hole filled. (Taken from
Ref. [6])

charge of +8 is consistent with the average charge found experimentally from photoionizing the K shell of xenon, as will be shown presently.)

To determine the extent of multiple ionization that results from inner shell vacancies, a specially-designed mass spectrometer was built that measured the charge spectra resulting from photoionizing gaseous atoms [7]. The x rays were obtained from line sources and cleansed from bremstrahlung background by use of selected metal foil filters. Ions were extracted in a sufficiently high electrostatic field to obviate any recoil energy effects and passed through a magnetic analyser. The design was an improvement over earlier time of flight measurements [8] because of the latter's dependence on the relative efficiency for electron collection.

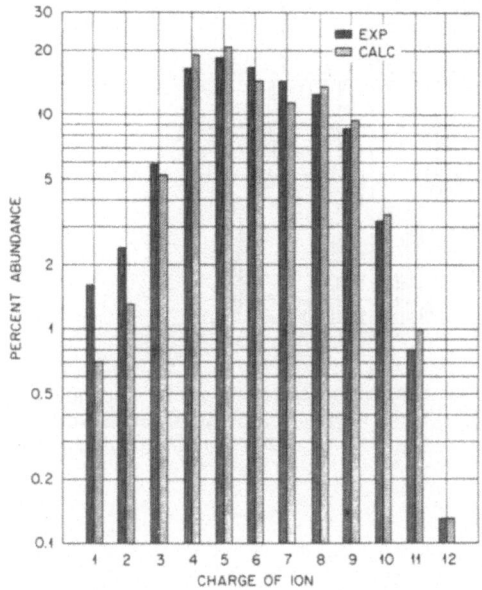

Fig.2. Distribution of Kr ions resulting from 5 vacancy cascades. The experimental values come from photoionization with 17.5 keV x rays, which result primarily (86.5%) in an initial vacancy for the K shell of Kr. The calculations are based on a Monte Carlo summing of Auger processes plus consideration of additional electron shake off. (cf. Ref.[9])

In Fig.2 are shown the results of one of these studies in which 90% of the primary vacancies are produced in the K shell of krypton [9]. Also included are results from a calculation of the charge spectrum. The calculation was based on a Monte Carlo approach in which the probability for each step in the vacancy cascade was determined by the relative Auger and radiative transition rates. Additional consideration was also made for electron shake off which can occur in the initial photoionization and in each subsequent Auger process. Considering the complexity of the total process, it is amazing how well one can predict the consequences of a vacancy cascade, which gives credance to the step wise description.

By choice of different x-ray energies and the knowledge of the relative photoelectric cross sections for the different rare gases, it was possible to extract charge spectra resulting from inner shell vacancies in the K, L, and M shells of the different rare gases (see Fig.3). These data, together with additional results on mercury vapor allowed the assignment of the average charge as a function of Z over the whole periodic table [10] (see Fig.4).

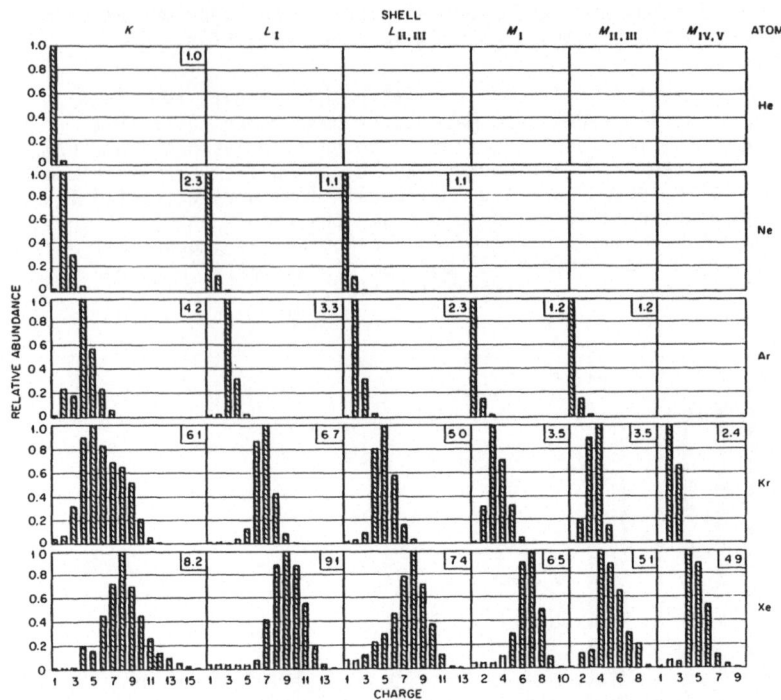

Fig. 3. The relative abundance of ions ound as a consequence of a vacancy in the K, L, and M shells of the rare gases, as derived from experimental data. The average charge is given for each spectrum in the upper right-hand corner (cf. Ref. [10])

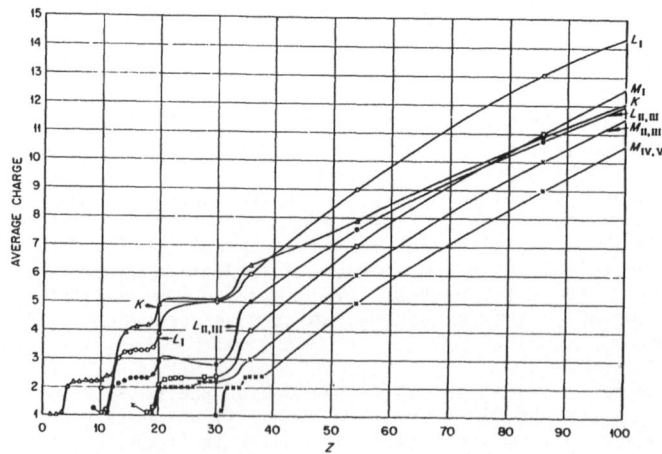

Fig. 4. Estimated average charge arising out of a sudden vacancy in the K, L and M shells as a function of the atomic number (cf. Ref. [10])

Though many details are lacking, the overall picture of the vacancy cascade is fairly clear. From the experimental data of charge spectroscopy, it is now possible to quantitatively predict the extent of multiple ionization resulting from vacancy cascades in most atomic situations. The next problem to discuss is what occurs when an atom undergoing a vacancy cascade is part of a molecule.

3. Coulomb Explosion in Molecules

The nature of a Coulomb explosion is graphically shown in Fig.5. Methyl iodide, chosen for our example, has been irradiated with x rays producing core vacancies primarily in the L shell of the iodine atom. A vacancy cascade takes place in iodine, and when the vacancies reach the valence shell, electrons are transferred from the other atoms leaving a collection of ions. (From experiment [3], the most likely combination is 4 H^+, C^{2+} and I^{5+}.) These ions are quite close to one another, roughly the inner nuclear distances measured for the neutral molecule. The ions then recoil from one another due to the Coulombic repulsion, gaining considerable recoil energy.

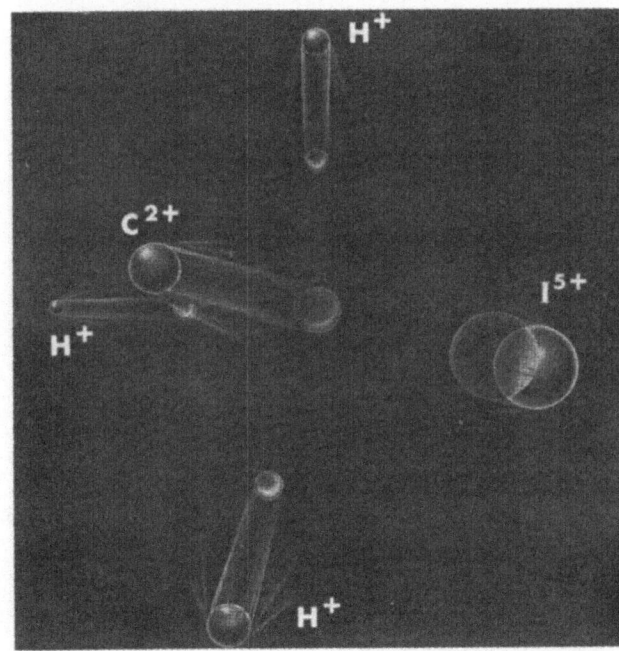

Fig. 5. A pictoral description of the Coulomb explosion of ion resulting from a vacancy in the L shell of iodine belonging to CH_3I (cf. Ref. [6])

To study these Coulomb explosions, experiments [3,4,5] were carried out with a charge spectrometer in which the fragment ions could be compared under conditions of equal efficiencies, and the average recoil energy for each fragment ion could be measured. This was accomplished by carrying out measurements as a function of extraction voltage. A summary of the results on four different molecules is given by the following equations:

$$HI + x\text{-ray} \quad \longrightarrow \quad 1.0 \ I^{6.5+} + 0.7 \ H^+ + 7.2 \ e^- \tag{1}$$

$$CH_3I + x\text{-ray} \quad \longrightarrow \quad 1.0 \ I^{4.7+} + 1.0 \ C^{2.0+} + 3.0 \ H^+ + 9.7 \ e^- \tag{2}$$

$$C_2H_5I + x\text{-ray} \quad \longrightarrow \quad 1.0 \ I^{3.0+} + 1.9 \ C^{1.6+} + 5.1 \ H^+$$
$$+ \ 0.15 \ C_2H_3^+ + 11.3 \ e^- \tag{3}$$

$$Pb(CH_3)_4 + x\text{-ray} \longrightarrow 1.0 \ Pb^{2.2+} + 3.7 \ C^{1.6+} + 10.4 \ H^+$$
$$+ \ 0.25 \ CH_n^+ + 18.7 \ e^- \tag{4}$$

For the first three molecules 98% of the initial vacancies are formed in the L shell of iodine. In the case of tetramethyl lead, most of the initial vacancies were created in either the L or M shell of lead. The measure intensities relative to 1.0 for the heavy ion are given before the symbol of the fragment ions. The average measured charge is given as a superscript. The number of electrons created in the average case is obtained by summing over the intensities times the average charge. Most of the ions found were the various charge states of the bare elements. Very few aggregate ions, made up of more than one element, were observed. The ratios of the observed ions were in reasonable agreement with the stochiometric formula, suggesting that few neutral or negative ions were formed. It is to be noted that the larger the molecule, the larger the total number of electrons ejected. It appears that when electrons are transferred from the lighter

Table 1. Recoil energy (eV) of fragment ions formed in a Coulomb explosion (Results taken from Ref.[3,4]

	CH_3I			$Pb(CH_3)_4$		
					Calculation[d]	
Ion	Exper.[a]	Calculation[b]	Ion	Exper.[c]	A(Sym)	B(Sym)
C^{2+}	40 ± 3	55	Pb^+	< 0.05	0.00	0.25
H^+	34 ± 2	61	C^+	13	11	10
I^{5+}	8.9 ± 1	**10.8**	H^+	29	52	44

[a] Initial vacancy created in L shell of iodine. Value represents peak of recoil energy spectrum.

[b] Based on assigning point charges to ions as listed.

[c] Initial vacancy created in L or M shell of Pb. Value represents peak of recoil energy spectrum.

[d] A(Sym) based on assigning single point charges to each atom in the molecules; B(Sym) based on assigning single point charges to each atom in the molecule, except for one CH_3 group, where the atoms remain neutral.

atoms to the highly charged heavy ion, further autoionization takes place. The efficiency of multiple ionization following a core vacancy is quite high. It was particularly impressive how nearly every atom in tetramethyl lead was found to be ionized.

As examples of the recoil energies measured in Coulomb explosions, results [3,4] are given in Table 1 for CH_3I and $Pb(CH_3)_4$. For comparison, calculated results are also given. The calculations are based on a model of point charges, separated by the internuclear distances found in the neutral molecules, together with knowledge of the atomic masses and the most likely charge as found experimentally for the different elemental ions. The collection of charges is allowed to recoil under small increments of time. After each increment the new positions, velocities, and kinetic energies are recalculated until the ions have separated to distances where the Coulombic repulsion is negligible, and the calculation is terminated. In general, the agreement between experimental recoil energies and the calculations is quite good, which justifies the model of the Coulomb explosion. On closer scrutiny, the calculated values of the recoil energy are, in general, slightly larger than those observed experimentally. This can be understood in terms of the time taken for the vacancy cascade.

To examine more unambiguously the effects of the timing of the vacancy cascade, studies were carried out [3] on HI and DI. It was found experimentally the deuterium ions received a greater recoil energy than hydrogen ions as the consequences of producing a vacancy from the L shell of iodine. This difference (47 eV versus 41 eV at the peak of the recoil energy curves) was shown to result from the fact that the molecule begins to separate during the vacancy cascade and thus does not receive the full recoil energy when the ionization is complete. Since deuterium is heavier than hydrogen, it moves more slowly and, thus, the interatomic distance at the end of the vacancy cascade is smaller for DI than for HI. It was possible, in fact, to estimate the time of the vacancy cascade in iodine from the difference in recoil energies. The value was about 3×10^{-15} sec in reasonable agreement with our knowledge of Auger rates.

The results [4] on tetramethyl lead give a good heuristic example of what happens when an atom undergoes an extensive vacancy cascade as part of a large molecule or in a condensed medium. The peripheral ions such as C^+ and H^+ have considerable recoil energy but the central ion, Pb^+, has essentially zero kinetic energy. This is due partly because lead is so heavy, but it is also due to the highly symmetric nature of the Coulomb explosion. For example, if we assume that one of the methyl groups remains uncharged, the Pb^+ would have received 0.25 eV of recoil energy or five times more than the observed experimental limit. In large molecules, condensed media, or in the case of molecules adsorbed on surfaces, it is not necessarily that the atom undergoing the vacancy cascade will receive the large recoil, but rather it might be some neighboring atom or group. The nature of the Coulombic explosions will depend on the effectiveness of the neutralization. It may be possible by the correct choice of the molecule adsorbed onto the surface, and the location of the inner shell vacancy determined by the photon energy, to engineer the expulsion of a particular ion.

Turning from heavy atoms to light atoms, Table 2 gives results on CO following vacancies in the K shells of carbon and oxygen. The results are typical of some 6 molecules studied [5], where the initial vacancies were produced in the K shells of carbon, oxygen, nitrogen or fluorine. One sees that the parent ion is found only in 5% of the cases and this probably arises from the small amount of valence shell photoionization present. Most of the

Table 2. Relative abundance and recoil energy of fragment ions resulting from radiation of CO by 930 eV x rays*

Ion	Abundance (%)	Recoil Energy (eV)
C^+	34 ± 2	10
C^{2+}	14 ± 3	24
C^{3+}	0.8 ± 0.4	60
O^+	40 ± 2	8
O^{2+}	4.4 ± 0.8	12
O^{3+}	1 ± 1	80
CO^+	5 ± 2	2
CO^{2+}	0.8 ± 0.2	4

*Estimated distribution of initial vacancies C(K) 15%; O(K) 80%; valence shell 5%. Results taken from Ref.[5]

time double ionization occurs (the K hole is filled by an Auger process better than 99% of the time). This leads generally to the Coulomb explosion into the fragment ions of C^+ and O^+. Only 0.8% of CO^{2+} remains intact long enough to be detected in our equipment. In some cases the Coulomb explosion is even more violent when additional ionization occurs (probably as the result of electron shake off). If CO were adsorbed on a surface, and neutralization were not a factor, carbon or oxygen ions woule be emitted from the surface as the result of a K vacancy, depending on the orientation of the molecule. The key factor in desorption from surface is however the role played by neutralization.

Some cases (N_2, O_2 and NO) were also studied [5] in which only the valence shell was affected. Compared with core vacancies, photoionization in the valence shell is rather benign even with 280 eV photons, which are sufficient to ionized deeper shell valence orbitals. Most of the ions are the single-charged parent ion, and when decomposition occurs the O^+ and N^+ appear with an average recoil of only about 1 eV. (This value compares favorably with the recoil energies as systematically studied by FRANKLIN for the consequences of ionization in the valence shell [12].) A small percentage of O^{2+} and N^{2+} (about 1%) was found with recoil energies of about 20 eV.

Since the survey of Coulomb explosion in molecules following photoionization was carried out at Oak Ridge [3,4,5], other work of relevant interest has appeared in the literature. VAN BRUNT et al. [11] have studied small molecules containing light elements using photon sources with energies of 0.28, 1.25 and 1.5 keV. Data on small angle inelastic scattering are relevant to photoionization experiments, since under the conditions of zero momentum transfer, electrons will impart energy as a dipole excitation and thus mimic the photoelectric effect. Studies have been carried out where the ion fragments have been measured for their intensity and recoil energies, and the data compare well with analogous photoionization experiments. Results of particular interest to the problem of Coulomb explosion can be found in Ref.[13]. The effects of Coulomb explosion are important to biological systems. There is the possibility of localizing radiation damage to a key position in a biological system by the proper choice of x-ray energy. Some reviews of such work are to be found in Ref. [14]. Finally, work on passing molecular ion beams through thin foils, whereupon the fragment ions Coulombically explode,has been used to elicit structural information on the initial molecule [15].

In order to pursue the study of Coulomb explosion further, it would be highly desirable to have photon sources that could be tuned to specific excited states and that could scan over the threshold for core shell electrons. Such sources have now become available in the form of laser and synchrotron radiation, and it will be the aim of the next section of this paper to discuss the usefulness of these new sources of radiation, not only to the study of Coulomb explosions but to the study of molecular decomposition, in general.

3. Experimental Studies in Molecular Decomposition

In the first portion of this paper we have tried to discuss in some detail one of the more violent modes of molecular decomposition, namely Coulomb explosion. Because of the high recoil energy generated in these explosions, this mode of decay should be of particular importance to the general field of desorption of ions from the surface. In this section of the paper a broader overview of molecular decomposition will be taken by examining recent experimental approaches to the subject.

Table 3. Advantages of photon sources

I.   Selective

   A. Photoelectric effect deposits full energy

   B. Cross section favors one orbital

   C. Well-defined resonances occur

II. Favorable peak-to-background

III. Low radiation damage

IV. Minimizes charging

Excitation of molecules may be initiated by several means. In addition to photons, interaction with ions and electrons are common sources for excitation and ionization. If the purpose of one's study is to clearly follow the physics of the consequence of the excitation process, it is desirable to identify the precise nature of the initial excitation state. The use of photons is particularly valuable in helping to establish this state. For convenience Table 3 lists a number of advantages inherent to the use of photons. The photoelectric effect permits a known quantity of energy to be taken up by the molecule. The number of possible processes are generally quite limited and highly selective. Background problems, radiation damage and charging

(when dealing with molecules adsorbed on surfaces) are minimized. This, in
no way, is a plea to use photons exclusively for the study of molecular decomposition. Charge particles have their own special advantages. The point
being made here is that photon sources are very valuable in studying molecular
decomposition, both for free molecules and for molecules adsorbed onto a
surface. A major problem in using photon sources is that they often are too
weak, or are not tunable over a wide range of energies. The recent develop-
ments in laser and synchrotron radiation, however, have helped correct these
problems. It therefore seems useful to evaluate the nature of these sources
and the types of research that can now be carried out effectively with them.

Table 4. Current new photon sources available for study of molecular
decomposition

I. Synchrotron radiation

    A. Intense

    B. Continuous from ionization threshold to region of x rays

    C. Polarized

    D. Pulsed

II. Laser

    A. Very intense

    B. Very narrow band width

    C. Tunable

    D. Polarized

    E. Pulsed

    F. Energy range

        1. recent development of sources < 1000 A°

        2. multiphoton excitation

    Table 4 lists some of the advantages of the new photon sources. The
special advantage of synchrotron radiation [16] is that synchrotron radiation
covers the full range of photon energies from the ionization threshold in the
valence shell to x-ray energies that can be used to photoionize core shell
electrons. Of particular value will be the ability to also scan over the
various resonances close to the different ionization thresholds. The value
of lasers, of course, lies in the enormous intensity of the sources and the
very high resolution. The study of well-defined excited states is possible,
even to the selection of specific vibrational and rotational states. Lasers
have limited tunability, but suitable lasers for a variety of energy ranges
can be found. Most lasers in general use have relatively low energy. To
affect molecular decomposition without ionization would not require a high-
energy laser. If one could monitor the decomposition of neutral fragments,
the use of lasers could be most fruitful both in the gas phase and for adsorbed molecules. For surface studies, one could alternate laser exposure
with typical surface analysis techniques such as Auger or photoelectron

spectroscopy. If one wants to study, however, ionization followed by molecular decomposition, higher energy photons are required. Recently, success has been achieved in developing lasers for wave lengths below 1000 Å [17]. However, the development is still in its infancy. Undergoing very fast growth is the use of multiphoton excitation and ionization [18]. Discussion of the use of this specific technique for molecular decomposition will be deferred until the next section.

Table 5 lists various possible experimental studies that can be carried out on molecular decomposition. Particular consideration is given the uses of laser and synchrotron radiation, and thought is given to the study of molecules adsorbed on surfaces as well as free molecules. An excellent review of the relationship of photoelectron spectroscopy and molecular decomposition has been written by ELAND [19]. This review discusses in detail low-energy excitation processes, so that it supplements the material covered in this paper on high-excitation processes (i.e. Coulomb explosion). Also useful in this regard is the book by BERKOWITZ [20].

Table 5. Types of experimental investigations used in the study of molecular decomposition

---

1. Photoabsorption cross sections

2. Photoelectron spectra

3. Mass spectroscopy

4. Photoelectron and Auger electron-ion coincidence

5. Angular distribution of fragment ions

6. Recoil energy of fragment ions

7. Fluorescence radiation from fragment ions

8. Threshold electron spectroscopy

9. Multi-photon ionization

A few random comments seem in order regarding Table 5. The various means of studying molecular decomposition may be divided under the categories of (1) cross sections for the initial excitation, (2) detection of the fragments involved in the decomposition, (3) coincidences between initial excited states and fragments, and (4) special characteristics of the fragment including its state of electronic excitation by means of its fluorescence spectrum, its recoil energy, and its angular distribution. Photoelectron spectra are useful in separating out the processes of excitation so that their cross sections can be determined independently. Mass spectroscopy makes the study of fragment ions easy. Unfortunately, neutral fragments are also very important, but their detection is much more difficult. Coincidences between the initial excitation and final fragments is an important feature in clarifying the precise mechanisms for decomposition. Recoil energy and angular distribution are useful in the study of Coulomb explosion. If a molecule is adsorbed on a surface, the angle at which a fragment ion is ejected relative to the surface plane could be helpful in determining the orientation of the adsorbed species.

Threshold electron spectroscopy [21] is based on the creation of low-energy electrons from autoionization states created just above the ionization threshold. The high efficiency of collecting these electrons makes possible a very sensitive probe for resonance cross sections for special ionization processes. Multiphoton excitation makes use of the high photon flux possible with lasers. By using a tunable laser it is possible to reach a given resonant state by a multiple sum of photon energies. Say, for example, it takes four photons to reach a given highly excited state, a resonance will take place when the laser reaches the energy of exactly 1/4 of the desired transition. A fifth photon can cause photoionization, and the usual means of photoelectron and mass spectroscopy may be employed to characterize the process [18]. This use of the laser allows one to create a very well resolved excited state from which it might be possible to initiate ion fragmentation.

4. Conclusion

In this paper the Coulomb explosion has been discussed in some detail. Of the different modes of molecular decomposition the high recoil energy experienced in Coulomb explosion gives this mode a specially good opportunity for ejecting fragment ions from the surface. One of the difficulties of utilizing the charge built up by the vacancy cascade is neutralization. It may be possible by choosing the correct molecule to adsorb on a surface to engineer an effective Coulomb explosion.

In general, molecular decomposition of free molecules is strongly related to desorption from surfaces. Molecules, particularly larger molecules adsorbed on surfaces, have an electronic structure that is very similar to free molecules except for slight perturbations. The perturbations are to be found primarily with the molecular orbitals involved directly with bonding to the surface. Those orbitals not involved with surface bonding behave very closely to those of the free molecules. The statements made above have been verified by the comparison of photoelectron spectra from free molecules and adsorbed molecules [22]. With regard to molecular decomposition there are, however, many very important differences between free molecules and molecules adsorbed on surfaces, in particular the capacity for neutralization that the surface offers, and the fixed orientated positions imposed on adsorbed molecules. However, there is much knowledge that can be transferred between the studies of free molecular decomposition and electronically excited desorption of surface adsorbed molecules. It is hoped that in the future close parallel investigations of the two phenomena will be pursued.

Research sponsored by the Division of Chemical Sciences, Office of Basic Energy Sciences, U. S. Department of Energy, under contract W-7405-eng-26 with the Union Carbide Corporation.

References

1. A. C. Wahl and N. A. Bonner, "Radioactivity Applied to Chemistry" (Wiley, New York 1951, p. 269).

2. S. Wexler and G. R. Anderson, J. Chem. Phys. 33, 850(1960); S. Wexler, J. Chem. Phys. 36, 1992(1962); T. A. Carlson and R. M. White, J. Chem. Phys. 38, 2930(1963).

3. T. A. Carlson and R. M. White, J. Chem. Phys. <u>44</u>, 4510(1966).

4. T. A. Carlson and R. M. White, J. Chem. Phys. <u>48</u>, 5191(1968).

5. T. A. Carlson and M. O. Krause, J. Chem. Phys. <u>56</u>, 3206(1972).

6. T. A. Carlson in *Chemical Effects of Nuclear Transformation in Inorganic Systems,* eds. G. Harbottle and A. G. Maddock (North Holland Pub. Co., Amsterday 1979, p. 13).

7. T. A. Carlson and M. O. Krause, Phys. Rev. <u>137</u>, A1655(1965).

8. M. O. Krause, M. L. Vestal, W. J. Johnston and T. A. Carlson, Phys. Rev. <u>133</u>, A385(1964).

9. M. O. Krause and T. A. Carlson, Phys. Rev. <u>158</u>, 18(1967).

10. T. A. Carlson, W. E. Hunt and M. O. Krause, Phys. Rev. <u>151</u>, 41(1966).

11. J. L. Franklin, *Energy Distribution in Unimolecular Decomposition of Ions in Gas Phase Ion Chemistry,* ed. by M. T. Bowen (Academic Press, New York, London 1979) p. 273).

12. R. J. VanBrunt, F. W. Powell, R. G. Hirsch and W. D. Whitehead, J. Chem. Phys. <u>57</u>, 3120(1972).

13. A. P. Hitchcock, C. E. Brion and M. J. van der Wiel, Chem. Phys. <u>45</u>, 461 (1980); A. P. Hitchcock, C. E. Brion and M. J. van der Wiel, Chem. Phys. Letters <u>66</u>, 213(1979); R. B. Kay, Ph. E. van der Leeuw and M. J. van der Wiel, J. Phys. <u>B10</u>, 2513(1977).

14. A. Halpern, in *Uses of Synchrotron Radiation in Biology,* ed. H. B. Stuhrmann (Academic Press, London 1982); A. Halpern and B. Mütze, Int. J. Radiat. Biol. <u>34</u>, 67(1978); A. Halpern and G. Stöcklin, Rad. Res. <u>58</u>, 329(1974).

15. D. S. Gemmell, Chem. Rev. <u>80</u>, 301(1980); S. T. Pratt and W. A. Chupka, Chem. Phys. <u>52</u>, 443(1980).

16. For a general discussion of the use and nature of synchrotron radiation, see *Synchrotron Radiation: Techniques and Application,* ed. by C. Kunz (Springer Verlag, Berlin, Heidelberg, New York 1979); *Synchrotron Radiation Research,* ed. by H. Winick and S. Doniach (Plenum Press, New York, London, 1980).

17. Topical Meeting on Laser Techniques for Extreme Ultraviolet Spectroscopy, March, 1982, Boulder, Colorado, Technical Report published by Optical Society of America.

18. L. Zandee and R. B. Bernstein, J. Chem. Phys. <u>71</u>, 1359(1979); C. D. Cooper, A. D. Williamson, J. C. Miller and R. M. Compton, J. Chem. Phys. <u>73</u>, 1527 (1980); J. C. Miller and R. M. Compton, J. Chem. Phys. <u>75</u>, 72(1981); <u>75</u> 2020(1981).

19. J. H. D. Eland, *Ion Fragmentation Mechanisms and Photoelectron Spectroscopy,* in *Electron Spectroscopy, Theory, Techniques and Applications, Vol. III,* ed. by C. R. Brundle and A. D. Baker (Academic Press, London, New York, 1979).

20. J. Berkowitz, *Photoabsorption, Photoionization and Photoelectron Spectroscopy,* (Academic Press, New York, London 1979).

21. P. M. Guyon, T. Baer, L. F. A. Ferreira, I. Nennes, A. Tabché -Fouhailé, R. Botter and T. R. Govers, J. Phys. B 11 L141(1978); P. T. Murray and T. Baer, Int. J. Mass Spectrons, Ion Phys. 30, 165(1979).

22. J. E. Demuth and D. E. Eastman, Phys. Rev. Lett. 32, 1123(1974); T. M. Thomas, F. A. Grimm, T. A. Carlson and P. A. Agron, J. Electron Spectrosc. 25, 159(1982),

Part 5

Ion-Stimulated Desorption

5.1 Desorption Stimulated by Ion Impact

P. Williams

Department of Chemistry, Arizona State University, Tempe, AZ 85287, USA

Abstract

The interactions of energetic ions with surfaces produce effects that have much in common with electron- and photon-stimulated desorption. Three distinct mechanisms for desorption stimulated by ion impact are reviewed here.

1. Introduction

Ion-stimulated desorption processes can conveniently be subdivided into three categories. The first two might be termed kinetic processes, by analogy with the terminology used in secondary electron emission, in that ionization is a consequence of the kinetic energy of the incident ion. Energetic light ions can act very much as electrons, that is, they can ionize by simple Coulombic perturbation. Ionization of an appropriate core level by proton or He^+ impact might therefore be expected to lead to electronic desorption through the mechanism described by KNOTEK and FEIBELMAN [1], while excitation of valence levels should give desorption according to the MENZEL-GOMER-REDHEAD model [2,3]. For slow (tens of keV) heavy ions, Coulombic excitation is inefficient. Instead, collisions involving such ions excite and ionize through inner (and outer) shell level-crossing processes; non-collisional desorption has been observed which appears to follow inner-shell ionization by heavy-ion impact. The final, and in many ways the most interesting, category can be called a potential process. Slow ions of high ionization-potential species can carry high-energy valence-shell holes into a surface, and there become neutralized through Auger processes. Secondary electron emission by such ions is independent of the ion energy at low energies, and is called potential ejection. It is to be expected, and has recently been shown, that ion ejection can also result from the Auger neutralization process, if the hole on the incident ion carries in enough energy to the surface.

2. Energetic Light Ion Impact: Ions Acting as Electrons

Because the theoretical treatment of ionization by fast light ions is identical to that for electron impact -- in essence, ionization results from the Coulombic perturbation experienced by the target atom during the rapid passage of a point charge -- it is not surprising that ionization cross sections for proton and electron impact have similar magnitude when the projectiles have the same velocity. It would therefore be expected that energetic proton or helium ion impact should stimulate desorption by mechanisms identical to those discussed elsewhere in these proceedings for

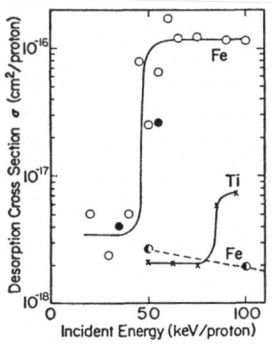

Fig. 1. Cross sections for desorption of oxygen adsorbed on Ti and Fe. Circles are for the Fe substrate; open circles for H_2^+ impact; closed circles for H_3^+ impact; half-filled circles for H^+ impact. The crosses are for H_2^+ impact on Ti. Data are shown as the cross section per proton as a function of energy per proton. The lines are meant only to guide the eye and do not represent measurements. From ref. [4]

electron or photon impact. Such an observation appeared to have been made by LEGG, WHALEY, and THOMAS who studied the desorption of adsorbed oxygen from iron and titanium by H^+, H_2^+, and H_3^+ impact [4]. They found a sharp increase in the (neutral) desorption cross section for projectile energies in excess of ∼ 50 keV/proton, at which energy the proton velocity corresponds to that of an electron with an energy equal to the oxygen 2s ionization threshold (Fig.1). LEGG et al. suggested that their observation might reflect a process of the KNOTEK-FEIBELMAN type. However, their data is puzzling, in that they see the effect only for the molecular projectiles, H_2^+ and H_3^+. For proton impact, they report that the desorption cross section does not deviate from that expected from sputtering theory (Fig.1). LEGG et al. ascribe the effect to ionization by the electron(s) carried along by the molecular projectiles which should also have a velocity corresponding to the oxygen threshold energy. However, the increase in desorption cross section is by a factor of thirty and it is not at all clear how the ionizing power of an electron and a proton could differ by such a large amount.

In fact, data for gas-phase ionization presents a very different picture. The threshold for electron impact ionization is at an energy corresponding to the ionization potential of the level in question, and the electron impact cross section is zero at threshold. The proton impact cross section is similarly zero at threshold, but the threshold in this case occurs when the center-of-mass energy of the projectile plus target exceeds the ionization potential, i.e., at a proton energy of some tens of eV in the present case. When the proton velocity is equal to that of an electron at the threshold energy, the proton cross section is a maximum. As Fig.2 shows for ionization of helium, which has an ionization potential comparable to that of the oxygen 2s shell, the electron-impact cross section is very much smaller than the proton impact cross section at velocities corresponding to the electron-impact threshold, and only begins to approach the proton cross section at energies far above threshold.

If indeed the LEGG data reflects a particularly efficient electron-impact ionization event, the data shown in Fig.1 shows an abnormally sharp threshold. Typically, electron-impact ionization cross sections are zero at threshold, rise roughly linearly with excess energy above the threshold, and maximize at energies approximately five times the threshold energy. Comparison of Figures 1 and 2 suggests that the threshold behavior is consistent neither with electron-impact nor proton-impact ionization events.

185

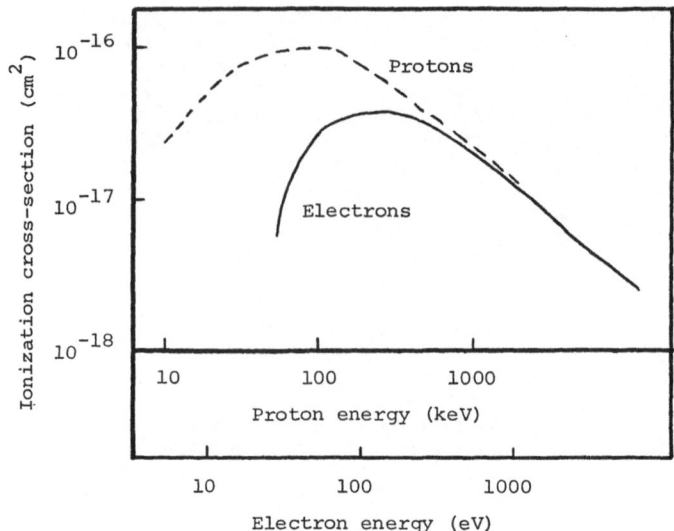

Fig. 2. Cross sections for electron- and proton-impact ionization of helium compared at the same impact velocities. From data collected in ref. [5]

In summary, the data of LEGG et al. is intriguing, but puzzling. The position and shape of their observed threshold is inconsistent with simple proton impact. For electron impact ionization of the level they propose, the threshold position is appropriate, but the shape is not understandable. Finally, given that protons of 50 keV energy should ionize the O 2s levels with a cross section near 10^{-16} cm^2, the apparent absence of a proton effect casts doubt on the idea that the authors' observations reflect the simple ionization-driven desorption process that they suggest.

3. Medium-Energy Heavy Ion Impact: Desorption Following Core-Hole Excitation

Excitation and ionization in heavy-ion collisions follows electron promotion which can occur when antibonding molecular orbitals in the colliding complex undergo pseudocrossings with bonding orbitals of higher principal quantum number [6]. Core ionization in such energetic collisions is the basis of a widely accepted model for the formation of multiply-charged ions. These ions are felt to result from the ejection of core-ionized atoms which subsequently undergo Auger decay, leading to further electron loss, in the gas phase [7]. It is reasonable to expect that core ionization could lead to desorption via Auger decay, independent of any momentum transfer in the collision. However, there is a significant experimental problem in looking for such an effect, because momentum transfer always does occur, so that any desorbed flux will inevitably be accompanied by an intense flux of directly sputtered ions. However, we know empirically that species of high ionization potential should have low sputtered positive ion yields (although there exists no quantitative model which would let us predict those yields), whereas high ionization potential species, being usually electronegative, tend to be efficiently electron-desorbed. In fact, it has been known for some time that hydrogen and, in particular, fluorine had positive ion yields very much higher than those expected from the systematics of ion yield vs. ionization potential [8].

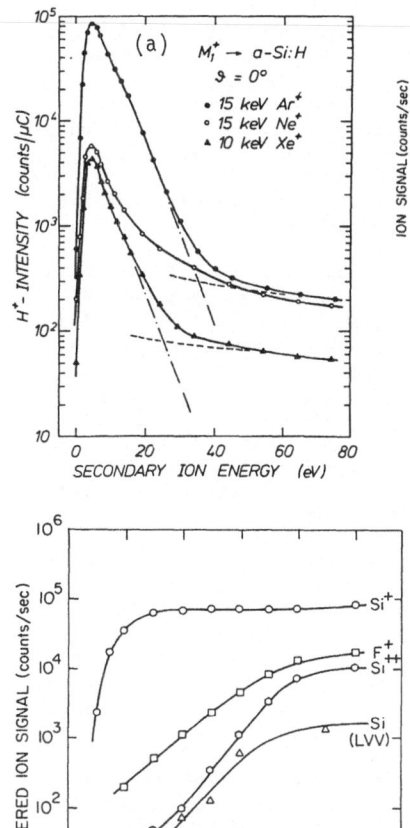

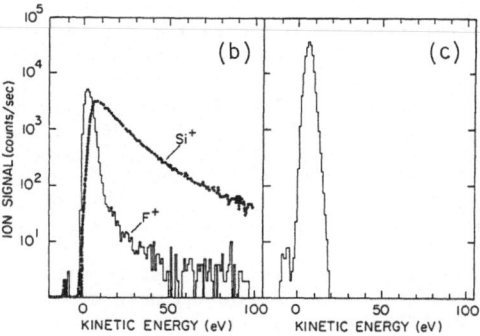

Fig. 3. (a) Energy distributions of H$^+$ ejected from hydrogenated silicon by inert-gas ion impact. From ref. [9]. (b) Energy distribution of F$^+$ ejected from fluorinated silicon by Ar$^+$ impact. From ref. [10]. (c) Energy distribution of F$^+$ ejected from fluorinated silicon by electron impact at 14 keV. Instrumental conditions identical to those of Fig. 3b. From ref. [10]

◁ **Fig. 4.** Variation of various ion yields with primary ion impact energy, shown together with the variation of the silicon ion-induced LVV Auger signal. From ref. [10]

In the last three years, models have appeared for H$^+$ and F$^+$ ejection from silicon (and for Cl$^+$ from aluminum) which involve collisional ionization of the substrate L-shell as the primary step. The subsequent mechanisms differ, although the experimental evidence in both cases is practically identical. For H$^+$ ejection from silicon by inert-gas ion bombardment, WITTMAACK [9] observed an ion energy distribution strongly peaked at low energy with an apparently distinct, and much less intense, component extending to higher energies (Fig.3a). WILLIAMS [10] observed a very similar distribution for F$^+$ ejection from silicon, shown in Fig.3b (this had earlier been observed by MORGAN and WERNER [8]), and, in addition, showed that the low-energy component was almost identical to the distribution for electron-desorbed F$^+$ obtained under identical instrumental conditions (Fig.3c). To identify the primary process, Wittmaack studied the dependence of the H$^+$ yield on the primary ion energy, and showed this to be identical to the behavior of Si^{++}. Adopting this approach, WILLIAMS

187

showed that the same was true for F^+, and further showed that data for the ion-excited Si LVV yield (taken from WITTMAACK [11]) was closely similar (Fig.4). The two papers diverge only in the mechanisms proposed. WITTMAACK suggests that hydrogen is sputtered bound to a core-ionized silicon atom, and that gas-phase Auger decay leads to further ionization of the molecule which dissociates in a "Coulomb explosion". Conversely, WILLIAMS suggests that F^+ is ejected directly from the solid in an ESD-like process, and further suggests that this might also be the mechanism for H^+ ejection. We will compare here the relative merits of the two models.

To begin, the data is inconsistent with any suggestion that F^+ ejection might result from the WITTMAACK mechanism. This can be seen from the F^+ energy distribution. If the F^+ resulted from gas-phase dissociation of a core-ionized SiF^+ ion, the F^+ would retain some 40% of the molecular ion energy, and would exhibit an energy spectrum resembling that of a collisionally-sputtered diatomic ion -- somewhat narrower than an atomic ion distribution, but still extending smoothly to energies of 100 eV or more. Conversely, hydrogen should retain only ~ 3% of the SiH^+ energy, and so the H^+ distribution is not inconsistent with the WITTMAACK mechanism. To explain the F^+ ejection, WILLIAMS suggested that the initial event was core ionization of a silicon atom somewhat distant from the fluorine to be ejected, the argument being that an energetic collision with a nearest neighbor to the fluorine could not avoid transferring kinetic energy to the fluorine. F^+ ejection was proposed to result from a true ESD process initiated by a silicon LVV Auger electron from the decay of the target atom. This process imposes certain constraints on the mechanism: the Si LVV electron cannot re-excite a silicon core level because it has insufficient energy. However, it can excite the F 2s level at 31 eV which KNOTEK and FEIBELMAN showed to be efficient in desorbing F^+. A comparable process could account for H^+ ejection. TRAUM [14] has shown that H^+ is electron-desorbed from silicon above a threshold energy of ~ 22 eV. Alternatively, it should be noted that collisional core ionization of a silicon nearest-neighbor to a hydrogen, followed by Auger decay, could result in K-F ejection of that hydrogen with the energy distribution observed by WITTMAACK, since again energy transfer to the hydrogen in a single collision is inefficient. This process is conceptually very similar to that suggested by WITTMAACK, the only difference being that the core-ionized silicon need not be ejected along with the hydrogen. It seems possible that such a process might in fact be more efficient than that discussed by WITTMAACK since the constraints for ejection of a bound molecular ion need no longer be obeyed.

4. Ions as Hole-Carriers: Ion-Neutralization Desorption

The possibility that the potential energy carried into the surface by the incident Ar^+ ion could initiate desorption was discounted in the study of ref. [10] due to the pronounced primary ion energy dependence of the F^+ yield. More recently, however, SCHULTZ and RABALAIS have looked at desorption of F^+ from alkali fluorides by inert gas impact at quite low energies [12]. They make the intriguing observation that the F^+ signal does not vanish as the incident ion energy approaches zero, which very strongly suggests that the potential, rather than kinetic, energy of the incident ion is responsible for the observed effect. The data is, in fact, strongly reminiscent of data on "potential" (i.e., Auger) ejection of secondary electrons which similarly shows a low-energy independence of incident ion energy for inert-gas impact [13]. Because these observations are reported in more detail by SCHULTZ and RABALAIS elsewhere in these proceedings, we will give no further discussion here, other than to point out that the process promises to bear a relationship to photon-stimulated desorption similar to that ion-neutralization spectroscopy bears to photoelectron spectroscopy. That is, the impact of a wide range of singly- and multiply-charged ions on surfaces offers an opportunity to excite surfaces by energies discretely tunable up to 54 eV (the second ionization potential of He). Further, the ejection of a previously adsorbed ion signals the neutralization of the incident ion at the surface, and so offers an interesting new probe of ion neutralization phenomena.

5. Conclusion

It appears unlikely that the observation of desorption processes initiated by energetic light or heavy ion impact will be of great fundamental importance to the understanding of stimulated desorption mechanisms, because the thresholds are less well defined than for electron or photon impact, and the processes can be complicated by other secondary processes, particularly sputtering processes. However, the availability of mature models for ESD/PSD does allow some discussion of the ion-induced desorption processes discussed here. The non-observation of proton-stimulated desorption in the work of ref.[4] remains to be explained, and it would be desirable to obtain electron-desorption cross sections under similar conditions. Given a measured cross section for electron-stimulated desorption of a particular species, it should be possible to predict a cross section for the analogous proton-stimulated process. Such predictions might, for example, be useful in estimating beam-damage effects in the use of energetic ion beams for surface physics studies. The fact that heavy-ion bombardment gives rise to an Auger-mediated desorption process in addition to sputtering is important to the understanding of sputtered ion emission, because it demonstrates that the "normal" systematic decrease in ion yield with ionization potential is not necessarily violated by the observation of anomalously high yields for H^+ and F^+.

The newly observed process of slow ion-induced desorption does appear to offer a new field of investigation, in that the effect is "clean" -- the energies used can be below the sputtering threshold for most solids -- and the input energies can be precisely specified. Because each ejected ion signals the neutralization of an incident ion, the process offers a probe of neutralization at low incident energies where detection of scattered neutral species would be extremely inefficient.

I thank Wayne Rabalais and Al Schultz for sharing with me information on their ion-neutralization desorption work prior to publication, Ed Thomas for discussions on his proton-impact studies, and Dwight Jennison and Mort Traum for information on ESD of H^+ from silicon.

References

1.   M. L. Knotek and P. J. Feibelman, Phys. Rev. Lett. 40 (1978) 969.

2.   P. A. Redhead, Can. J. Phys. 42 (1964) 886.

3.   D. Menzel and R. Gomer, J. Chem. Phys. 41 (1964) 3311.

4. K. O. Legg, R. Whaley and E. W. Thomas, Surf. Sci. 109 (1981) 11.

5. H. S. W. Massey, E. H. S. Burhop and H. B. Gilbody, "Electronic and Ionic Impact Phenomena", Vol. IV. Clarendon, Oxford, 1974.

6. U. Fano and W. Lichten, Phys. Rev. Lett. 14 (1965) 627.

7. P. Joyes, J. Phys. C5 (1972) 2192.

8. A. E. Morgan and H. W. Werner, J. Chem. Phys. 68 (1978) 3900.

9. K. Wittmaack, Phys. Rev. Lett. 43 (1979) 872.

10. P. Williams, Phys. Rev. B23 (1981) 6187.

11. K. Wittmaack, Phys. Lett. 74A (1979) 197.

12. J. W. Rabalais and A. Schultz. These proceedings.

13. M. Kaminsky, "Electronic and Ionic Impact Phenomena on Metal Surfaces", Springer, Berlin 1965.

14. M. M. Traum, Bell Labs. Private communication.

5.2 F$^+$ Ejection from LiF Surfaces by Ion Bombardment

J.A. Schultz, P.T. Murray, R. Kumar, Hsin-Kuei Hu, and J.W. Rabalais

Department of Chemistry, University of Houston
Houston, TX 77004, USA

INTRODUCTION

Desorption or decomposition of surfaces can be induced by electron, photon
or ion stimulated electronic transitions; this contribution is concerned with
ion neutralization stimulated decomposition of surfaces. The energy supplied
by the neutralization of an ion near the surface of an insulating solid can
cause decomposition of the solid [1-3]. For example, the yields of both positive and negative secondary ions ejected by 100 - 1500 eV Ar^{+n} and Kr^{+n}(n=1-5)
bombardment of alkali halides are almost totally determined by the charge on
the primary ion [4,5]. Auger-induced desorption of fluorine has been postulated [6] as a reason for the metallization of LiF surfaces following 10 - 400
eV He$^+$ impact and Auger electron yields of $\sim$ 0.5 electrons/ion from 10 - 200
eV He$^+$ bombardment of LiF have been measured [7]. The formation of halogen
cations by Auger core-hole decay and their subsequent movement within alkali
halide crystals has been treated theoretically [8,9]. Recent investigations
of ion bombardment of silicon [10] and of fluorine-doped silicon [11] show a
correlation between Si^{+2} and F$^+$ secondary ion yields and the ejection of
Si-L$_{2,3}$MM Auger electrons by ion impact. In this report we present data demonstrating the long-predicted production of F$^+$ secondary ions during ion
bombardment of LiF.

EXPERIMENTAL

The experiments were performed on evaporated LiF films using the apparatus
shown in Fig. 1. The primary ions were formed by 200 eV electron impact in
a commercial Varian ion gun and mass selection of the primary ions was accomplished by a Wien filter of our own design. Primary ion beam current densities were < 10^{-9} A/cm^2 and total ion doses on the sample were less than 10^{15}
ion/cm^2 during the acquisition of any one data curve. Positive charging of
the sample was compensated by using a low-energy (< 7 eV) electron flood.
Electron stimulated desorption of F$^+$ was observed when the electron energy was
increased to ca. 32 eV or greater, as shown in Fig. 2. Mass selection of secondary ions ejected from the surface was performed by an Extranuclear quadrupole mass filter. The secondary ions were not energy analyzed when obtaining
the data in Figs. 1 and 3. However, an energy filter was later used to demonstrate that sample charging does not affect the conclusions drawn from the
data presented.

RESULTS

Fig. 1A shows the results from He$^+$ bombardment with the Wien filter removed.
The x's in the figure result from ion bombardment by residual gases at current densities of 1 x 10^{-10} A/cm^2 (P = 8 x 10^{-10} torr). The x's were obtain-

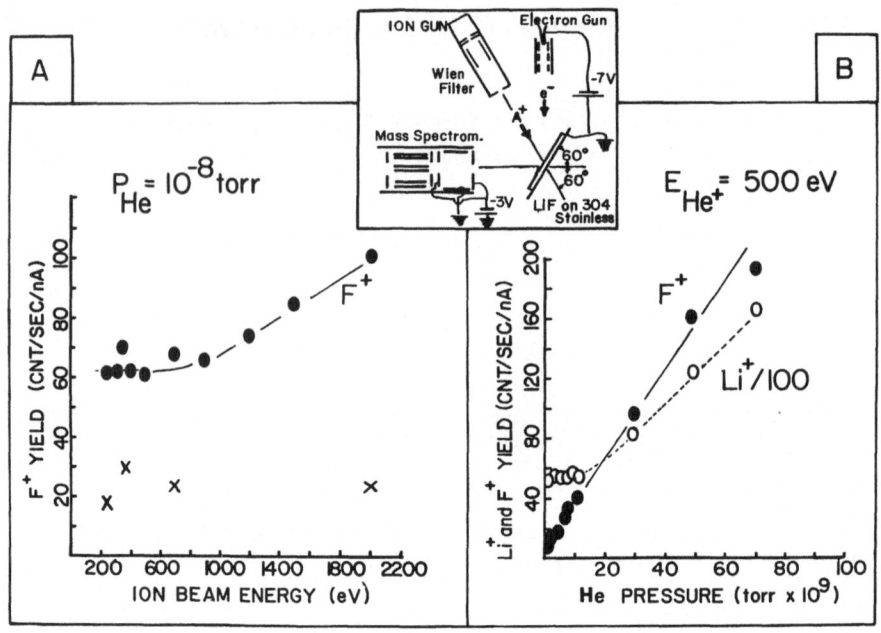

Fig. 1A. Yield of F^+ as a function of beam energy during bombardment of LiF(s) with He^+ and ionized residual gases. The chamber was backfilled with 2×10^{-8} torr He at all beam energies. Fig. 1B. Yield of Li^+ and F^+ as the ion current to the sample is increased by increasing the pressure of the backfilled He. Notice the large component of sputtering of Li^+ by residual gases alone. The ion beam energy was constant at 500 eV for all pressures of He. A schematic diagram of the experimental setup is shown in the inset

ed by normalizing the F^+ signal observed in the absence of He^+ to the ion current of 1×10^{-9} A/cm^2 produced when 2×10^{-8} torr He was leaked into the chamber. At this pressure the residual gas peaks at m/e = 18 and 28 increased by < 10%. The pressure of 2×10^{-8} torr He was constant for the data in Fig. 1A. Thus the yield of F^+ can be seen to be constant as a function of He^+ beam energy below 800 eV, although a small but significant contribution comes from the residual gases. It is seen in Fig. 1B that the residual gas ions sputter Li^+ many times more efficiently than does He^+. For example, when no helium is in the chamber, residual gas sputtering gives an intensity of Li^+ of 5400 c/s while raising the chamber pressure with He from 10^{-9} to 10^{-8} torr increases the Li^+ count rate to only 5800 c/s. This result clearly points to the danger of attempting low current He^+ ion bombardment studies without mass selection of the He^+ ion beam. Nevertheless, the dependence of the F^+ yield with helium pressure in Fig. 1B shows that F^+ is being produced almost exclusively by the He^+ primary ion.

The data in Fig. 3 were obtained using the experimental geometry and focusing potentials shown in Fig. 1. Relative sputtering yields of F^+ by the four primary ions can, therefore, be determined if sample charging is adequately controlled by the electron flood. This was shown by reproducing the experimental conditions of Fig. 3 and measuring the Li^+ secondary ion energy distributions. The uncertainty in the charge compensation was estimated by

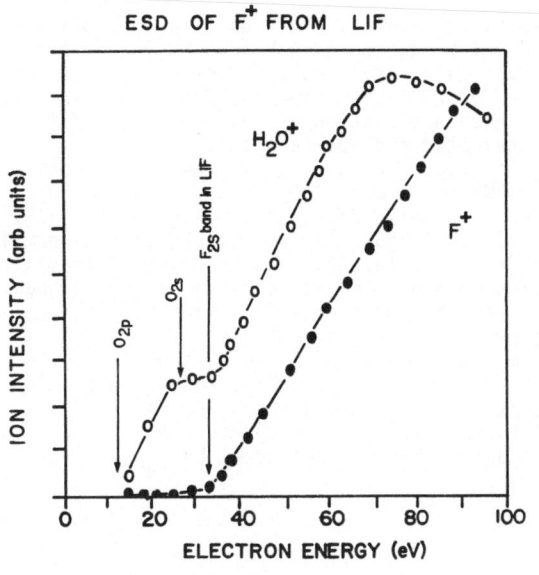

ESD OF F$^+$ FROM LIF

Fig. 2. Electron-stimulated decomposition of lithium fluoride. The ionization curve for gaseous H_2O is included for comparison of the inner valence levels

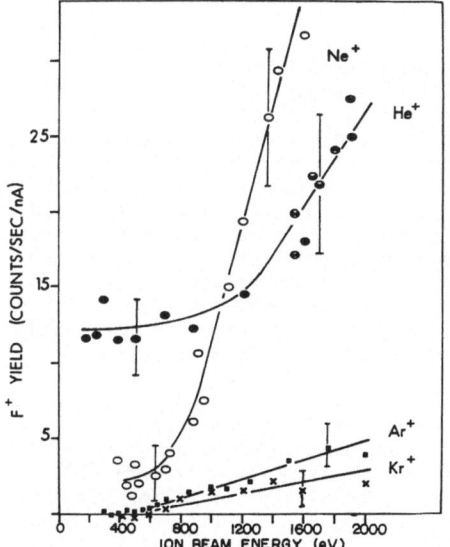

Fig. 3. Relative yields of F$^+$ from $\overline{LiF(s)}$ as a function of bombardment energy for four velocity selected singly charged rare gas ions

biasing the sample so that the peak in the secondary Li$^+$ distribution shifted by ± 5 volts. This was greater than the energy differences between the peaks in the Li$^+$ energy distributions produced by the four primary ions and resulted in a reduction of the Li$^+$ signal by about 50%. This was used to assign error bars in Fig. 3 for the Ne$^+$, Ar$^+$, and Kr$^+$, and for He$^+$ above 1200 eV.

Because the velocity selected He$^+$ intensity was inadequate below 1200 eV, the data from Fig. 1A is transferred to Fig. 3 by normalizing to make the

193

points above 1200 eV in the two experiments match. This gives an approximate qualitative comparison of the yields of F^+ at energies below 600 eV for the four primary ions. The slope of each yield curve above 600 eV was reproduced within experimental error from three separate LiF samples. Error bars shown in Fig. 3 for He^+ data below 800 eV have been assigned from the uncertainties in matching the slopes of the mass selected and non-mass selected data for He^+ above 1200 eV. The data for F^+ in Fig. 3 below 600 eV beam energy shows that the yields ($\gamma_{He^+} > \gamma_{Ne^+} > \gamma_{Ar^+} \approx \gamma_{Kr^+} \approx 0$) correlate with the potential energy contained in the primary ion hole (in eV, He^+ = 24.5, Ne^+ = 21.6, Ar^+ = 15.8, Kr^+ = 14.0). This hierarchy of F^+ yields is qualitative because of the normalization of the low energy He^+ data from Fig. 1A. All data contain at least 100 total counts in the determination both of background signal and F^+ signal. The negative points for Kr^+ and Ar^+ result from statistical fluctuations in the background count rate produced by fast ions and neutrals.

The data for F^+ production by ion bombardment below 600 eV can be classified into two groups of He^+, Ne^+ and Ar^+, Kr^+. In the He^+, Ne^+ group an Auger electron can be ejected when the ions approach the surface, whereas the Ar^+, Kr^+ group does not have sufficient potential energy to eject an Auger electron from LiF. This is shown in Rxns. 1 and 2 where we represent $Li_n F_n$ solid by two F^-'s in the following reaction.

Two electrons removed from LiF:
$$2F^- + He^+ \rightarrow 2F^0 + He^0 + e^-(KE) \tag{1a}$$

$$F^- + F^+ + He^0 + e^-(KE') . \tag{1b}$$

One electron removed from LiF:
$$2F^- + Kr^+ \rightarrow F^- + F^0 + excess\ energy . \tag{2}$$

In Rxn. 1 the relative partitioning of the products between ($Li^+ F^-_{n-2} + 2F^0$) and ($Li^+ F^-_{n-1} + F^+$) is unknown. The yield of F^+ depends not only on its probability of initial formation which is likely very high, but also on its probability of survival and escape [13]. The F^+ would be thought to be efficiently neutralized by its next nearest neighbors F^-; however, theoretical estimates show that the relaxation time for this neutralization in maximal valency solids is sufficiently slow to allow F^+ to desorb from the surface layer [13]. We propose that the F^+ escape is limited to those anions ionized in the first atomic layer. Although F^+ is probably produced in the bulk, particularly at higher beam energies where recoil sequences extend deep into the crystal [14], the escape probability from the interior as F^+ is certainly low [13]. Ultimate sputtering as a neutral, which for electron bombardment has a threshold at 5 eV [15], is one of the possible ways of disposing of this electronic excitation of the bulk [16].

Ionization of sputtered F^0 or F^- by a second primary ion could also explain our results. A correlation has been observed between F^+ and Cl^+ production (from impurities in an Al-Mg matrix) and the potential energy in He^+, Ar^+, and Xe^+ for primary current densities of 100 mA/cm^2 (4 - 12 keV) [17]. At these high current densities a significant probability exists for bimolecular collisions between incoming primaries and secondary particles sputtered by a previous event. This has also been seen in O_2^+ (10 mA/cm^2; 12 keV) sputtering of an alloy of Cd, Hg, and Te [18]. With the low current density and dose (< 10^{15} ions/cm2) employed herein, we do not expect bimolecular reactions between the primary ions and either sputtered secondary particles or surface defects.

Yields of F^+ are dependent on the beam energy above 600 eV. This is illustrated in Table 1 where the slopes of the F^+ yield curves are tabulated along with the fraction of the initial primary ion energy transferred to a target fluorine atom by an elastic collision, i.e., $4M_1 M_F/(M_1 + M_F)^2$, where M_1 is the mass of the primary ion and M_F is the mass of fluorine. For Ne^+ and He^+ bombardment, in which Auger neutralization is possible, ratios have been taken

Table 1. F^+ yields and collisional energy transferred during bombardment

	F^+ yield slope	Ratio	Fractional energy transferred	Ratio
Ne^+	0.032		0.99	
He^+	0.018	1.8 ± 0.3	0.58	1.7
Ar^+	0.0025		0.87	
Kr^+	0.0017	1.5 ± 0.5	0.60	1.4

of the yield slopes and of the fractional energy transferred by the two ions. The two ratios are equal within experimental error. A similar comparison has been performed in Table 1 for the data involving Ar^+ and Kr^+. The rate of change of the F^+ yield is seen to be about an order of magnitude greater for the case of He^+ and Ne^+ compared to Ar^+ and Kr^+ bombardment. These data show that although the fluorine recoil is important, the other factor determining the yield of F^+ may be the number of electrons removed by the potential energy neutralization process. If two electrons have already been removed before collision as shown in Rxn. 1, this apparently increases the probability that F^+ will escape without neutralization compared to the case for one electron removed from the initial state as shown in Rxn. 2. Intuitively, one would expect that the escape probability of F^+ would be increased as the number of electrons within the collision site is decreased.

The F^+ ion could come either from the surface or from the decomposition of a sputtered cluster. If the F^+ comes from cluster decomposition, an upper limit of 100 nsec for the cluster lifetime has been experimentally determined. This was obtained by bombarding the sample with a 300 nsec 2keV He^+ pulse (also Ne^+) and observing the time of flight of the secondary ions through the mass spectrometer to an on-axis channeltron detector. A tungsten grid located 1 mm in front of the sample provided a ground plane so that the secondary ions could be accelerated by biasing the sample to a selected voltage (in this case + 100 V). The slowest of any particular secondary ion has a unique flight time determined by the sample bias. Sample charging is directly measured by observing a uniform shifting of all secondary ions to shorter flight times (i.e., to a flight time of the ion determined by the sample bias plus the sample charging). The F^+ was observed at the flight time appropriate to its having been formed at the surface and thus being accelerated through the entire sample bias of 100 volts. The mass spectrometer could be tuned to pass only F^+ (plus fast scattered ions and neutrals which are temporally separated). From this it was determined that no F^+ is produced by decomposition of clusters on their way to the detector. The combination of time-of-flight and quadrupole mass selection allows simultaneous measurements of elastic scattering for surface analysis [19] and determination of the secondary ion yields while monitoring and controlling surface charging of insulators. We plan a more refined study of the lithium fluoride system as well as other insulating materials using this experimental technique.

CONCLUSIONS

1. The production of F^+ from ion bombardment of LiF below 600 eV correlates with the potential energy of the primary ion. We have tentatively attributed this to the removal of bonding electrons from fluorine with subsequent desorption of F^+ from the solid.

2. The F^+ yield above 600 eV correlates with the fraction of the primary
beam energy transferred to fluorine recoil. The yields from Ne^+ and He^+ are
an order of magnitude higher than from Kr^+ and Ar^+ bombardment, indicating
that the potential energy of the ion may be important even at high bombardment energies.
3. F^+ can be produced from LiF by electron stimulated decomposition. Although this effect was predictable, its observation is important as an illustration of a general artifact in ion bombardment of ionic solids. If the
region around the sample is not field free, secondary electrons are accelerated
and interact with the sample or areas coated with sputtered sample.

ACKNOWLEDGEMENT

This material is based upon work supported by the National Science Foundation
under Grant No. CHE-7915177. We would also like to thank Prof. E. Bauer and
a referee for their germane comments.

REFERENCES

1. S.P. Wolsky and E.J. Zdanuk, Phys. Rev., 121 (1961) 374.
2. E.S. Parilis in: Proceedings of the International Conference on Atomic
    Collision Phenomena in Solids, North Holland, Amsterdam (1970) 324.
3. D.V. McCaughan and V.T. Murphy, IEEE Trans Nucl. Sci., 19 (1972) 249;
    D.V. McCaughan, C.W. White, R.A. Kushner and D.L. Simms, Appl. Phys.
    Lett., 35 (1979) 405.
4. Sh.S. Radzhabov, R.R. Rakhimov, and D. Abdusalomov, Izv. Akad. Nauk SSSR,
    Ser. Fiz., 40 (1976) 2543.
5. S.N. Morozov, D.D. Gruich, and T.U. Arifov, Izv. Akad. Nauk SSSR, Ser.
    Fiz., 43 (1979) 612. (Figure captions 2 and 3 are reversed in the English
    translation.)
6. V.G. Matsevich and G.K. Zyryanov, Izv. Akad. Nauk SSSR, Seriya Fiz., 38
    (1974) 244.
7. R. Rakhimov and S. Gaipov, Izv. Akad. Nauk SSSR, Ser. Fiz., 43 (1979)
    1894.
8. J.H.O. Varley, J. Nuclear Energy, 1 (1954) 130.
9. R.E. Howard, S. Vosko, R. Smoluchowski, Phys. Rev., 122 (1961) 1406.
10. K. Wittmaack, Nucl. Instr. Meth., 170 (1980) 565.
11. P. Williams, Phys. Rev. B, 23 (1981) 6187.
12. M.L. Knotek and P.J. Feibelman, Surf. Sci., 90 (1979) 78.
13. P.J. Feibelman, Surf. Sci. Lett., 102 (1981) L51.
14. J.P. Biersack and E. Santner, Nucl. Instr. Meth., 132 (1976) 229.
15. P.D. Townsend, Surf. Sci., 90 (1979) 256.
16. M. Szymonski, A.E. deVries, Rad. Eff., 54 (1981) 135.
17. R.F.K. Herzog, W.P. Poschenrieder, and F.G. Satkiewicz, Rad. Effects, 18
    (1973) 199.
18. K. Wittmaack, Adv. Mass. Spec., 8 (1980) 503.
19. Y.S. Chen, G.L. Miller, D.A.H. Robinson, G.H. Wheatley, and T.M. Buck,
    Surface Sci., 62 (1977) 133.

5.3 Similarities in the Relative Populations of Excited States Produced by Sputtering and by Electron Impact

R. Kelly

IBM Thomas J. Watson Research Center, Yorktown Heights, NY 10598, USA

Abstract

In both sputtering and medium-energy electron-atom impact, the inelastic energy transfer, ΔE_e, should have a similar distribution, $dN/d(\Delta E_e) \propto 1/(\Delta E_e)^2$. This suggests that the relative populations of excited states may be similar for each type of experiment and, in order to test this idea, previously published results for the formation of SiI, SiII, and SiIII in sputtering and for the formation of RbI, RbII, and RbIII in electron-atom impact are compared with theory based on an inverse square distribution of ΔE_e. Agreement between experiment and theory is found to be good over the entire 55-60 eV span of excitation energies.

1. Excited States Formed by Sputtering

When solids, especially those with electronic band gaps, are bombarded with medium-energy (1-100 keV) ions of medium to high mass (e.g., Ne^+, Ar^+, Kr^+, Xe^+), easily detected quantities of atoms and ions are found to be sputtered in excited states. For example, count rates of up to 10^6 s^{-1} are obtained for low-lying neutral excited states of group I-IV elements provided the surfaces are flooded with oxygen. We have previously discussed the formation of such excited states in terms of an electron-pickup model [1] as well as in terms of a thermal model [2]. The thermal model was concluded to be inadequate [2] but the electron-pickup model [1] is still advocated, especially for metallic targets [3]. More recently we have shown that both excited atoms [4,5] and excited ions [6] can be understood, qualitatively and quantitatively, in terms of a statistical apportioning of inelastic energy transfer. The model was then extended to describe excitation in a low-pressure R.F. plasma (0.3-3 Pa) [7].

The present work will carry the argument one step further by comparing excitation in the sputtering of Si (data from [6]) with that arising in medium-energy electron bombardment of gaseous Rb (data from [8-10]). Since the comparison will be made for excited atoms, excited single ions, and excited double ions, particularly wide spans of excitation energies will be involved.

The basis of the statistical model is easily outlined. (a) There is an inverse square kinetic-energy (E) distribution of cascade atoms, i.e., the number of such atoms per energy interval is $dN/dE \propto 1/E^2$. (b) The cascade atoms expel surface atoms, the latter receiving inelastic energy, ΔE_e, of average amount ΔE_e $\approx KE$, where K is calculable [11,12] in terms of the motion of electrons during the encounter of Thomas-Fermi atoms (Fig.1). An approximately linear relation, $\Delta E_e \propto E$, is also borne out experimentally [13,14]. (c) The inelastic energy is, as a consequence of (a) and (b), distributed as

$$dN/d(\Delta E_e) \propto 1/(\Delta E_e)^2, \tag{1}$$

where we recognize an <u>inverse-square</u> distribution. (d) The inelastic energy then populates accessible excited states in single events having two steps [6]. First of all, for ΔE_e greater than the first ionization potential, the appropriate number of electrons is expelled. Then, the accessible excited states, i.e., those with internal (excitation) energy, ε, less than ΔE_e, are postulated to populate in proportion to their degeneracies. This is equivalent to postulating equal probability of forma-tion for each accessible J level and is easily shown [4,5] to be preferable to the alternative postulate that only the excited state lying immediately below ΔE_e is formed. Singly and doubly excited species, as well as species with different spins, are treated in the same way.

An important part of the argument is the assumption that excitation occurs in single events having two steps. For example, SiIII is assumed to form in events of the type

$$\text{energetic cascade atom} \; + \; Si = Si^{**} = Si^{++*} + 2e^-, \tag{2}$$

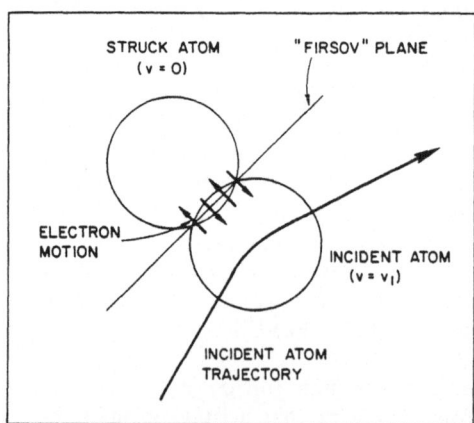

Fig. 1. Sketch relating to the Firsov [11,12] model for inelastic energy transfer in collisions between medium or heavy atoms. As the one atom passes the other, a plane can be regarded to occur at the internuclear midpoint through which electrons from each atom pass in accordance with the radial distributions. The fluxes are evaluated throughout the entire classical trajectory and do not cancel out since the one atom is moving

where Si^{**} represents an autoionizing intermediary with ΔE_e exceeding the sum of the first and second ionization potentials. There is not a stepwise ionization, a conclusion which has been reached previously by ZAPESOCHNYI and SHIMON [15] in connection with the formation of NaII from Na vapor by electron impact in a demonstrably single-collision regime.

It is not too difficult to show [4-6] that a direct consequence of (a)-(d) is that excited atoms or ions of type j form according to the unnormalized distribution function, P_j:

$$P_j \text{ (unnormalized)} = \sum_{n=0}^{\infty} \frac{g_j}{g_0 + g_1 + \cdots + g_{j+n}} \{\varepsilon_{j+n}^{-1} - \varepsilon_{j+n+1}^{-1}\}, \qquad (3)$$

where g is the degeneracy, $(2S + 1)(2L + 1)$. As will be seen from Figs. 3 and 4 (to follow), (1) is approximately of the form $g_j \exp(-\varepsilon_j/\varepsilon^*)$, where ε^* is a parameter with units of energy. Such a result should probably be expected for any statistical or near-statistical partitioning of energy. The normalized expression corresponding to (3) was shown [4,5] to be

$$P_j \text{ (normalized)} = 2UK \times \text{Eq. (3)}, \qquad (4)$$

where U is the surface binding energy and K is defined by $\Delta E_e \approx KE$. The reasonableness of (4) is shown by the factor-of-two agreement with the absolute excited-state yields of TSONG and YUSUF [16].

Had the postulate of each accessible J level having an equal probability of formation not been made, (3) would be replaced with

$$P_j \text{ (incorrect)} = \int_{\varepsilon_j/K}^{\varepsilon_{j+1}/K} E^{-2} dE \ \alpha \ \{\varepsilon_j^{-1} - \varepsilon_{j+1}^{-1}\}.$$

This relation is unacceptable as it varies erratically rather than monotonically as ε_j increases.

As will be discussed in Section 4 it is possible that the arguments leading to (3) are incomplete. There is an indication, based on the population of RbII and RbIII being excessive, that (3) should contain as a factor the number of equivalent outer electrons in the ground state of the species in question. This factor is 6 for RbII and 5 for RbIII.

2. Excited States Formed by Electron Impact

A similar formalism can be devised, based on work by SEATON [17], for describing excited-state production due to the impact of medium-energy ($\sim >$ 100 eV) electrons of energy E_1 on gas atoms. This is discussed elsewhere [7], the final result being

$$\Delta E_e \approx \frac{4I_H^2 a_o^2}{p^2 E_1} \times \sum f_{fi}, \tag{5}$$

where f means "final", i means "initial", I_H is 13.60 eV, a_o is 0.5292 Å, and f_{fi} is the usual optical oscillator strength.

An alternative and highly instructive approach to electron-atom collisions is to treat the collisions as classical Coulomb scattering (Fig.2). (a) The energy transferred from an incident electron (mass M_1) to an effective number, Z_{eff}, of atomic electrons (mass $M_2 = Z_{eff}M_1$) will be identified with ΔE_e, so that one can use the very general result,

$$\Delta E_e = \frac{4M_1 M_2 E_1}{(M_1 + M_2)^2} \times \sin^2 \frac{\phi}{2} \approx \frac{4E_1}{Z_{eff}} \times \sin^2 \frac{\phi}{2}.$$

Here ϕ is the center-of-mass scattering angle. (b) The angle ϕ can be eliminated in favor of the impact parameter p by introducing the general relation for Coulomb scattering [18]:

$$\tan \frac{\phi}{2} = \frac{Z_1 Z_{eff} e^2 (M_1 + M_2)}{4\pi\varepsilon_o \times 2pE_1 M_2} \approx \frac{Z_{eff} e^2}{4\pi\varepsilon_o \times 2pE_1},$$

where $Z_1 = 1$ and $4\pi\varepsilon_o$ is the usual factor when S.I. units are used. The result is

$$\Delta E_e = \frac{4I_H^2 a_o^2}{p^2 E_1} \times Z_{eff}, \tag{6}$$

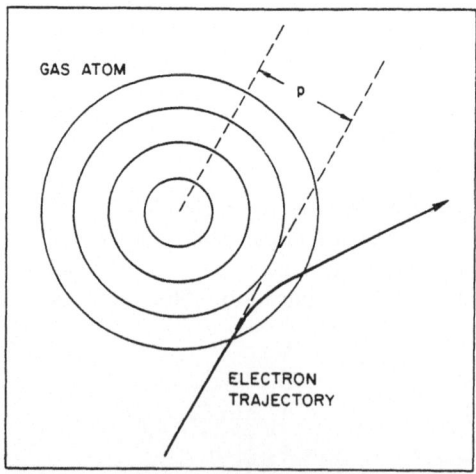

Fig. 2. Sketch of the "impact parameter rings" of area $2\pi pdp$, p being the impact parameter, which are associated with a gas atom which is being bombarded randomly by electrons. According to (5) or (6), the extent of inelastic energy transfer to the gas atom for a given p scales as $1/p^2$.

a relation which is closely similar to (5) but with the problem of evaluating the sum of optical oscillator strengths, Σf_{fi}, replaced by the comparably intractable problem of evaluating Z_{eff}. We will circumvent both problems by deducing only <u>relative</u> populations.

The final step in the description of electron-atom impact follows from the form of $dN/d(\Delta E_e)$. Since for random binary collisions (Fig.2) we have $dN/dp \propto 2\pi p \propto p$ and since from either (5) or (6) we have $d(\Delta E_e)/dp \propto p^{-3}$, it follows that $dN/d(\Delta E_e)$ is of precisely the same form as (1). It is therefore reasonable that ΔE_e be assumed to populate the accessible excited states in terms of the same distribution function, (3), as for a sputtering geometry. For ΔE_e greater than the sum of the first and second ionization potentials, this again means that a two-step process occurs as in (2) in which the first step involves the expulsion of the appropriate number of electrons.

We thus obtain in (3) a formalism for describing electron-atom impact which is of such a type that, provided only relative populations are sought, is readily evaluated for any state of ionization of any element. The only restriction is that reasonably complete tabulations of degeneracies (g) and excitation energies (ε) be available and it is noted, in this regard, that the revised N.B.S. tabulations have now reached Ni [19].

For completeness, it is worth indicating that the expressions for ΔE_e given by (5) or (6) are readily recast in the form of a cross section, σ, by using, as starting point, the general relation, $d\sigma = 2\pi p dp$. We thus obtain

$$d\sigma \text{ (for a given } \Delta E_e) = |2\pi p dp| = \frac{4\pi I_H^2 a_o^2 \Sigma f_{fi}}{E_1 (\Delta E_2)^2} d(\Delta E_e),$$

so that the total cross section to form a given level is

$$\sigma\text{(for a given level j)} = \sum_{n=0}^{\infty} \frac{g_j}{g_0 + g_1 + \ldots + g_{j+n}} \int_{\varepsilon_{j+n}}^{\varepsilon_{j+n+1}} d\sigma$$

$$= 4\pi I_H^2 a_o^2 \Sigma f_{fi} P_j / E_1.$$

Here P_j is given by (3) and Σf_{fi} could be replaced with Z_{eff}.

3. Similarity

A basis for expecting a similarity between sputtering and medium-energy electron impact has been demonstrated. As seen in Section 1, the distribution of ΔE_e amongst sputtered particles follows from the kinetic-energy distribution, $dN/dE \propto 1/E^2$, as in (1),

$$dN/d(\Delta E_e) \propto 1/(\Delta E_e)^2. \tag{1}$$

With medium-energy electron impact, the impact parameter distribution is $dN/dp \propto 2\pi p$, so that using either (5) or (6) the distribution of ΔE_e takes on the same form. The dimensionalities of the distributions of ΔE_e are thus the same and a full similarity of the final relative excited-state populations follows from the postulate that (3) can be used.

It has been emphasized at this point that sputtering and medium-energy electron impact have a similar dimensionality. It is not necessary for what follows to take the argument to the next logical step and ask whether the atomistic processes are similar. Nevertheless, it is interesting that the basis of the FIRSOV [11,12] model for inelastic energy transfer in atom-atom collisions is, as seen in Fig.1, basically electron-atom collisions as in Fig.2. A limited degree of similarity in the atomistic processes therefore appears to exist.

4.   Experiments to Compare Sputtering with Electron Impact

To confirm whether a similarity in relative excited-state populations for sputtering and medium-energy electron impact occurs, the relative populations for 30 keV Ar^+ impact on solid Si as reported in [6] will be compared with those from medium-energy electron impact on gaseous Rb as reported in [8-10]. The experiments with solid Si involved bombarding an elemental target with mass-separated 30 keV Ar^+ in an O_2 atmosphere of 3×10^{-3} Pa and detecting the sputtered excited states with a 0.5 m Spex monochromator followed by an R.C.A. type C31034-02 photomultiplier tube. Si, which is a group IV element, was chosen in view of the likelihood that excited ions would be prominent, and indeed a reasonable number of SiII and SiIII was found. It is unclear why SiIV was not found. The presence of O_2 or an equivalent gas in such an experiment is essential to insure that, by virtue of a combination of thermal oxidation and recoil implantation [20], the surface of the target has an electronic band gap and radiationless deexcitation or neutralization of sputtered excited atoms and ions therefore cannot occur. Such radiationless processes are peculiar to situations involving solid (especially metallic) surfaces and have no analog in electron-atom impact.

The experiments with gaseous Rb, carried out initially by ZAPESOCHNYI and SHIMON [8] and more recently by SMIRNOV and SHAPOCHKIN [9,10], involved crossing a beam of electrons (with energies up to 200 eV) with a beam of Rb atoms from a furnace. These data are not necessarily of the best quality: those of [21] for Na are, for example, probably better. They are, however, nearly unique with electron-impact work in that information on three different ionization states, RbI, RbII, and RbIII, is available. An important part of the present argument is that a proper test of (3) can be made only if a sufficiently wide range of excitation energies is investigated. It is not satisfactory just to consider excited neutrals, as has normally been done in the past [4,5], as the data points tend to form a nearly vertical array so that agreement with theory, whether in a relative or absolute sense, always tends to look favorable.

In both cases the observed count-rates or cross sections ("signal") were converted to level populations by division by the branching ratio (b_{fi}) and were made to be in accordance with (3) by a further division by the initial-state degeneracy (g_i). This could be called a normalized population:

normalized population = signal/$g_i b_{fi}$.

Division by the transition probability (A_{fi}) is, as we have demonstrated earlier [22], not correct for conditions of low pressure unless the excited species has a long life. Convenient sources of branching ratios for RbI, SiII, and SiIII are the extensive compilations of LINDGARD [23,24] of calculated transition probabilities, A_{fi}, with branching ratios following from the relation

$$b_{fi} = \frac{A_{fi}}{\Sigma_f A_{fi}}.$$

In addition, [9] gives b_{fi} values for RbII. Information on SiI and RbIII is rather less accessible.

An important point is that we are using (3) to determine the level populations rather than optical (or generalized) oscillator strengths as in quantum-mechanical approaches. This means that, thanks to the availability of extensive tabulations of the relevant quantities, degeneracies (g) and excitation energies (ε), for the lowest spectra of all light-weight elements, it is possible to explore the populations of quantum-mechanically difficult species such as SiI, RbII, or RbIII.

The results of the two experiments, namely ion bombardment of elemental Si and electron bombardment of gaseous Rb, are compared with theory in Figs.3 and 4. A single normalization is necessary in each case, as (3) is being used in a relative rather than absolute sense. It is seen that, with this single normalization, the populations of SiI, SiII, and SiIII follow theory to within a factor-of-two scatter. This scatter is regarded as acceptable in view of the particularly wide span of excitation energies (55 eV). In addition, we note that there is a severe lack of experimental data for branching ratios, b_{fi}, and that cascade corrections have not been made.

With Rb, the populations of RbI, RbII, and RbIII follow theory less well in that the points for RbII and RbIII are all high. This is, of course, equivalent to stating that the points for RbI are low but we prefer Fig.4 in the form given. The reason is that RbII and RbIII differ from most other systems that have been studied in that the ground states have several equivalent outer electrons. It is possible that (3) should contain this number, so that (as shown in Fig.4) the theoretical curve for RbII should be raised by ln6 and that for RbIII should be raised by ln5. It is unfortunate, in this regard, that good quality data like those for Na [21] are not available with Rb.

It is concluded that the relative excited-state populations are indeed similar for sputtering and medium-energy electron impact over rather wide spans of excitation energies, 55 eV for Si and 60 eV for Rb. The similarity is adequately understood as being due to ΔE_e being in both cases distributed as $1/(\Delta E_e)^2$ and

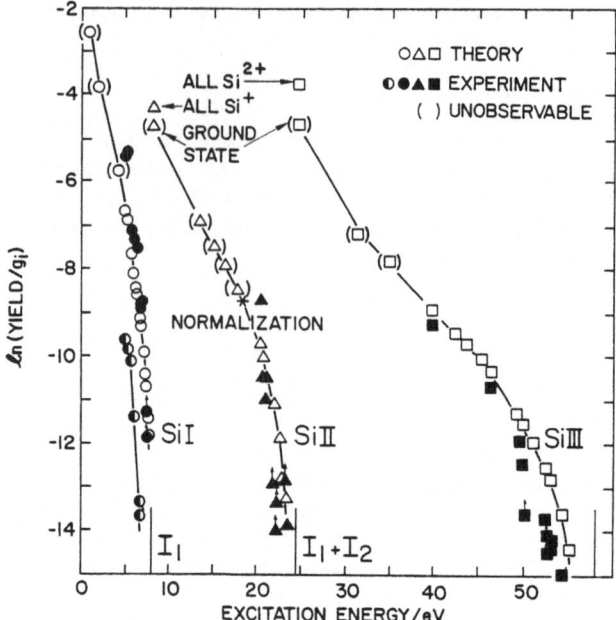

Fig.3. Comparison of theory, (3), with experiment for the relative populations of SiI, SiII, and SiIII. I_1 and I_2 are the first and second ionization potentials, with the short line at the bottom right representing $I_1 + I_2 + I_3$. The experiments, reported in [6], involved bombarding solid Si with a 30 keV Ar^+ beam in the presence of O_2 and detecting the sputtered excited states with a 0.5 m monochromator followed by a photomultiplier tube. Normalization was done <u>only once</u> for the entire ensemble of populations. Theory and experiment will be recognized to agree well over a 55 eV span of excitation energies

to (3) being in both cases a reasonable description of how ΔE_e is partitioned amongst the accessible excited states. We reiterate that an important part of the present argument rests on the fact that (3) is easily evaluated even for quantum-mechanically difficult species such as SiI, RbII, and RbIII. In fact, the present work is one of the few in which an attempt has been made to explain the populations of such species, thence to study wide spans of excitation energies.

In a preceding paper [7] it is shown how (3) can be used as an effective means for analyzing plasmas. The particular plasma was set up in Kr and sputtered an Al cathode. As a result the light from the plasma had a component of AlI, AlII, and AlIII, the relative populations of which could be used to show whether the plasma was more nearly of a collisional type (i.e. in accordance with (5) or (6)) or a thermal type (i.e. in accordance with a Boltzmann factor, $g_j \exp{(-\varepsilon_j/\varepsilon^*)}$). The key was again the possibility of using (3) for quantum-mechanically difficult species, thence to have access to wide spans of excitation energies.

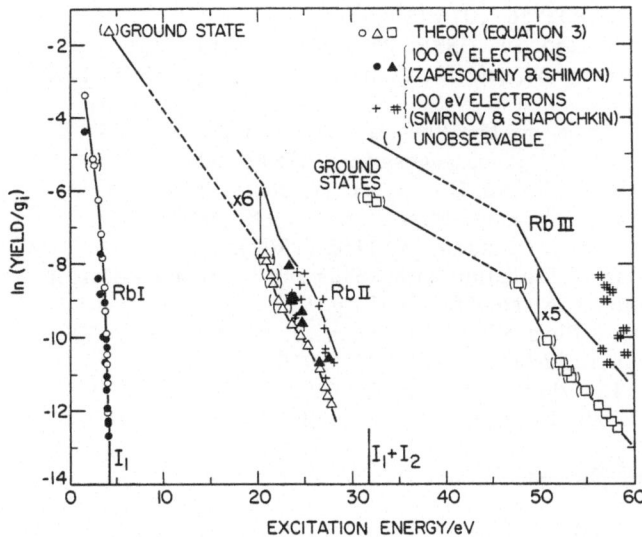

Fig.4. Comparison of theory, (3), with experiment for the relative populations of RbI, RbII, and RbIII. I_1 and I_2 are the first and second ionization potentials, with the short line at the bottom right representing $I_1 + I_2 + I_3$. The experiments, reported in [8-10], involved bombarding gaseous Rb with a 0-200 eV electron beam in the absence of O_2. Normalization was done <u>only once</u> for the entire ensemble of populations. Theory and experiment agree less well than in Fig.3 in that the points for RbII and RbIII are all high. This is discussed in terms of whether (3) should contain a factor giving the number of equivalent outer electrons in the ground state of the species of interest. Inclusion of such a factor raises the curves for RbII and RbIII as shown and leads to reasonable agreement over a 60 eV span of excitation energies

References

1. R. Kelly and C. B. Kerkdijk, Surface Sci. <u>46</u>, 537 (1974).
2. C. J. Good-Zamin, M. T. Shehata, D. B. Squires, and R. Kelly, Rad. Effects <u>35</u>, 139 (1978).
3. E. Veje, Surface Sci. <u>110</u>, 533 (1981).
4. R. Kelly, <u>in</u>: Inelastic Particle-Surface Collisions, eds. E. Taglauer and W. Heiland (Springer Verlag, Berlin, 1981) p. 292.
5. R. Kelly, Phys. Rev. B<u>25</u>, 700 (1982).
6. R. Kelly, Phys. Rev. Lett. (submitted).
7. R. Kelly, S. A. Shivashankar, and J. J. Cuomo, J. Vac. Sci. Technol. (in press).
8. I. P. Zapesochnyi and L. L. Shimon, Opt. Spect. <u>20</u>, 525 (1966).
9. Yu. M. Smirnov and M. B. Shapochkin, Opt. Spect. <u>47</u>, 243 (1979).
10. Yu. M. Smirnov and M. B. Shapochkin, Opt. Spect. <u>47</u>, 131 (1979)

11. O. B. Firsov, Sov. Phys.-JETP 36, 1076 (1959).
12. L. M. Kishinevskii, Bull. Acad. Sci. USSR, Phys. Ser. 20, 1433 (1962).
13. W. Eckstein, V. A. Molchanov, and H. Verbeek, Nucl. Instr. Meth. 149, 599 (1978).
14. W. Heiland and E. Taglauer, in: The Physics of Ionized Gases, ed. R. K. Janev (Institute of Physics, Beograd, Yugoslavia, 1978) p. 239.
15. I. P. Zapesochnyi and L. L. Shimon, Opt. Spect. 19, 268 (1965).
16. I. S. T. Tsong and N. A. Yusuf, Appl. Phys. Lett. 33, 999 (1978).
17. M. J. Seaton, Proc. Phys. Soc. (London) 79, 1105 (1962).
18. C. Lehmann, Interaction of Radiation with Solids and Elementary Defect Production (North-Holland, Amsterdam, 1977) p. 58.
19. C. Corliss and J. Sugar, J. Phys. Chem. Ref. Data 10, 197 (1981).
20. R. Kelly and J. B. Sanders, Surface Sci. 57, 143 (1976).
21. J. O. Phelps and C. C. Lin, Phys. Rev. A24, 1299 (1981).
22. M. R. Morrow, O. Auciello, S. Dzioba, and R. Kelly, Surface Sci. 97, 243 (1980).
23. A. Lindgård and S. E. Nielsen, Atomic Data and Nucl. Data Tables 19, 533 (1977).
24. A. Lindgård (H. C. Ørsted Institute, Copenhagen) (personal communication, 1982).

5.4 Erosion of Dielectric Solids by High-Energy Ions

T.A. Tombrello

W.K. Kellogg Radiation Laboratory, California Institute of Technology
Pasadena, CA 91125, USA

Abstract

A fast ion passing through matter loses energy mainly by its interaction
with electrons along its path. In an insulator the resulting electronic
excitation may be sufficiently long lived that some of this energy is
transferred to motion of the atomic nuclei, which leads to track formation
and to the ejection of atoms from surfaces where the ion enters or leaves
the material. The resulting erosion involves sputtering yields that are
comparable to those observed at much lower bombarding energies where nuclear
recoil processes dominate the energy loss of the incident ion. This paper
briefly summarizes our experimental and theoretical work that concerns this
electronic excitation erosion mechanism. The solids in which we have observed such effects include SiO_2, CaF_2, UF_4, Al_2O_3, and $LiNbO_3$.

Introduction

In contrast with the interactions of photons or electrons with solids, where
the mechanisms are relatively well defined and the excitations are separated
and well localized, a fast ion can produce a continuous chain of excited
atoms with a deposited linear energy density of keV per lattice spacing.
With this much energy available it is clear that not only are a wider variety
of erosion mechanisms possible but also that along the ion's path the properties of the excited atoms are not necessarily well described by those of
the unexcited bulk material. For all types of incident radiation, it is the
transfer of a portion of the electronic excitation to atomic motion that
leads to erosion; however, the nature of these processes can be quite different. For light ions, photons, or electrons the excitation may tend to be
stored in individual defects which are transported to the surface by focused
replacement sequences leading to ejection or by thermal migration leading to
molecule formation and evaporation [1]. In the case of heavier ions there
is continuous damage that is seen as a well-defined track [2], and atoms
ejected from the surface can be characterized by high effective temperatures [3,4].

The work described herein concentrates on situations which involve bombardment by heavy ions such as ^{19}F and [35]Cl at energies (MeV/amu) in the
vicinity of the peak in the electronic part of the stopping power. Some of
the erosion yields observed are large — comparable to those obtained for low-
energy sputtering of these dielectric materials but peaking with the electronic stopping power. However, other materials like Si and UO_2, which in
the experiments performed have low resistivities, the sputtering yields are
smaller and do not have the same dependence on the bombarding energy. Our
data are most extensive for UF_4 where we have observed the yields as a

function of bombarding energy, ion type, and incident charge state. We also have measured velocity spectra of the sputtered atoms [4,5]. For other materials we have data only for the erosion yield versus bombarding energy.

In the last section I will present a description of a new model whose predictions are in reasonable agreement with the data [6]. This model is based on the concept that the intense electronic excitation modifies the effective interatomic (intermolecular) lattice potential transiently but significantly in a small region about the bombarding ion's path. The acceleration of the nuclei in this modified potential provides the energy transfer mechanism that leads to track production and/or erosion. The erosion yields are calculated without the introduction of adjustable parameters or normalization.

Experimental Data

1. UF_4

For the experiments described here we collect the material sputtered from targets containing enriched ^{235}U.

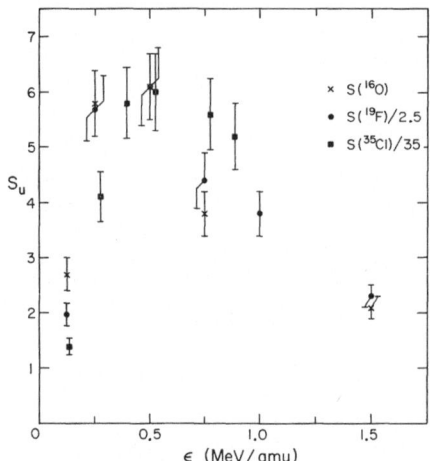

Fig.1. The sputtering yield (U atoms/incident ion) versus bombarding energy (MeV/amu) for ion beams of ^{16}O, ^{19}F, and ^{35}Cl. The beams were passed through a carbon foil so that the ions would be in charge state equilibrium when they hit the UF_4 target. The data and their associated uncertainties were taken from Ref.[5]

The amount of deposited uanium is determined using mica detectors for the thermal-neutron induced fission tracks. Erosion yields (given as ^{235}U atoms per incident ion) haven been obtained for ion beams from ^4He to ^{35}Cl; however, the yields for ^4He are consistent with those expected from the nuclear colli- sion cascade mechanism [7]. As we shall show, the sputtering yield observed depends very strongly on the charge state of the incident ion, which indicates that the ejected atoms arise from a near surface region in which the bombard- ing ion has not reached charge state equilibrium [4,5]. In Fig. 1 we show UF_4 yield measurements versus bombarding energy for beams of ^{16}O, ^{19}F, and ^{35}Cl in which the beam is brought to charge state equilibrium in a thin carbon foil before it strikes the UF_4 target [5]. The yields all have the same general shape and peak at the same bombarding energy - somewhat lower than the peak

208

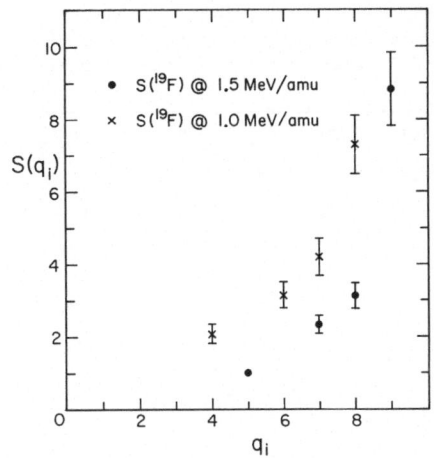

Fig.2. The sputtering yield (U atoms/incident ion) versus incident charge state for ^{19}F ions incident upon a UF$_4$ target at bombarding energies of 1.0 and 1.5 MeV/amu. These data come from Ref.[5]

S(^{19}F) @ 1.5 MeV/amu

S(^{19}F) @ 1.0 MeV/amu

in dE/dx [Ref. 8]. The shapes are, however, not identical; the yield from ^{35}Cl rises more steeply at low energies and decreases somewhat more slowly above the peak.

In Fig. 2 are shown the sputtering yields of UF$_4$ versus incident charge state for ^{19}F beams incident at energies of 1.0 and 1.5 MeV/amu [4,5]. These data are very similar to those observed in the ion yields from ^{16}O incident upon targets of CsI, glycylglycine, and ergosterol [9]. Because of this charge state dependence (and the possibility of different yields from different projectile excited states [4]) one must be very careful in comparing total yields obtained under different experimental conditions.

The velocity distribution of sputtered uranium from 13 MeV ^{35}Cl on UF$_4$ is shown in Fig. 3. The solid curve corresponds to a Maxwellian emission spectrum for T = 524°K [4,5]. For 4.74 MeV ^{19}F bombardment the same functional form is observed with a temperature of 3620°K [3,4,7]. The comparison of yield and

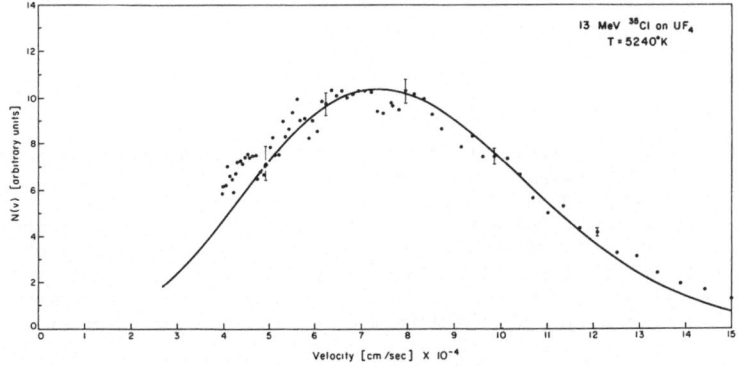

Fig. 3. The velocity spectrum of sputtered uranium for 13 MeV ^{35}Cl ions incident on UF$_4$. The curve corresponds to a Maxwellian emission spectrum for T = 5240°K (Refs. [4,5])

Table 1. The comparison of experimental data and the calculated results
[Ref. 6] for the electronic erosion of several materials. In all cases the
ion beam was in charge state equilibrium. kT refers to the emitted particle
spectrum

Target	Ion Beam	S_{exp}	S_{calc}	kT_{exp}(eV)	kT_{calc}(eV)	Ref.(exp)
UF_4	4.74 MeV, ^{19}F	14.5	63	0.31	0.60	4,5
UF_4	13 MeV, ^{35}Cl	200	250	0.45	0.66	4,5
SiO_2	20 MeV, ^{35}Cl	2	8	—	0.67	12
CaF_2	20 MeV, ^{35}Cl	1	133[†]	—	0.99	12
H_2O	0.25 MeV, ^4He	18	22	—	0.11	*
H_2O	6 MeV, ^{19}F	1400	1400	—	0.11	4
SO_2	7.5 MeV, ^{19}F	6500	2.3×10^4	—	0.062	**

[†]A band gap of 10 eV was assumed for the single crystal CaF_2; however, the
bombardment probably made the material amorphous. A decrease of the band
gap by 3 eV is sufficient to lower S_{calc} to 8.

*W. L. Brown, L. J. Lanzerotti, J. M. Poate, and W. M. Augustyniak, Phys.
Rev. Lett. 40, 1027 (1979).

**D. J. LePoire, B. H. Cooper, C. L. Melcher, and T. A. Tombrello, submitted
to Radiat. Eff. (1982).

velocity spectrum data with the model described in the next section is given
in Table 1. As you see, the predictions are in excellent agreement with the
data.

2. Al_2O_3 and $LiNbO_3$

We have obtained erosion yield data for Al and Nb from these materials as a
function of ^{35}Cl bombarding energy [10]. The sputtered material was collected on a thin carbon foil through which the beam passed before hitting
the target; thus, the incident ions were in charge equilibrium. The quantity
of deposited material was measured by forward angle Rutherford scattering of
a ^{19}F beam from the Al and Nb atoms [10]. Both erosion yield curves have approximately the same variation with bombarding energy – approximately that of
the electronic part of dE/dx. The Nb yield from ^{35}Cl on $LiNbO_3$ is shown in
Fig. 4.

Tracks have been observed in $LiNbO_3$ but not in Al_2O_3 [Ref. 11]. We note
that the magnitudes of the erosion yields are quite comparable ($\sim .2$/incident
ion) despite the very different physical properties of these materials [10].

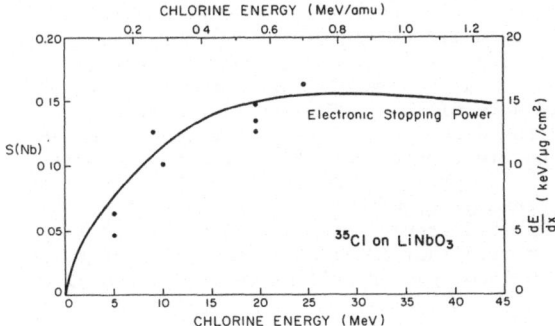

Fig. 4. The sputtering yield (Nb atoms/incident ion) versus bombarding energy (MeV) for a ^{35}Cl ion beam incident on a $LiNbO_3$ target. The solid curve is dE/dx [Ref. 10]

3. UO_2

It would be easy to suppose that the erosion of most materials might look like those cases described above. However, we have not seen erosion of Si by ^{35}Cl ions at levels above those predicted for collision cascade sputtering [12]. UO_2, the resistivity of which is easily reduced by small impurity levels, was bombarded by ^{35}Cl ions; the yield, which is given by Fig. 5, falls monotonically with increasing beam energy. Thus, one cannot be completely sure whether the erosion of this material is controlled either purely by nuclear collision processes or by a combination of nuclear and electronic mechanisms 5 .

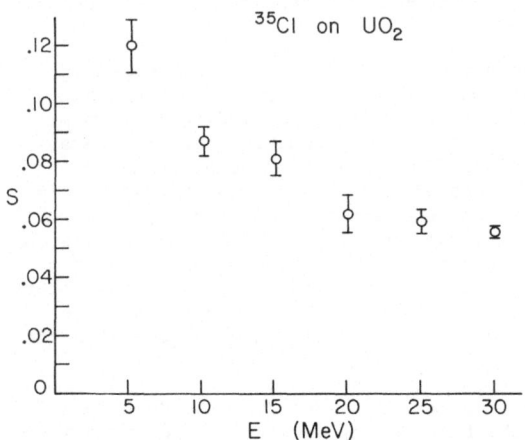

Fig. 5. The sputtering yield (U atoms/incident ion) versus bombarding energy (MeV) for a ^{35}Cl ion beam incident on a UO_2 target [Ref. 5]

4. SiO_2

The erosion yield for quartz was observed using the same technique as that for $LiNbO_3$ and Al_2O_3 [Ref. 10]. The shape of the yield versus bombarding energy is essentially that observed for those materials, but the magnitude of the yield is an order of magnitude larger ($\sim$ 2 Si atoms per incident ^{35}Cl ion). No difference was observed for the yields from amorphous and crystalline SiO_2 targets; however, the beam fluences employed were probably sufficient to amorphize the crystalline material [10].

Because the erosion yield is so much larger for SiO_2, we were able to observe the resputtering of the material deposited on the carbon catcher foil. This resputtering did not take place at low levels of areal coverage, but required an areal density equivalent to approximately three atomic layers of SiO_2 [Ref. 12]. Since we are not sure of the uniformity of the microscopic coverage, the exact structure of the deposited SiO_2 is quite uncertain. However, these data indicate that the electronic excitation erosion mechanism does, as one would expect, require the presence of a "coherent" material at the microscopic level and cannot take place with individual molecules. This, of course, emphasizes the essential difference between nuclear collision cascade sputtering which can occur with individual atoms and the electronic excitation erosion mechanism which involves the collective properties of the material.

Examples of erosion yield data for the dielectric substances listed above as well as for CaF_2 [Ref. 12] are summarized in Table 1 together with predictions from the model that is described in the next section.

A Model of the Electronic Erosion Process

The model described here is developed in detail elsewhere and only a schematic description will be given here [6]. In general, our goal is to emphasize the universal properties of the processes that lead to erosion in a variety of very different materials and not to develop a detailed model that applies only to a particular substance. Our yield data for electronic erosion for solids and condensed gases cover four orders of magnitude; the model reduces this spread to less than a factor of ten.

The electronic erosion of solids may be divided into several distinct stages. The first is the deposition of the ion's energy into electronic excitation as it slows down in the material. The spatial distribution of the energy given by the ion to the electrons is calculated from the δ-ray (recoil electron) distribution — this defines the track volume. Although the overall energy loss (dE/dx) is accessible to measurement, it is not in general possible to split dE/dx into the different forms that may be more appropriate for describing how the excitation energy is transferred to atomic displacement in the second stage of the process. Our model proposes that this transfer is most efficiently achieved by the modification of the interatomic (or intermolecular) potential. In effect, the heating of the electrons causes a shift in the minimum of the potential, which accelerates the atoms.

The effectiveness of the energy transfer is governed by the duration of the electronic excitation. Thus, the third stage involves the dissipation of the energy that was deposited in the electrons. One is in essence interested in the electrical component of the thermal conductivity. If the electrons can cool quickly, there will be little energy transferred and therefore there is little atomic motion that would lead to erosion. Thus, this is the most critical stage of the process, and the length of this stage is probably the controlling factor in preventing electronic erosion or tracks in metals and most semiconductors. We estimate the electrical dissipation on the basis of external Auger transitions between electronically "hot" and "cold" atoms at the boundary of the track. Radiative transitions, which are much slower, are neglected. A key parameter here is the electronic band gap because energy can be transferred only for collisions that result in a cold electron being promoted to a state above the band gap.

212

So far we have not tried to calculate in detail the spectrum of the moving atoms but have used the average energy transferred to determine their effective temperature. This will, of course, lead to a Maxwellian distribution of velocities, which was what we observed for UF_4 [Ref. 4]. The comparison of these temperatures is also given in Table 1.

Once energy has been transferred to atomic or molecular motion it may also be dissipated by atomic or molecular collisions. For most of the solids discussed in this paper the time scale for cooling by these collisions is long compared to the electronic dissipation lifetime; however, for the frozen volatile materials discussed elsewhere in this volume [13], the molecular collision cooling is fast and the molecules can "freeze" into the configuration required by the modified lattice potential and then relax very gradually.

Although at this stage this model has many limitations, the comparisons given in Table 1 are sufficiently promising that more detailed calculations seem warranted.

Conclusions

We have seen that there is a powerful ion erosion mechanism of dielectric materials that is associated with the deposition of the ion's energy into electronic excitation of the solid. This erosion process takes place in a variety of insulating materials and is comparable in magnitude to ordinary low-energy sputtering; however, the systematic behavior of the effect is distinctly different. A model has been briefly described that allows estimates of electronic erosion yields to be made with accuracies of better than an order of magnitude, even though the erosion yields vary by about four orders of magnitude.

Acknowledgments

The author thanks his collaborators, B. H. Cooper, J. E. Griffith, C. K. Meins, M. H. Mendenhall, Y. Qiu, L. E. Seiberling, C. C. Watson, and R. A. Weller, for their permission to quote liberally (and hopefully accurately) from our published and unpublished work. The research described herein was supported in part by NASA [NAGW-202 and -148] and by the NSF [CHE81-13273 and PHY79-23638].

References

1. J. P. Biersack and E. Suntuev, Nucl. Instrum. Methods 132, 229 (1976); M. Szymoński, H. Overeijnder, and A. E. deVries, Radiat. Eff. 36, 189 (1978).
2. R. L. Fleischer, P. B. Price, and R. M. Walker, Nuclear Tracks in Solids (Univ. of Calif. Press, Los Angeles, 1975).
3. L. E. Seiberling, J. E. Griffith, and T. A. Tombrello, Radiat. Eff. 52, 201 (1980).
4. L. E. Seiberling, C. K. Meins, B. H. Cooper, J. E. Griffith, M. H. Mendenhall, and T. A. Tombrello, Nucl. Instrum. Methods; in press (1982).
5. C. K. Meins, Ph.D. thesis (California Institute of Technology, 1982).
6. C. C. Watson and T. A. Tombrello, Proc. Lunar Planet Sci. Conf. 13th, 845 (1982); and submitted to Radiat. Eff. (1982).
7. J. E. Griffith, R. A. Weller, L. E. Seiberling, and T. A. Tombrello, Radiat. Eff. 51, 223 (1980).

8. J. F. Zeigler, <u>Stopping Cross Sections for Energetic Ions in All Elements</u> (Pergamon Press, New York, 1980).
9. P. Hakansson, E. Jayasinge, A. Johansson, I. Kamensky, and B. Sundqvist, Phys. Rev. Lett. <u>47</u>, 1227 (1981).
10. Y. Qiu, J. E. Griffith, and T. A. Tombrello, Radiat. Eff., in press (1982).
11. A. Sigrist and R. Balzer, Helv. Phys. Acta <u>50</u>, 49 (1977).
12. J. E. Griffith, Y. Qiu, M. H. Mendenhall, and T. A. Tombrello, submitted to Radiat. Eff. (1982).
13. T. A. Tombrello, a contribution to these proceedings in the session on condensed gases.

Part 6

Electronic Erosion

6.1 Sputtering of Alkali Halides by Electrons

M. Szymoński*

Arizona State University, Department of Physics, Tempe, AZ 85287, USA

A.E. de Vries

FOM-Institute for Atomic and Molecular Physics, Kruislaan 407
1o98 SJ Amsterdam, The Netherlands

1. Introduction

It has been established that incident ionizing radiation on alkalihalides mainly activates the halogen sublattice. Potential energy, stored in localized excited states, formed by this radiation, can be converted into kinetic energy of F and H centers, leading to their separation along a closely packed direction of the crystal. This is known as the Pooley-Hersh mechanism of defect formation [1,2] with some further refinements [3]. It is believed that high mobility of the H-centre (in fact an interstitial halogen atom), which may easily migrate to the surface, is responsible for sputtering of alkali halides with photons or electrons.

There are different views, however, on details of this process. From one side it is assumed that the exciton can diffuse towards the surface and subsequently deexcite radiationless, producing the H-centre transposed in a short-range focused replacement sequence [4]. Only those excitons which deexcite near the surface can contribute to sputtering since the range of the H-centre does not exceed a few lattice sites. One can expect, therefore, that halogens emitted in this process should have kinetic energies of the order of a few eV. On the other hand it is argued that the exciton is quickly selftrapped and its radiationless decay produces the H-centre in an excited state [3,5], which has a much lower activation energy for migration than a ground-state H centre. Thus, it is likely that the range of the excited H-centre is significantly larger [3,5,6].

Recent experiments on electron- and photon-induced sputtering of excited, neutral alkali atoms [7,8] and ion-beam induced enhanced sputtering of insulators [9] have posed the additional question: how far can the electron sputtering mechanism be relevant for description of these new experimental data?

In this short contribution we would like to discuss some of our experimental results and theoretical calculations which might be helpful for answering the questions stated above.

2. Energy distributions of sputtered atoms

The time of flight spectra of halogen atoms generally have a double structure. An example is shown in Figure 1 for Br atoms sputtered from KBr at 100°C

---

* Permanent address: Instytut Fizyki, Uniwersytet Jagielloński, Reymonta 4, Pl. 30-059, Kraków, Poland

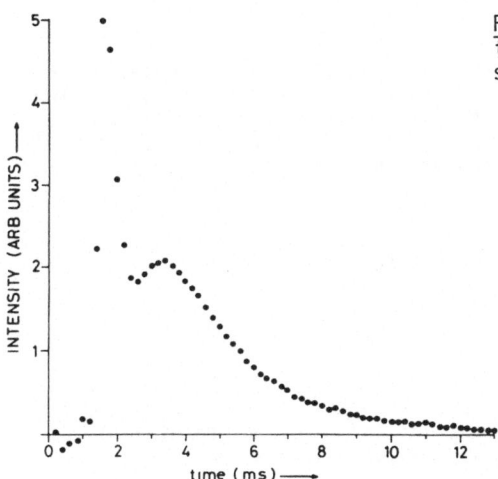

Fig. 1. The time of flight distribution of Br atoms sputtered from a KBr sample at 100°C by 540 eV electrons

with 540 eV electrons [10]. At the high-energy side of the spectrum (0.5- 0.1 eV) a very sharp peak is visible, followed by a broad thermal distribution, well described by a Maxwellian form with the sample temperature. This finding, observed for poly and single crystal samples, has very distinct properties, leading to consequences for the mechanisms involved. These properties are:

a) the nonthermal contribution to the energy spectrum has always a well-defined cut-off energy below 1 eV. It implies that neither the surface exciton itself nor the H-centre formed just below the surface would explain the emission of such low-energy particles.

b) The ratio S_{nth}/S_{th} (S_{nth} and S_{th} denote nonthermal and thermal halogen sputtering yields respectively) is larger, at a given target temperature, for samples with a larger interionic space along the <110> direction of the crystal. This supports the proposed symmetry of the defects involved in sputtering and the assumption that the focused replacement sequences are directed along the <110> row of the crystal [1-4].

c) For single crystals the S_{nth}/S_{th} ratio varies with the observation angle [11]. The magnitude of these variations is strongly dependent on the target temperature and the electron current density. This might be caused by the directional emission of nonthermal atoms being affected by the alkali atom enrichment on the surface. At low surface temperatures and for high electron densities the sample surface can be covered with an extra layer of alkali atoms because alkali vaporization process may not be fast enough for compensation of the rapid halogen emission.

d) Measurements on the time lapse between the incident electron and the sputtering have also been performed. These measurements have shown no delay for nonthermal particles but for thermal particles delay times are of the order of $10^{-4} - 10^{-3}$ s [12].

e) The energy distributions of thermal particles (alkali and halogen) are very well described by the Maxwell-Boltzmann formula with the target temperature.

217

3. Model calculations

Basing on the above experimental results we have followed the concept of a
long range directional movement of the H-centre along <110> directions. Thus
we have assumed that the H-centre taking part in this focused replacement
sequence might cross the surface, giving rise to the sputtering of nonthermal
halogen atoms. If the range distribution of the replacement sequence is [5]:

$$\lambda(x) = \frac{1}{\lambda} \exp(-\frac{x}{\lambda}) ,$$

where λ is the mean range of the replacement sequence, one can express the
nonthermal sputtering yield by:

$$S_{nth} = \frac{a}{\varepsilon} \int_0^\infty dx H(x) \exp(-\frac{x}{\lambda}) ,$$

where $H(x)$ is the energy deposited per unit length at depth x, ε is the energy required to produce a replacement sequence and a is the geometrical factor describing how many replacement sequences generated at depth x are oriented towards the surface.

Furthermore it was assumed [6] that all H-centres, which were primarily
transported via the focused replacement sequences and terminated at a certain
distance ξ from the surface, can continue their migration by random diffusion.
The total probability $p(\xi)$ that the randomly moving centre can reach the
surface at the distance ξ equals [13]:

$$p(\xi) = \exp(-\frac{\xi}{\lambda_{th}}) ,$$

where λ_{th} denotes the random diffusion length for H-centre migration. So,
the thermal sputtering yield can be approximated as:

$$S_{th} = \frac{a}{\varepsilon} \int_0^\infty dx H(x) \frac{1}{\lambda} \int_0^x dy \exp(-\frac{x-y}{\lambda}) \cdot \exp(-\frac{y}{\lambda_{th}}) .$$

Using this approach we were able to reproduce the experimental dependences S_{nth}/S_{th} on the electron beam energy and the target temperature, the
S_{total} dependence on the electron beam energy and the S_{x_2}/S_x dependence on
the target temperature for the investigated samples.

4. Electronic sputtering in ion-bombarded alkali halides

It was reported previously that for ion-bombarded halides the energy spectra
of sputtered atoms contain energetic and thermal particles [14,15]. An example is shown in Figure 2 where the energy distribution of I atoms sputtered
from LiI with a 6 keV Kr[+] beam is plotted [15]. The low-energy part of the
spectrum is described, like in the case of sputtering with electrons, by a
Maxwellian distribution with the target temperature. The ratio of this thermal part to the energetic (cascade) one varies for different projectiles,
following the ratio of electronic to nuclear energy loss of the projectile
[15]. This indicates that ionization and excitation of the halogen sublattice
by the incident ion beam cause defect migration and sputtering quite similar
to that induced by electrons. It seems plausible, therefore, that at high
projectile energies (in the MeV range) a similar type of the energy conversion process, from electronic excitation into nuclear motion, may occur,
causing rapid sputtering of condensed gases and other insulators.

218

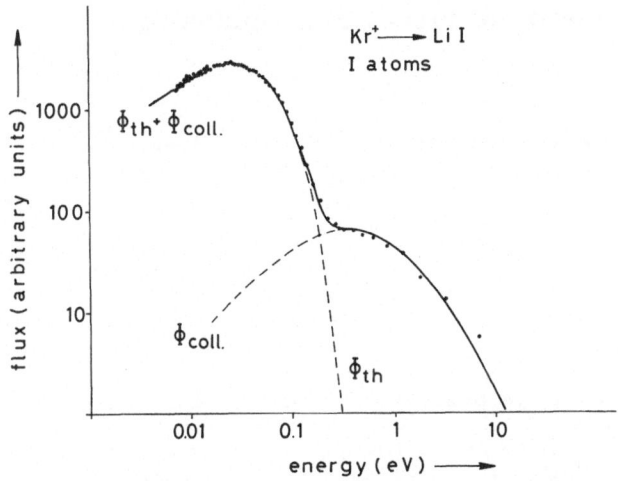

Fig.2. The energy distribution of I atoms sputtered from LiI with a 6 keV Kr$^+$ beam. The theoretical curve (solid line) represents a sum of the cascade distribution $\Phi_{col} \sim E/(E+E_b)^3$ (E_b is the binding energy) and the Maxwellian distribution $\Phi_{th} \sim E \cdot Exp(-E/kT)$ (T is the target temperature)

References

[1] D. Pooley, Proc.Phys.Soc. 87, 245 (1966).
[2] H.N. Hersh, Phys.Rev. 148, 928 (1966).
[3] N. Itoh, Nucl.Instr.Meth. 132, 201 (1976).
[4] P.D. Townsend, These proceedings.
[5] N. Itoh, These proceedings.
[6] M. Szymoński, Rad.Effects 52, 9 (1980).
[7] N.H. Tolk, M.M. Traum, J.S. Kraus, T.R. Piau, W.E. Collins, N.G. Stoffel and G. Margaritondo, Phys.Rev.Lett., in press.
[8] N.H. Tolk, L.C. Feldman, J.S. Kraus, R.J. Morris, M.M. Traum and J.C. Tully, Phys.Rev.Lett. 46, 134 (1981).
[9] T.A. Tombrello, These proceedings.
[10] H. Overeijnder, M. Szymoński, A. Haring and A.E. de Vries, Rad.Effects 36, 63 (1978).
[11] A. Haring and A.E. de Vries, unpublished data.
[12] H. Overeijnder, R.R. Tol and A.E. de Vries, Surf.Sci. 90, 265 (1979).
[13] R. Kelly, Surf.Sci. 90 (1978).
[14] M. Szymoński, H. Overeijnder and A.E. de Vries, Rad.Effects 36, 189 (1978);
H. Overeijnder, A. Haring and A.E. de Vries, Rad.Effects 37, 205 (1978).
[15] M. Szymoński and A.E. de Vries, Rad.Effects 54, 135 (1981).

6.2 The Contribution of Electronic Processes in Sputtering

P.D. Townsend and F. Lama

School of Mathematical and Physical Sciences, University of Sussex
Brighton, BN1 9QH, Great Britain

Abstract

Many materials decompose when exposed to ionising radiation. For some, such as the alkali halides, the threshold energy for defect formation is merely that for exciton production. Because of the long history of defect studies in alkali halides, a detailed model has evolved for defect generation and vacancy-interstitial separation. Key steps are exciton diffusion, exciton induced halogen relaxation and a halogen replacement collision sequence. The model successfully predicts sputtering of material from the surface with directionality in the halogen yield, plus complex impurity, temperature, and lifetime effects. Time of flight measurements reveal component species with energies up to a few eV. All these effects are compatible with the basic model.

Sputtering by electronic processes is noted for many other insulators and they often show a preferential loss of one element until the non-sputtered component forms a surface barrier. Some guide lines are considered to help predict which materials may decompose by electronic excitation. The mechanisms differ from those in alkali halides but excitons play an important role in several of the cases studied.

The details of the excitonic process in alkali halides will be discussed together with the more speculative charge explosion models.

1. Introduction

In any discussion of defect formation in insulators one first considers the alkali halides. They form a convenient starting point as there are a series of simple components with the same lattice structure and chemistry which may be used to test models involving ionic or defect sizes and bond properties. Lattice defects in alkali halides are well identified and within the halogen sub-lattice the F and H colour centres are the basic vacancy and interstitial sites. The F centre is merely an electron trapped at a halogen ion vacancy. The H centre is more complex as it is a crowdion interstitial formed along a <110> halogen row in which four ions occupy three lattice sites. Charge compensation is achieved by capture of a hole.

From the wealth of experimental data for alkali halides we know F and H centres are the primary defects under pulse irradiation. Such defects are not necessarily stable and diffuse, anneal, combine into complexes with impurities or other intrinsic defects (e.g. F_2, F_3, H_2, etc.).

Defect formation close to the surface can lead to ejection of halogen atoms, diffusion and evaporation of interstitials or excess metal. This decomposition under electron or photon excitation is an example of electronic sputtering.

By contrast with the extensive literature on colour centres in insulators, there is a shortage of data specifically describing sputtering as a consequence of electronic excitation. In part this is a semantic problem as "sputtering" is normally used in the context of energetic ion bombardment. For incident ion energies of several keV the ejected target particles are also energetic and have energies of up to a few hundred eV. For the faster particles the energy distribution has an E^{-2} dependence. Electronic processes commence with energy quanta of at most a few tens of eV and sputtered particles are inevitably in this same low-energy regime. Many of the sputtered atoms will have a thermal velocity distribution. Induced excess energy can only arise from chemical processes so in electronic sputtering ions will be limited to some 10 eV. Within this broader low-energy definition of sputtering one may select a wide range of examples from materials that decompose or explode during excitation. Surface desorption and decomposition of molecular solids will not be considered here.

Many of the examples of electronic sputtering identified so far, for example the alkali halides, silica, ZnO, PbO, CdS, cause the preferential ejection of one element. Therefore the rate of sputtering is time dependent as the surface layer becomes depleted in the volatile component. The residual material acts like a containment layer and inhibits further sputtering. No such restriction occurs in ion beam sputtering as all elements will be removed to some extent. For a correct understanding of the ion beam sputtering of insulators one must separate out the contributions from nuclear collisions and electronic excitation. For these continuously 'cleaned' surfaces the electronic term may play a major role (e.g. as with UF_4).

This review will briefly summarise the established mechanism of defect formation in the alkali halides and demonstrate that it describes some, but not all, of the features of the electron or photon sputtered surfaces for the alkali halides.

More speculative models of multiple ionisation and charge explosion will be mentioned. Whilst these seemed inappropriate for alkali halide defects they are still under discussion for other materials (e.g. SiO_2, UF_4, etc.) both for sputtering, bulk defect formation and fission track formation. However, without the sophistication of the alkali halide studies it is difficult to decide between alternative proposals.

2. Defect Formation in Alkali Halides

Displacement of lattice atoms or sputtering as a result of electron or photon irradiation requires both energy and momentum. Since the incident radiation lacks the requisite momentum a two-stage process must be involved. Thus one can generalise that the major steps include a localisation of energy at a single lattice site, some rearrangement of the local structure and a displacement event. The currently accepted steps for alkali halides fit this general pattern and are:- [1-7]

(i) exciton production by the electron or photon,
(ii) exciton migration to preferred sites such as impurities, defects or the surface,

(iii) relaxation of the lattice during the excited state of the
trapped exciton,
(iv) decay either by luminescence or by a non-radiative step in
which the energy causes halogen displacement,
(v) separation of the vacancy - interstitial pair by a
replacement collision sequence.

Detailed studies of optical absorption, luminescence and sputtering support the detailed model. Pulse irradiation experiments have simplied the otherwise complex set of absorption features [8-10]. From these important experiments one notes that the F and H centres are formed as the primary defects at all temperatures, the F centre is created in the ground state and can be observed within pico-or nanoseconds after the primary pulse of energy. More recent experiments with polarized light also reveal that the orientation of the H centre is the same as the primary excitation [11].

Points of particular note for each of the stages are as follows.

(i) Threshold energy corresponds to exciton production.
(ii) There are a disproportionate number of interstitial centres
linked to impurities (H_A) and the ratio of H to H_A centres
may be controlled by the size of the impurity ions [12,13].
(iii) Identical lifetime components are found in both the
luminescence and sputtering signals which emphasises that
they have the same precursor state [14].
(iv) Luminescence and defect formation, including sputtering,
are alternatives. The signal intensities anti-correlate
with temperature and have the same activation energies [14].
(v) Both in the original POOLEY [15,16] or HERSH [17] excitonic
models, and in the later refinements [5,11,18-20], the F-H
separation involves movement of an interstitial along the
<110> halogen row by a replacement sequence. The
efficiency of such a process is linked to the space available
between the halogen ions along the row (S) and the diameter
of the moving entity. This is effectively a neutral halogen
of diameter (D). Overall the ratio S/D strongly influences
the efficiency of F centre production and the possibility
of <110> halogen sputtering [21].

Theoretical estimates of the energy scheme for excited excitons and H centres confirm that the process is feasible. They also show that the F centre is formed directly in the ground state, as observed, and stable F-H centre separation is possible. Minimisation of energy losses during the collision sequence are suggested by polarised orbitals of the moving H centre [18,19] so that the momentum transfer occurs with a small neutral core and charge balance is preserved by a separate parallel electronic charge transfer.

Key ideas in the defect formation of alkali halides are summarised in Figs. 1 to 4. Figure 1 shows a simplified configuration coordinate diagram in which the exciton states rearrange during lattice relaxation. Decay paths are either by luminescence or a direct transition between the states which leaves the kinetic energy available for halogen displacement. Figure 2 shows a more complex energy scheme for the excited exciton. There are a range of states which could decay and act as precursors for the formation of F and H centres. Figure 3 gives an example of luminescence and sputtering data showing how these are alternative uses of the exciton energy and Fig. 4 gives a sketch of the F and H centre models together with a prediction of <110> halogen ejection at the end of a replacement collision sequence.

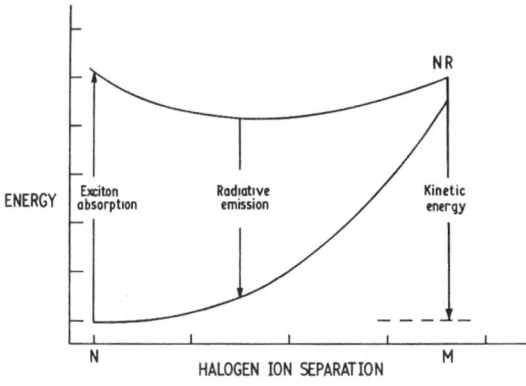

Fig. 1. The effect of trapping an exciton at a pair of halogen ions, after POOLEY [15,16]. Lattice relaxation allows decay of the excited state via luminescence or, at NR, by a non-radiative transition. The ions may relax from the normal spacing, N, to a molecular-like state, M. The kinetic energy in the non-radiative decay is used for atomic displacement of halogens

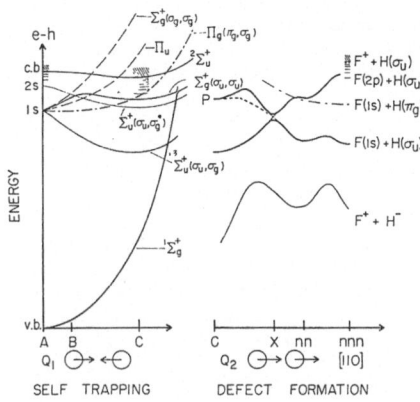

Fig. 2

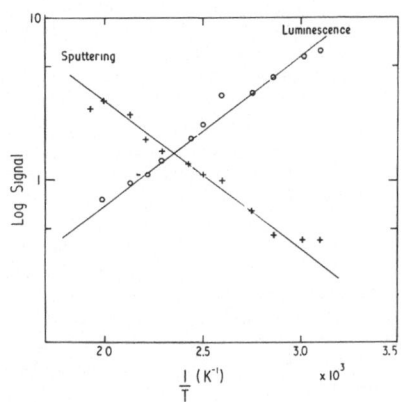

Fig. 3

Fig. 2. An example of a detailed calculation of exciton and (F,H) energy states for relaxation of the alkali halide lattice along a <110> direction, after WILLIAMS [5]. A precursor state for defect formation is indicated at P leading to F centres in the ground states

Fig. 3. A comparison of luminescence and sputtering data for NaCl [14] . Note the two processes have the same activation energy resulting from the same precursor state, see text

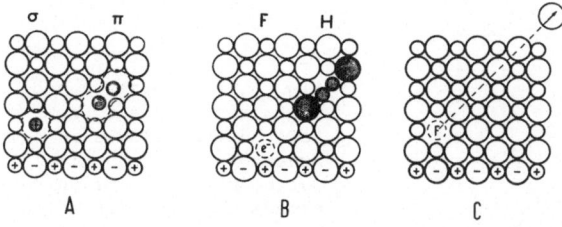

Fig. 4. Exciton trapping in the alkali halide lattice can form either σ or π states, Fig. A. The π example leads via a halogen replacement collision sequence to vacancy (F) and interstitial centres (H), Fig. B. If the sequence intersects the surface there is a directional halogen emission, Fig. C

223

3. Observations of alkali halide sputtering

The energy available for the replacement collision sequence is a few eV and so the range of the sequence is short. This is not a major limitation for the sputtering as the initial stage of exciton diffusion can transport energy to the surface layers. However the low energy is insufficient to sputter contaminant layers overlaying the alkali halides. Consequently alkali halide sputtering is only observable from atomically clean surfaces. The apparent sputtering rate may be reduced by the build-up of a metal rich layer. Evaporation of the metal can then appear to control the total sputtering rate [22].

The previously noted correlation between the decay of the exciton energy into sputtering or luminescence was shown in Fig. 3. Whilst it is clear that the anti-correlation of luminescence and sputtering implies alternative decay routes from the same energy state the lifetimes involved are up to 10^{-3} sec whereas pulse experiments of F centre formations imply a nanosecond time scale. These features may not be totally in conflict as the colour centre studies involve defect formation in the interior of the crystal with very high exciton energy densities. The sputtering is inherently from near surface levels and experiments have been made at modest power levels. A delay between excitation and sputtering can be explained by charge capture which delays electron-hole recombination into exciton states. One also notes the luminescence spectra taken during sputtering have the normal exciton wavelength dependence but, as with ion bombarded samples, are observed at higher temperatures than normally expected from bulk crystal studies.

Detailed analyses reveal several lifetime components which in fact are identical in the luminescence and sputtering. However the fractional values of the components may be altered by selective excitation of electron-hole pairs from inner shell states [14]. In the original publication [14] the electron energy dependence of the signal recorded in stage II (Fig. 5) was suggested to be evidence for a resonance in yield when the electron energy matched the NaK_α transition energy. It has been pointed out by P. WILLIAMS [23] that as the yield rises at sub-threshold energies and the cross-section continues to increase, one should propose an alternative mechanism in which the state II features are suppressed after the onset of K_α transitions.

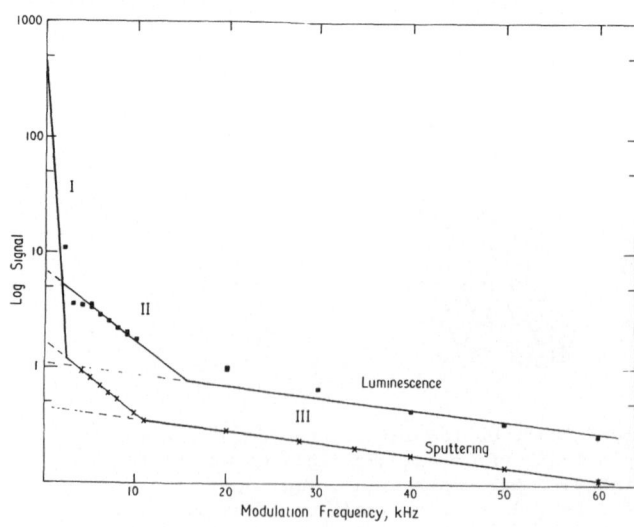

Fig. 5. Luminescence and sputtering signals from NaCl have similar size component features in time resolved measurements which emphasise the correlation between them. The activation energies for stages II and III are the same but differ from the value for stage I [14]

Sputtering patterns from electron [14] and photon [24] excitation are slightly different but both clearly show directional features for the halogen emission [7]. In the simple theory for defects in the interior of the crystal only <110> sequences were considered. Other halogen directions such as <211> would not lead to stable F-H separation. This limitation does not apply to sputtering and indeed strong <211> spot patterns are evident [14,24].

Measurements of the energy spectra confirm that metal atoms are evaporated with thermal velocities whereas the halogen emission has a hyperthermal component [25,26]. The hyperthermal component depends on the lattice structure (i.e. the S/D ratio) and was detected for the more open lattices. There is also a crystal temperature effect. The remainder of the halogen emission results from either interstitial diffusion to the surface or halogen molecular dissociation at the surface. Experiments at FOM also reveal both atomic and molecular fragments and their ratio (X/X_2) further correlates with crystal temperature, etc. Initial interpretations of the data vary slightly but all lie in the broad framework sketched above [7,27,28]. Although hyperthermal halogen emission is noted along <110> axes, there is also an energetic component perpendicular to the (100) face. This is only apparent below 100°C.

At present no explanation has been offered to account for the differences between the electron and photon sputter pattern, nor for the presence of energetic <100> emission. Experimental data are sparse so suggestions, such as additional sputtering process by ionisation of surface atoms, have not been tested. For alkali halides the multiple ionisation model of defect formation proposed by VARLEY [29] seems inappropriate for defects in the interior of the crystal but it might still apply at the surface. The related mechanism for surface desorption [30] should also be considered. However much of the sputtering is consistent with the excitonic route to halogen displacement.

To summarise, a detailed mechanism has been identified for the conversion of the energy of an exciton to produce a displaced halogen atom. A feature of the model is the halogen replacement collision sequence and this is directly viewable in the sputtering pattern. The wide range of ionic sizes in the various alkali halides acts as a testing ground for models by their effects of the (S/D) ratio, alternative decays of luminescence or sputtering, and the changes of (X/X_2) ratio with lattice and temperature.

4. Other Examples of Electronic Sputtering

Few experiments have been performed which set out to search for electronically induced sputtering and, as with the alkali halides, without a deliberate effort to start with clean surfaces accidental observation of the effect is difficult. However, experiments with LEED or AES have the requisite surface conditions and hence some examples have been detected. To list examples one can commence with two systems which have similarities with the alkali halides. These are the alkaline earth halides and the azides. Of the former a deliberate search [31] revealed that $ZnBr_2$, $AgBr$, CaI_2 and PbI_2 all show halogen emission during electron irradiation. In each case the rate of emission was retarded or stopped by the development of a metal rich surface layer. Experiments with these materials have not yet been made at a sufficiently high temperature to re-establish the normal stoichiometry.

The second group of related materials are the alkali and alkaline earth azides [32]. In these the replacement of a halogen ion by an N_3^- group alters the unit cell from a cubic to a tetragonal form. Photon irradiation of the crystals NaN_3, KN_3, PbN_6, BaN_6, etc. causes an efficient process of decomposition. The lattice destruction is exothermic and experimentally difficult for PbN_6 or pure BaN_6 as both materials are excellent explosives. In the more controlled experiments with NaN_3 and KN_3 a number of parallels with the alkali halides have been noted. Both have V_K-like colour centres (i.e. self trapped exciton defects) and both have temperature dependent luminescence signals associated with the exciton states [33,34].

As a first step to a model of decomposition of the azides one can suggest exciton formation and lattice relaxation of the nitrogen atoms followed by gas release. The exothermic nature of the process also evaporates the excess metal.

In most of the preceding examples the instabilities in the lattice had first been identified by ionisation induced colour centres with bulk material. One may extend the range of study by considering materials which have displacement energies comparable with the band gap. Although displacement energies are determined by dynamic events during electron bombardment, whereas in low-energy processes there may be thermal lattice relaxation, the presence of a low displacement energy is a useful guide. For example many II-VI compounds have low displacement energies [35] and during light production in light emitting diodes are degraded [36].

A similarity between alkali halide and II-VI decomposition is provided by studies of CdS. Using high intensity laser light [37] ($10^8 Wcm^{-2}$) the luminescence spectrum at 77K is modified. This is interpreted in terms of exciton states bound to donor centres. The laser damage at 77K causes the formation and movement of sulphur vacancies and interstitials. At high temperatures above 680°C, the normally stable CdS can be made to decompose by illumination with band gap light. Laser intensity is not required and powers of only 10^3 μWcm^{-2} are sufficient for sputtering [38]. The decay rate varies with time of illumination and may again reflect a preferential loss of one element followed by thermal evaporation of a second. This is confirmed by experiments with non-stoichiometric CdS where excess sulphur enhances the optical effect.

Other examples of II-VI sputtering occur for ZnO. Both electron or band gap energy photons cause oxygen release [39,40]. There are many oxides which give evidence for electronic sputtering and numerous examples have been reviewed from AES literature [41]. Materials include ZnO [11,42], SiO_2 [42], $LiNbO_3$ and $LiTaO_3$ [43], SnO_2 [44], PbO and molybdenum oxides [45].

One should note for silica there is again a similarity between E_{gap} and E_d even though in this case both numbers are rather large ($\sim$ 9 eV). Electron irradiation of silica produces a blue fluorescence [e.g. 46,47] during the decay of excitons and this same luminescence has been anti-correlated with the rate of damage formation over the temperature range 77 to 300 K [48]. From experience with alkali halides an excitonic process of lattice relaxation to produce defects and sputtering would seem feasible.

This short list of examples indicates that electronic sputtering is detectable in a wide range of insulators. Replacement collision sequences may not be important outside of the alkali halides but there is clear evidence in many of the examples that exciton transitions are linked to the atomic ejection. The higher charge state mechanisms [29,30] may also be important but acceptance, or rejection, awaits the detailed threshold, pulse and lifetime measurements that have been so successful in alkali halides. Some such experiments are in progress and for example [49] in CdSe, GaP, ZnO and TiO_2

laser pulses cause sputtering via a non-linear process, by contrast with sputtering yields of alkali halides.

5. Conclusion

Whilst electronic processes contribute to the sputtering from a range of materials it is likely that no single mechanism describes all the examples. In the alkali halides a subtle lattice relaxation event caused by exciton decay is clearly an appropriate model.

In general few crystalline insulators have received a serious study for electronic sputtering features and until more detailed experiments are made a further discussion of mechanisms is premature. The success with the alkali halides suggests that there is no *a priori* reason why similar progress should not be made with other materials.

References

1. M. Saidoh and P.D. Townsend, Rad. Eff. 27, 1 (1975).
2. A.M. Stoneham, "Theory of Defects in Solids", Oxford (1975).
3. N. Itoh, J.de Physique 37, C7-207 (1976).
4. N. Itoh, Nucl. Inst. Methods 132, 201 (1976).
5. R.T. Williams, Semicond. and Insulators 3, 251 (1978).
6. P.D. Townsend and F. Agullo-Lopez, J. de Physique Colloque C6 (Supp. 7, vol. 41), 279 (1980).
7. "Sputtering by Electrons and Photons", in Vol. 2 of "Sputtering by particle bombardment", ed. R. Behrisch, Springer Verlag, to be published 1982.
8. H. Hirai, Y. Kondo, T. Yoshinari and M. Ueta, J. Phys. Soc., Japan 30, 440 (1971).
9. T. Karasawa and M. Hirai, J. Phys. Soc. Japan 34, 276 (1973).
10. I.N. Bradford, R.T. Williams and W.L. Faust, Phys. Rev. Lett. 35, 300 (1975).
11. N. Itoh (private communication).
12. G. Giuliani, Phys. Rev. B2, 464 (1970).
13. M. Saidoh, J. Hoshi and N. Itoh, J. Phys. Soc. Japan 39, 155 (1975).
14. P.D. Townsend, R. Browning, D.J. Garlant, J.C. Kelly, A. Mahjoobi, A.J. Michael and M. Saidoh, Rad. Eff. 30, 55 (1976).
15. D. Pooley, Solid State Commun. 3, 241 (1965).
16. D. Pooley, Proc. Phys. Soc. 87, 145 (1966); ibid. 87, 257 (1966).
17. H.N. Hersh, Phys. Rev. 148, 328 (1966).
18. R. Smoluchowski, O.W. Lazareth, R.D. Hatcher and G.J. Dienes, Phys. Rev. Lett. 27, 1288 (1971).
19. N. Itoh and M. Saidoh, J. Physique 34, C-9, 101 (1973).
20. Y. Toyozawa, J. Phys. Soc. Japan 44, 488 (1978).
21. P.D. Townsend, J. Phys. C. 6, 961 (1973).
22. Y. Al Yammal and P.D. Townsend, J. Phys. C 6, 955 (1973).
23. P. Williams (discussion during DIET-1, 1982).
24. A. Schmid, P. Braunlich and P.K. Rol, Phys. Rev. Lett. 35, 1382 (1975).
25. M. Szymonski, H. Overeijnder and A.E. de Vries, Surf. Sci. 90, 274 (1979).
26. H. Overeijnder, R.R. Tol and A.E. de Vries, Surf. Sci. 90, 265 (1979).
27. F. Agullo-Lopez and P.D. Townsend, Phys. Stat. Sol. b 97, 575 (1980).
28. M. Szymonski, Proc. Symp. on Sputtering, Vienna, 1980, 761; ibid. Rad. Eff. 52, 9 (1980).

29. J.H.O. Varley, Nucl. Energy $\underline{1}$, 130 (1954).
30. M.L. Knotek and P.J. Feibelman, Phys. Rev. $\underline{40}$, 964 (1978).
31. H. Overeijnder, M. Szymonski, A. Haring and A.D. de Vries, Rad. Eff. $\underline{36}$, 63 (1978).
32. F.J. Owens, Rad. Eff. $\underline{21}$, 1 (1974).
33. A.M. Stoneham and R.H. Bartram, Sol. State Elect. $\underline{21}$, 1325 (1978).
34. P.J. Kemmey, R.H. Bartram, A.R. Rossi and P.W. Levy, J. Chem. Phys. $\underline{70}$, 538 (1979).
35. T. Taguchi and B. Ray, "Prog. in Crystal Growth and Characteristics", in press.
36. N. Ettengerb and C.J. Nuese, J. Appl. Phys. $\underline{46}$, 2137 (1975).
37. M.S. Brodin, N.A. Davydova and I. Yu Shablii, Sov. Phys. Semicond. $\underline{10}$, 375 (1976).
38. G.A. Somorjai, Surf. Sci. $\underline{2}$, 298 (1964).
39. G. Heiland, Z. Physik $\underline{132}$, 354 and 367 (1952).
40. N. Itoh (to be published).
41. C.G. Pantano and T.E. Madey, Appl. Surf. Sci. $\underline{7}$, 115 (1981).
42. B. Carriere and B. Lang, Surf. Sci. $\underline{64}$, 209 (1977).
43. V.H. Ritz and V.M. Bermudez, Phys. Rev. B $\underline{24}$, 5559 (1981).
44. Y. Shapira, J. Appl. Phys. $\underline{52}$, 5696 (1981).
45. T.T. Lin and D. Lichtman, J. Vac. Sci. Techn. $\underline{15}$, 1689 (1978).
46. G.N. Greaves, Phil. Mag. $\underline{37}$, 447 (1978).
47. A.N. Trukhin and A.E. Plaudis, Sov. Phys. Sol. State $\underline{21}$, 644 (1979).
48. F. Jaque and P.D. Townsend, Nucl. Inst. Meth. $\underline{182/183}$, 781 (1981).
49. T. Nakayama, N. Itoh, T. Kawai, K. Hashimoto and T. Sakata, Rad. Eff. Letters, in press.

6.3 Mechanisms for Defect Creation in Alkali Halides

N. Itoh

Department of Crystalline Materials Science, Faculty of Engineering
Nagoya University, Furo-cho, Chikusa-ku, Nagoya 464, Japan

1. Introduction

A number of mechanisms of defect creation in alkali halides have been suggested and their survey [1] is useful to give a general view of the atomic processes induced by electronic transitions. Table 1 gives a summary of the mechanisms proposed so far. In this table the mechanisms are categorized in two ways: (1) how the excitation wave function is localized and (2) how the energy possessed by the electronic system is transferred to the atomic system, causing defect migration. Both of these processes are necessary for the defect migration to take place since the elementary excitation in solids produces electrons, holes and excitons, the wave functions of which are delocalized.

2. Discussion

In crystals, such as alkali halides, in which self-trapping of excitons takes place, the creation of an exciton is now known to be able to cause defect formation. Thus for alkali halides only those mechanisms in Table 1 under the heading of self-trapping are up to date. (These are called Pooley-Hersh or excitonic mechanisms). In many semiconducting and insulating crystals, a core excitation or a double excitation of valence electrons induced by an Auger process creates desorption (FEIBELMAN-KNOTEK mechanism [2]). Double excitation is not necessarily localized but here we are concerned with the localization only within the time constant of the inverse of the lattice characteristic frequency. Double excitation of valence electrons has been suggested to be the cause of defect creation in alkali halides. However, because the yield of defect creation in alkali halides is much higher than that of double excitation, this latter process cannot be the main cause of defect creation. Most probably, however, it may account for some of the defects created by ionizing irradiation. Localization by defects may induce defect migration in some cases (e.g. the mechanisms proposed by LANG and KIMERLING [3] for semiconductors), but the defect creation in alkali halides is an intrinsic process not related to any existing defects in crystals. (A few atomic migration processes induced by excitation of defect electronic states are known to occur in alkali halides).

The mechanisms of energy transfer from the electronic system to the atomic system have been divided into two categories [4]. In the local heating, the non-radiative transition excites local lattice vibrations (often called the accepting mode), which is different from the mode (reaction mode) that causes the reaction. On the other hand the local excitation mechanism assumes that the reaction mode and the accepting mode are the same, that is: a certain excited state of the self-trapped exciton has a (repulsive) adiabatic potential curve leading to the primary reaction products (the F-H pair in alkali halides) without any or with only a small barrier. This concept is rather similar to the MENZEL-GOMER-REDHEAD [5] mechanism of desorption. Local excitation is more efficient for electron-lattice energy transfer than local heating. It is generally accepted that the cause of defect creation in alkali halides is excitation to a certain excited state of the self-trapped exciton at which the energy is transferred to the lattice through the local excitation process. At this stage, we remark that much effort has been made to identify the excited state leading to defect creation [1].

Table 1. Suggested Mechanisms of Defect Formation in Alkali Halides. The mechanisms are categorized in two ways: (1) the localization of the excitation wave function, and (2) the mechanisms of electron-lattice coupling

		Process of localization of the exciton wave function		
		Self-trapping	Multiple excitation	Trapping by defect
Electron-lattice coupling	Local heating (phonon-kicking)	Pooley (1965) Hersh (1965) Smoluchowski-Lazarus -Dienes (1971)		Seitz (1954)
	Local excitation	Kabler (1973) Itoh-Saidoh (1973) Toyozawa (1974) Kabler-Williams (1978) [Menzel-Gomer-Redhead (1964)]	Varley (1954) Howard-Vosko -Smoluchowski (1961) Klick (1960) Williams (1962) [Feibelman-Knotek (1978)]	

References to Table 1

F. Seitz, Rev. Mod. Phys. 26, 7 (1954).
J. H. O. Varley, Nature 174, 886 (1954a).
J. H. O. Varley, J. Nucl. Energy 1, 130 (1954b).
C. C. Klick, Phys. Rev. 120, 760 (1960).
R. E. Howard, S. Vosko and R. Smoluchowski, Phys. Rev. 122, 1406 (1961).
F. E. Williams, Phys. Rev. 126, 70 (1962).
D. Menzel and R. Gomer, J. Chem. Phys. 41, 3311 (1964).
P. A. Redhead, Can. J. Phys. 42, 886 (1964).
D. Pooley, Solid St. Commun. 3, 241 (1965).
H. N. Hersh, Phys. Rev. 148, 928 (1966).
R. Smoluchowski, O. W. Lazareth, R. D. Hatcher and G. J. Dienes, Phys. Rev. Lett. 27, 1288 (1971).
N. Itoh and Saidoh, J. de Physique 34, C9-101 (1973).
Y. Toyozawa, Vacuum Ultraviolet Radiation Physics, edited by E. E. Koch, R. Haensel and C. Kunz, (Braunschweig: Pergamon Vieweg) (1974).
M. N. Kabler, Proceedings for NATO Advanced Study Institute on Radiation Damage Processes in Materials, Corsica, edited by C. H. S. Duput (Leyden: Noordhoff International), p. 171 (1975).
M. N. Kabler and R. T. Williams, Phys. Rev. B18, 1948 (1978).
P. J. Feibelman and M. L. Knotek, Phys. Rev. B18, 6531 (1978).
E. Kotomin and A. Shluger, Solid State Commun. 40, 669 (1981).

230

References

1. N. Itoh, Defects in Insulating Crystals, eds. V. M. Tuchkevich and K. K. Shvarts (Springer-Verlag, Berlin 1981) and Radiation Effects (in print).

2. P. J. Feibelman and M. L. Knotek, Phys. Rev. *18*, 6531 (1978).

3. D. V. Lang and L. C. Kimerling, Phys. Rev. Lett. *33*, 489 (1974).

4. A. M. Stoneham, Advances in Physics *28*, 457 (1979).

5. D. Menzel and R. Gomer, J. Chem. Phys. *41*, 3311 (1964); P. A. Redhead, Can. J. Phys. *42*, 886 (1964).

Part 7

Condensed Gas Desorption

7.1 The Non-Linear Erosion Yield of Condensed Gas Solids Electronically Excited by Fast Light Ions

W.L. Brown, W.M. Augustyniak, L.J. Lanzerotti, K.J. Marcantonio
Bell Laboratories Murray Hill, NJ 07974, USA

R.E. Johnson
Department of Nuclear Engineering and Engineering Physics, University of Virginia, Charlottesville, VA 22901, USA

1. Introduction

The erosion of rare gas and simple molecular gas solid films (i.e., H_2O [1-4], Xe [5], Ar [6], CO_2 [3], SO_2 [7,8], CO) by fast ions is a special case of desorption due to electronic excitation. It is special in several respects. First, the films studied are not monolayers as in most desorption work, but are typically ∿1000 Å thick, essentially bulk material. Second, the surface binding energies of the condensed gas species are low ($\lesssim$0.5 eV), characteristic of VanderWaals bonded atoms and molecules or, in the case of water, hydrogen bonded molecules. Third, all of the films are insulators. Fourth, the use of MeV ions allows large changes in the density of electronic excitation along well-defined linear paths by choosing the nuclear charge and velocity of the incident ions. Fifth, the erosion or desorption yields are large, from 0.1 to 1000 molecules desorbed per incident ion even for H and He ions.

The focus of this paper is to draw attention to the strong non-linearity in the erosion (desorption) yield with electronic excitation density that has been observed in every condensed gas that has been studied to date. The recent availability of both electron and ion-induced erosion yields for solid CO films indicates that the non-linearity extends to ionization densities even lower than those encountered in many electron-stimulated desorption experiments. The paper also discusses possible erosion mechanisms.

2. Experimental Approach

In all cases the films that we have studied have been condensed from the gas phase on a cold beryllium substrate, usually at a temperature of about 10K [9]. The principal experiments have been carried out using Rutherford backscattering of MeV helium ions to provide a measurement of the average molecular thickness of the films, as grown and at various stages in the erosion process. In many cases backscattering also provides the average atomic composition of the films as they are eroded, for example, by giving independent measurements of the column density of C and O atoms in a CO films. Collimation of the eroding ion beam and current measurements at the target permit absolute determination of erosion (desorption) yields, i.e., the number of molecules or atoms lost per incident ion. Observation that the erosion yield is independent of current density over an order of magnitude in current density gives assurance that the phenomena we are studying occur on a per-particle basis and are not due, for example, to macroscopic heating effects [9].

We focus our attention on erosion produced by MeV He and H ions for which the overwhelmingly dominant energy loss process in the films is electronic excitation and ionization. We have used $(dE/dx)_e$, the electronic stopping power or the energy loss per unit path length of the incident ion as a measure of the initiating effect, since the relative importance of individual excitation events is not yet understood. This regime is in contrast to the collisional stopping regime at keV energies in which the dominant energy loss of the incident ion is in collisions with the nuclei of the solid. It is the latter regime which leads to conventional sputtering [10].

3. Results and Discussion

The erosion yield Y for H_2O ice is shown in Fig. 1 as a function of $(dE/dx)_e$ [9]. The yield is highly non-linear in $(dE/dx)_e$; the plotted line in the figure has a slope of 2. The overall trend of the data indicates the process is roughly proportional to $(dE/dx)_e^2$. Deviations from the trend, particularly the high points for He ions at low $(dE/dx)_e$, arise from a combination of non-equilibrium charge state of the incident ions, velocity dependent track radius and collision cascade contributions [11]. A striking feature of Fig. 1 is the close to quadratic dependence on $(dE/dx)_e$ for hydrogen ions, for which these three effects are all small.

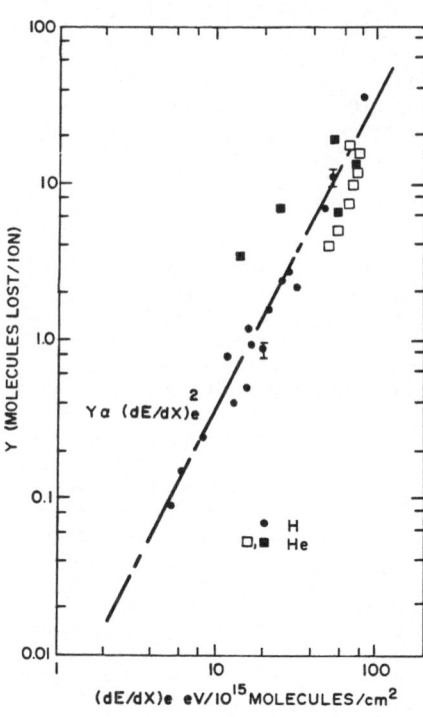

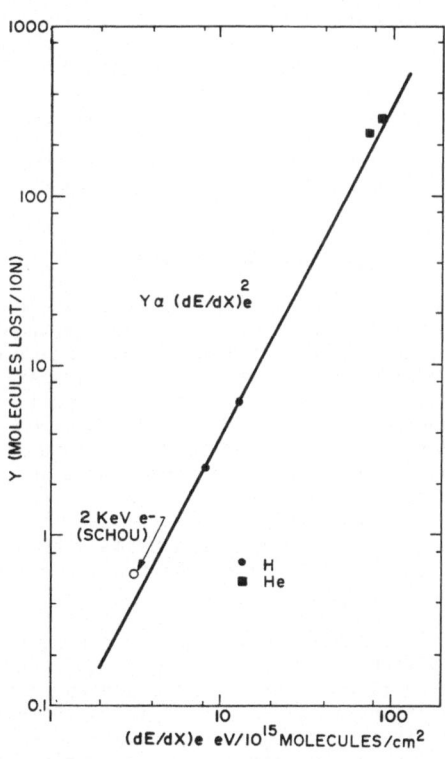

Fig. 1. Erosion yield Y of H_2O ice at low temperature plotted as a function of $(dE/dx)_e$ for H and He ions. The He points plotted as open squares are from [2]

Fig. 2. Erosion yield Y for solid CO at 10 K plotted as a function of $(dE/dx)_e$ for H and He ions and for 2 keV electrons. The electron data point is due to SCHOU [12]

235

Approximately the same non-linearity is found for Ar [6], CO_2 [3], SO_2 [7] and CO. Recent measurements in our laboratory for the case of solid CO are shown in Fig. 2. The data is much less extensive than for H_2O, but the same approximate $(dE/dx)_e^2$ trend is evident. The absolute yields are much higher: for 1.5 MeV He ions, Y for CO is approximately 220 compared with about 7 for H_2O. This is the direction that would be expected from the difference in intermolecular binding for the two cases (0.55 eV/molecule for H_2O; 0.08 eV/molecule for CO). However, the ratio of yields depends more strongly than linearly on the inverse of the binding energy [11,6]. One important question is whether the yield becomes linear or remains non-linear at low $(dE/dx)_e$, where the average separation between individual ionization events becomes large. This effect can be studied for ions by using higher energy MeV protons. Alternatively, it can be studied with keV electrons which have the same ionization probabilities as protons of the same velocity.

The erosion of frozen gases with keV electron bombardment is under investigation by SCHOU and his associates at Riso, Denmark [12]. Preliminary, as yet unpublished, work suggests that for solid hydrogen targets the erosion is linear in $(dE/dx)_e$. The group has also determined a value of Y = 0.6 for 2 keV electrons on CO. This value has been placed on Fig. 2 at the appropriate $(dE/dx)_e$ for such electrons (equivalent to $(dE/dx)_e$ for 3.7 MeV protons). Considering the small number of points and the very different measuring techniques, the agreement of the electron value with the ion results is good. Nevertheless, the electron point lies above the $(dE/dx)_e^2$ line drawn through the ion data by more than the 10-20% uncertainty the experiments are believed to have. This suggests that a transition to a linear dependence may occur at $(dE/dx)_e \lesssim 4$ eV/10^{15} molecules/cm^2. It is important to extend the $(dE/dx)_e$ measurement range to lower values in order to follow this transition in detail if indeed it does occur.

4. The Question of Mechanism

A major issue in these results, as in the case of electron or photon stimulated desorption, is how electronic excitation is transferred to the atomic motion required for release or ejection of atoms or molecules. There are only two general means of generating nuclear motion from electronic excitation and ionization produced in the material. The first is 'free electron' heating of the molecules of the solid [13] and we do not consider it here. The second is due to repulsive forces between neighboring atoms and molecules. These forces must occur in response to the change in the local electronic state of the solid. In an insulating material such forces become important because the excitations remain localized. The differences in possible models (mechanisms) therefore reduce to a consideration of the spatial extent (along the track and radially) of the effect of these repulsive forces [11].

Two models, both of which can account for erosion yields that vary non-linearly with $(dE/dx)_e$ have frequently been discussed in the ion erosion literature: thermal spike and coulomb explosion. In a thermal spike one envisions an extremely rapid ($\lesssim 10^{-12}$ sec) transfer of energy from the electronic system to random atomic motion (heat) in a narrow cylinder (20-40 Å radius) around the ion track. Where this cylinder meets the surface, a transient in sublimation takes place as the cylinder is very rapidly heated and then very rapidly cooled by radial heat conduction to the surrounding material. Such a transient will give a net sublimation yield which is non-linear in the thermal source strength [14]. Under appropriate conditions the non-linearity can be quadratic. The crucial step in this picture is not really the transient sublimation, which may or may not be treated correctly as a local Maxwellian process subject to thermal diffusion, but rather the rapid transfer of energy from excitation and ionization in the electronic system. Both the transfer time and the fraction of the energy which is rapidly transferred are critical to the results [11]. For the thermal spike formulation to give a quadratic result the individual sites at which electronic-to-atomic energy transfer takes place must be close enough together temporily and spatially so that the ion track can be treated as a very hot cylinder. Individual hot spheres will give a sublimation yield that depends linearly on $(dE/dx)_e$. At a $(dE/dx)_e$ of 3 eV/10_o^{15} molecules/cm^2 in CO the individual ionization events are spaced about 40 A apart on the average along the electron path. Using Eq. 3a of ref [11], if rapid transformation of 3-4 eV of electronic excitation to atomic motion occurs at each of these ionization locations, the local "hot" spots will just overlap. This is the critical condition for which a transition between a linear and a quadratic dependence on $(dE/dx)_e$ would be expected to occur.

In the coulomb explosion picture, electrons are radially separated from positive ions in the ionization process along an ion track. If the resulting space charge persists, the ions can be set in motion, similar to the mechanism proposed for formation of etchable tracks in room temperature insulating solids [15]. At the surface, the fields and forces are such as to promote ejection from the solid and the process has a quadratic dependence on the charge density and hence also on $(dE/dx)_e$ [16,9]. Preserving the space charge long enough for this to happen seems very improbable [11, 12]. Extensions of the coulomb explosion model have been proposed in which the coulombic force along the track is incompletely screened by the excited electrons [11,18]. That is, even though the free electrons return rapidly to the vicinity of the track they may remain in an excited state long enough to allow the generation of nuclear motion by the repulsive ion cores. Even at low $(dE/dx)_e$ a coulomb repulsion model can retain a quadratic dependence. That is, the probability of one ion being formed close enough to the surface to be ejected and that of a second ion being formed close enough to the first to provide a force for ejection of the first are both linear in the electronic stopping power, giving a yield with a quadratic dependence [11].

A large scale collective coulomb explosion seems unlikely to explain the erosion process. At the low ionization densities of interest in this discussion, atomic scale coulomb explosions seem much more likely to provide a means for rapid transfer of electronic excitation to atomic motion. For rare gas solids, such a possibility has been treated by JOHNSON and INOKUTI [17]. If a transition from quadratic to linear erosion with electronic stopping power is confirmed at low stopping power, a picture of the collective aspects of erosion as being due to overlapping "hot" spots produced by atomic scale explosions will be relatively convincing. The quantitative aspects of the energy transfers will remain the main issue.

5. Summary

The erosion yields of condensed gas solids by fast light ions are non-linear and approximately quadratic in their dependence on the excitation density down to quite low densities for H_2O, CO_2, SO_2, Ar and CO. Recent measurements of the erosion of CO with keV electrons confirm the conclusion of the MeV ion results that the important processes of excitation are electronic, not collisional. However, the electron results show a possible transition from a quadratic dependence toward a linear dependence at very low $(dE/dx)_e$.

At low excitation density the observation of a non-linear dependence of the erosion yield on excitation density demands one of two types of cooperative processes by which electronic energy is transferred to atomic motion: 1. Rapid electron-atom energy transfers initiated by individual excitation or ionization events are closely enough spaced for them to overlap in a hot cylinder. Such a model would be related to the MGR model of desorption [19]. 2. Two ionization events are produced close together by passage of a single particle. These two events might be on neighboring molecules giving rise to mutual repulsion related to the KF model of desorption [20]. Such events might also be related to a double ionization model proposed for laser stimulated desorption [21]. For the case of solid CO, the indications of the most recent results with electrons and ions suggest that process 1 is dominant. In the case of solid H_2O, this picture still poses some quantitative difficulties.

References

1. W. L. Brown, W. M. Augustyniak, L. J. Lanzerotti, R. E. Johnson and R. Evatt, Phys. Rev. Lett. 45, 1632 (1980).
2. J. Bottiger, J. A. Davies, J. L'Ecuyer, H. K. Haugen, N. Matsunami and R. Ollerhead, Proc. Conf. on Ion Beam Mod. of Materials, Budapest, 1978: ed. J. Gyulai, J. Lohner, E. Pasztor (Cent. Res. Inst. for Physics, Budapest) p. 1521 (1979).
3. W. L. Brown, et al., Nuc. Inst. and Meth., 197, (1982).
4. B. H. Cooper, Ph.D. Thesis, Cal Tech (1982), L. E. Seiberling, C. K. Meins, B. H. Cooper, J. E. Griffith, M. H. Mendenhall and T. A. Tombrello, Nuc. Inst. and Meth., to be published (1982).
5. R. Ollerhead, J. Bottiger, J. A. Davis, J. L. E'cuyer, H. J. Haugen and M. Matsunami, Rad. Effects 49, 203 (1980).
6. F. Besenbacher, J. Bottiger, O. Graversen, J. L. Hansen, and H. Sorenson, Nuc. Inst. and Meth., 191, 221 (1982).
7. L. J. Lanzerotti, W. L. Brown, W. M. Augustyniak, R. E. Johnson and T. P. Armstrong, Astrophys. J., in press (1982).
8. D. J. LePoire, B. H. Cooper, C. L. Melcher and T. A. Tombrello, submitted for publication.
9. W. L. Brown, W. M. Augustyniak, E. Brody, L. J. Lanzerotti, A. L. Ramirez, R. Evatt and R. F. Johnson, Nuc. Inst. Meth., 170, 321 (1980).
10. P. Sigmund, Phys. Rev. 184, 383 (1969); P. D. Townsend, J. C. Kelley and N. E. W. Hartley, "Ion-Implantation, Sputtering and Their Applications", Academic press, N.Y. (1976); R. Behrisch, ed. "Sputtering by Particle Bombardment", Springer, Berlin (1981).
11. R. E. Johnson and W. L. Brown, Nuc. Inst. and Meth., 197, in press (1982).
12. J. Schou, Private Communications.
13. C. Claussen and R. H. Ritchie, Nuc. Inst. and Meth., 197, in press (1982).
14. G. H. Vineyard, Rad. Effects. 29, 245 (1976). R. E. Johnson and R. Evatt, Rad. Effects 52, 187 (1980).
15. R. L. Fleischer, P. B. Price and R. M. Walker, Nuclear Tracks in Solids, Principles and Applications, U. Calif. Press, Berkeley (1975).
16. P. K. Haff, App. Phys. Lett. 29, 473 (1976). J. O. Stiegler and T. S. Noggle, J. App. Phys. 33, 1894 (1962).
17. R. E. Johnson and M. Inokuti, Nuc. Inst. & Meth., submitted (1982).
18. C. C. Watson and T. A. Tombrello, Lunar Planet. Sci. XIII (Lunar and Planetary Institute, Houston, 1982) p. 845.
19. D. Menzel and R. Gomer, J. Chem. Phys. 41, 3311 (1964); P. A. Redhead, Can. J. Phy., 42, 886 (1964).
20. M. L. Knotek and P. J. Feibelman, Phys. Rev. Lett., 40, 964 (1978).
21. T. Nakayama and N. Itoh, Rad. Effects, 67, 129 (1982).

7.2 Desorption of Condensed Gases and Organic Molecules by Electronic Processes

T.A. Tombrello

W.K. Kellogg Radiation Laboratory, California Institute of Technology
Pasadena, CA 91125, USA

Abstract

It has been demonstrated that condensed gases and heavy organic molecules
are desorbed very efficiently by energetic charged particles. This occurs
even at high velocities where the particle loses energy in the material by
its interaction with atomic electrons. Although there are a variety of
effects observed, the overall pattern that has emerged shows many similarities for electronic excitation erosion in dielectric solids, condensed
gases, and organic materials. This paper emphasizes these similarities in
the belief that there may be a common underlying desorption mechanism.

Introduction

The presence of frozen gases on the surfaces of the moons of Jupiter and
Saturn has provided a major impetus for studying how such materials are
affected under bombardment by energetic neutral and charged particles that
are trapped in the planetary gravitational and electromagnetic fields. The
bombarding particles have energies that range from eV to MeV; thus, one
expects a variety of erosion mechanisms to be involved. Material that is
removed from the satellite surfaces can itself be trapped and re-impact the
satellites, making the system regenerative. Trapped ions, neutral atoms,
and even excited atoms are observed. For example, there is a toroidal cloud
of Na in the orbit of Io that shines with yellow Na(D) light.

It is likely that a large part of the erosion comes from the intense low-
energy bombardment. The desorption mechanism is ordinary collision-cascade
sputtering which produces a hard spectrum of ejected atoms and molecules
that can overcome the gravitational binding (a few eV) of the large Galilean
satellites [1]. However, in a number of situations (e.g., the small H_2O
covered moons of Saturn) the less intense, high-energy bombardment could be
of comparable importance. For such cases, it is especially important to
have a better understanding of the electron desorption mechanisms that are
the subject of this conference [2].

The systematic behavior of the electronic desorption of condensed gases
does not seem to be very different from that observed for other dielectric
solids [3], but what is perhaps more surprising is the similarity with ion-
induced desorption of organic molecules [3,4]. In the latter case not only
the masses (10^2-10^4 amu) but also the dimensions of the molecules (10-100 Å)
can be especially large. This probably is a reflection of the size of the
region affected by the excitation mechanism.

In this paper I shall emphasize the similarities in charged particle induced desorption phenomena that seem to imply a common mechanism. Although each substance exhibits its individual peculiarities, the differences between condensed gases and organic and inorganic solids are no more pronounced than the differences among different condensed gases.

Observations

The data of BROWN et al. for the proton and helium ion erosion of frozen H_2O provide a superb example of the departure of the behavior of the sputtering yield versus bombarding energy from what would have been expected on the basis of collision-cascade sputtering [5]. These data (shown in Fig. 1) peak at bombarding energies close to the maximum in the ion's energy loss to electronic excitations. In Fig. 2 measurements of the neutral species desorbed indicate that at low temperature the yield is virtually all H_2O molecules [6]. Only at much higher temperatures are appreciable numbers of hydrogen and oxygen molecules desorbed.

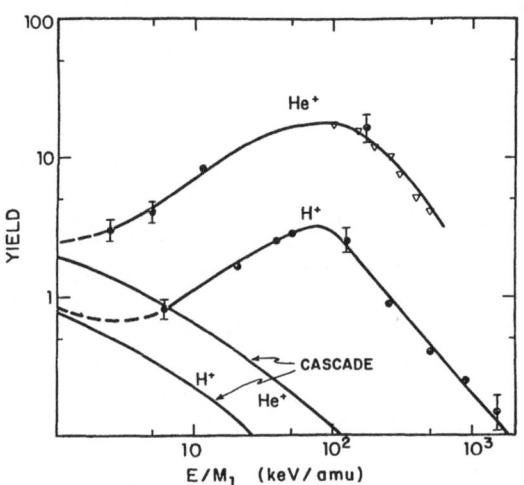

Fig.1. The sputtering yield of H_2O molecules per incident ion are shown for the bombardment of frozen H_2O (at $10^\circ K$) by beams of protons and $^4He^+$ ions. The calculated collision cascade predictions are shown. The solid lines through the data are intended only as a guide to the eye. (These data come from [5])

Although the general trend of the H_2O yield data tends to match that of the square of the electronic part of the energy loss, $(dE/dx)_e^2$, the H_2O yield curve is actually double valued when plotted versus $(dE/dx)_e$ [7]. This same behavior is shown in Fig. 3 for the Cs^+ yield from CsI, where we see that although the overall trend is consistent with a power of $(dE/dx)_e$, the detailed shape of the individual yield curves is somewhat more complicated [8]. Nature, however, may be somewhat simpler than our theories. In Fig.4 are shown the desorption yields for UF_4 (a crystalline, dielectric solid) and two condensed gases (H_2O and SO_2) under bombardment by ^{19}F ions. Although the magnitudes of the yields differ by a factor of 10^3, the shapes of the yield versus ^{19}F energy are identical [9]. In fact, the yield curves are far more alike than the energy losses calculated for ^{19}F ions in these dissimilar materials [10].

The fact that ion-induced desorption approximately follows the square of the energy loss can be obtained either from ion-explosion models [11,7] or thermal spike models [12] of the desorption process. In the former one should use $(dJ/dx)_e^2$ instead of $(dE/dx)_e^2$ because dJ/dx is the ionization

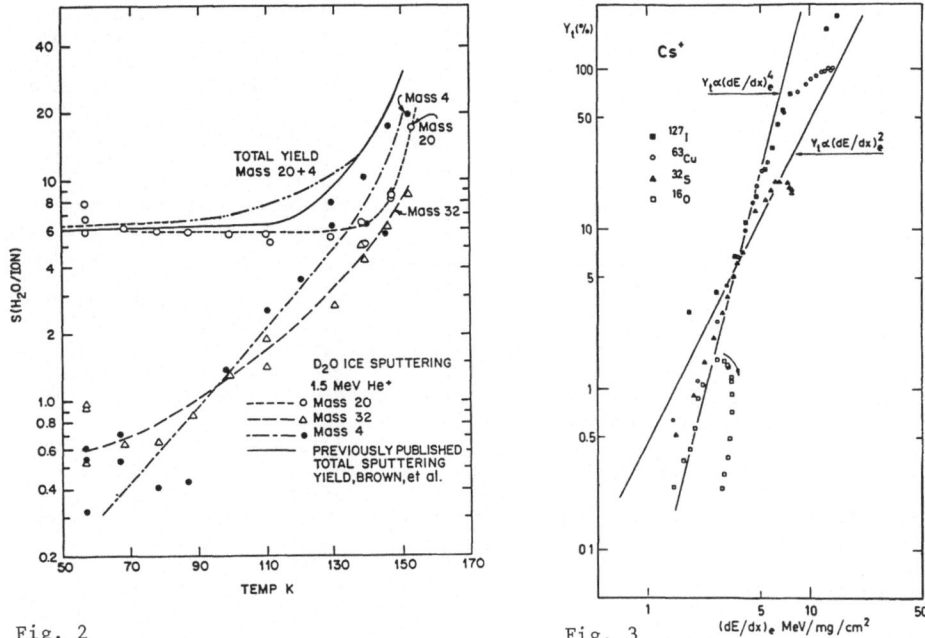

Fig. 2

Fig. 3

Fig. 2. The neutral components observed in the sputtering of H_2O ice at various temperatures by a beam of 1.5 MeV $^4He^+$ ions. (These data come from [6])

Fig. 3. The Cs^+ ion yield from the bombardement of CsI by various ion beams is plotted versus the electronic part of the ion's energy loss $(dE/dx)_e$ in the material. Note the double-valued nature of the yield curves for the ^{16}O and ^{32}S ion beams. (These data are from [8])

produced per unit length of ion path. Unfortunately, dJ/dx is not easily or unambiguously obtained [13], and it has a form that is not the same as $(dE/dx)_e$ [Ref. 14]. Both of these models also neglect the fact that the density of deposited energy is governed to a large extent by the recoil electron (delta ray) distribution. The maximum projected range of the delta rays is proportional to the square of the ion's velocity; thus, the energy is deposited in a larger volume as the ion's energy increases. This is probably the reason that the observed yields peak a bit earlier than $(dE/dx)_e$ and fall off much faster at higher bombarding energies [14]. The track etching data of DIAMOND show this behavior in Fig.5 [15]. In this experiment the ratio of the track etching rate to the etching rate of the bulk material was obtained as a function of the ion beam energy in an organic material (Makrofol). The track is etched faster because of the damage; thus, the data should follow the density of the ion's deposited energy.

In situations where the energy loss is low, the excitation tends to be concentrated on isolated defects, which lead to desorption by focused colli-sion sequences or by thermal migration leading to molecule formation and evaporation [16]. In this regime the number of defects, and hence the de-sorption yield, are proportional to $(dE/dx)_e$. A fine example of this is provided by the data of BØRGESEN and SØRENSEN for the erosion of frozen D_2 by keV electrons [17]. A portion of these data presented in this session by

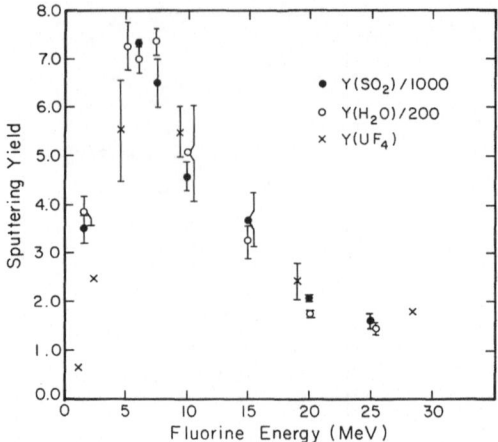

Fig.4. The sputtering yields for SO_2, H_2O, and UF_4 per incident ^{19}F ion are given versus ^{19}F bombarding energy. The data points for each substance were taken with projectiles in the same charge states at each energy. (These data are from [9])

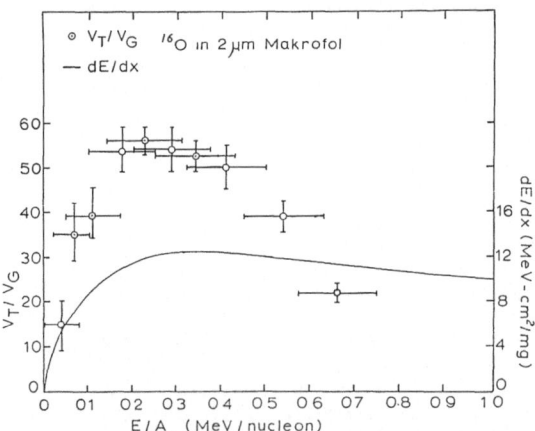

Fig.5. The ratio of the track etching rate to that of the bulk material for Makrofol (an organic track detecting material) bombarded by [16]O ions. The solid curve is the electronic part of the energy loss of the ions in the material. (These data are from [15])

SCHOU show that the desorption yield is proportional to $(dE/dx)_e$ rather than to its square [18]. For thick D_2 targets the yield per 2 keV electron is ~ 8; however, for targets thinner than 500 Å the yield increases and is approximately proportional to the inverse of the target thickness [17].

The erosion yield for helium ions (0.1-3 MeV) on frozen [40]Ar also depends on the target thickness [19]. However, in this case (as shown in Fig.6) the yield is suppressed for thin targets. Here, the yield increases approximately linearly with target thickness up to about 750 Å. Neither for this case nor for the electron erosion of D_2 is the thickness dependence understood — particularly for thicknesses as great as these.

In Table 1 are given the desorption yields per incident 1.5 MeV $^4He^+$ ion for a number of frozen gases [20]. These data show that the yield tends to decrease with increasing surface binding energy of the molecule, but there is no detailed correlation of the two quantities. One notices immediately that the range of yields is at least as large as those shown in [21] or in Fig.4 between dielectric solids and condensed gases.

The desorption of large organic molecules by energetic heavy ions shows many of the trends just discussed for condensed gases and for dielectric solids. The yield curves have the same shapes, i.e., they peak earlier than $(dE/dx)_e$ and decrease more rapidly [8]. There is a general tendency for yields to scale as $(dE/dx)_e^2$ and the data look very much like those in Fig.3 [8].

Table 1. Erosion yields of various condensed gases bombarded by 1.5 MeV He[+] (as summarized in [20])

Gas	Surface Binding Energy (eV)	Yield (per incident ion)	
Ne	0.02	2000	
Ar	0.08	50	
Kr	0.12	20	
Xe	0.16	20	
CO_2	0.27	120	
SO_2	0.38	16	(Y = 50, Ref. [9])
H_2O	0.52	8	

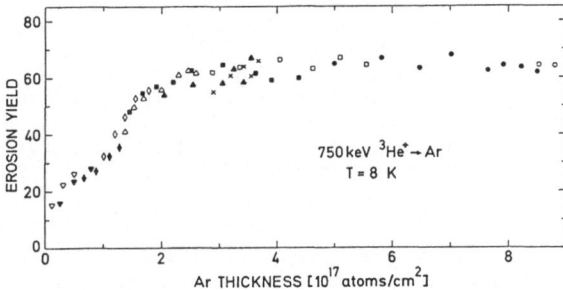

750 keV ^{3}He* → Ar
T = 8 K

Fig.6. The erosion yield of frozen ^{40}Ar per incident 750 keV ^{3}He$^+$ ion as a function of the target thickness. (These data come from [19])

The yield versus energy for a valine target has been measured for oxygen ions whose energies span the range between predominantly nuclear energy loss (ordinary collisional sputtering) and exclusively electronic excitation energy loss. These data are shown in Fig.7 and clearly exhibit the two regimes in which large molecules are desorbed by quite different processes [22]. Usually, there is an appreciable enhancement of the yield with high-energy ions as compared with similar ions at low energy. This enhancement increases with the mass of the target molecule; the ratio of yields is 1.5 for CsI (133 amu), 14 for glycylglycine (132 amu), and 196 for trinucleoside diphosphate (1884 amu) [4]. The high value of the ratio for organic molecules probably reflects the larger surface region affected by the electronic excitation mechanism, allowing a large molecule to be ejected intact.

Generally, the spectra are cleaner in the region of the full mass peak for high energy bombardment [4]. For example, a portion of the mass spectrum is shown in Fig.8 for molecular ions from a chemically blocked deoxyoligonucleotide that were desorbed by ^{252}Cf-fission fragments [23]. Above the spectrum is a structural diagram of the molecule, which is 90 Å long.

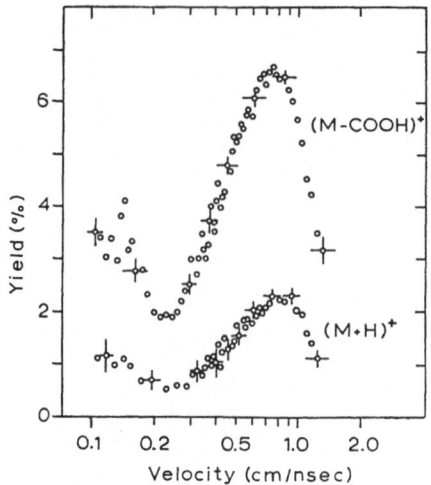

The desorption of such molecules by photons from lasers has also been observed [23]. In this case the heavy part of the positive molecular ion spectrum is dominated by molecule-alkali ion complexes which seem to arise from ion-molecule reactions in the vapor that is produced thermally from the target material [24]. Thus, photon-induced desorption of heavy organic molecular ions does not primarily involve an electronic excitation mechanism.

Fig. 7. The yield of (valine - COOH)$^+$ and (valine + H)$^+$ molecular ions from the ^{16}O ion bombardment of valine. The two peaks shown correspond to the regimes where nuclear energy loss processes dominate (low velocities) and where electronic energy loss processes (high velocities) dominate. (These data are from [22])

It is worth noting that data exist which indicate that collective effects of the electronic excitation mechanism are present for these large molecules. If a mixture of different organic molecules is bombarded, then although fragments are observed, there are few whole molecules in the spectrum [25]. This result, which is similar to that observed for thin layers of SiO_2 bombarded by ^{35}Cl ions [25], tends to confirm the result that in the absence of a coherent piece of material the desorption mechanism is strongly inhibited. These results may be related to the dependence of the erosion yield on target thickness that was described above for frozen D_2 [17] and ^{40}Ar [19].

Conclusions

Observations in a variety of materials exhibit comparable behavior. The shape of the desorption yield for a given ion versus ion beam energy is similar for many substances. Generally, the yield varies as $(dE/dx)_e^2$, with

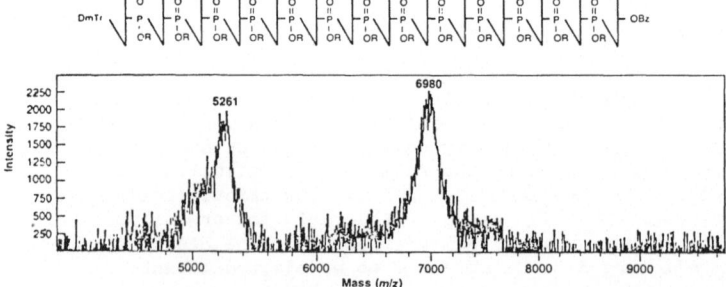

Fig. 8. A portion of the molecular ion mass spectrum arising from the ^{252}Cf-fission fragment bombardement of a chemically blocked deoxyoligonucleotide. The structure of the molecule is given above. The vertical lines represent the deoxyribose moiety. DmTr = dimethoxytrityl, R = p-chlorophenyl, Iso = isobutyl, bz = benzyol. (Taken from [22])

typical departures from that behavior that probably reflect the changing spatial density of deposited energy; however, the data for different target materials are more alike than their $(dE/dx)_e$'s. As indicated in my contribution to the preceding session of this conference [21], the magnitudes of the observed yields can be reproduced by a simple model that depends upon only a few of the properties of the materials [14]. The success of this calculation is probably a consequence of the fact that the extraordinary amount of energy deposited along the ion track (keV/Å) is sufficient to excite many electrons into the conduction band. Thus, we are dealing with a plasma whose properties are nothing like those of the bulk material. In the model the nature of the target material only affects the process through the interaction of the cooling plasma with the surrounding target volume.

Although the yield data, indeed, exhibit many similarities, it is clear that additional data (e.g., the energy spectra of the desorbed molecular species) will be important in determining if a common desorption mechanism is involved. It is also especially important to determine the role played by the minimum thickness or purity of a material that exhibits the electronic erosion mechanism [24,25].

Acknowledgments

I want to thank those at Aarhus, Bell Labs, Caltech, Erlangen, FOM, Risø, Schlumberger, Texas A & M, University of Virginia, and Uppsala who provided their recent data which is referred to here. The work associated with the preparation of this paper was supported in part by NASA [NAGW-202 and -148] and by the NSF [CHE81-13273 and PHY79-23638].

References

1. P. K. Haff, C. C. Watson, and Y. Yung, J. Geophys. Res. 86, 6933 (1981).
2. T. A. Tombrello, Radiat. Eff., in press (1982).
3. L. E. Seiberling, C. K. Meins, B. H. Cooper, J. E. Griffith, M. H. Mendenhall, and T. A. Tombrello, Nucl. Instrum. Methods, in press (1982).
4. B. Sundqvist, Invited talk at the Symposium on "Role of Physics in Development," Dacca, Bangladesh, Jan. 10-24, 1982.
5. W. L. Brown, L. J. Lanzerotti, J. M. Poate, and W. M. Augustyniak, Phys. Rev. Lett. 40, 1027 (1979).
6. W. L. Brown, W. M. Augustyniak, E. Simmons, K. J. Marcantonio, L. J. Lanzerotti, R. E. Johnson, J. W. Boring, C. T. Reimann, G. Foti, and V. Pironello, Nucl. Instrum. Methods, in press (1982).
7. W. L. Brown, W. M. Augustyniak, E. Brody, B. H. Cooper, L. J. Lanzerotti, A. Ramirez, R. Evatt, and R. E. Johnson, Nucl. Instrum. Methods 170, 321 (1980).
8. P. Håkansson and B. Sundqvist, Radiat. Eff. 61, 179 (1982).
9. D. J. LePoire, B. H. Cooper, C. L. Melcher, and T. A. Tombrello, submitted to Radiat. Eff. (1982).
10. J. F. Zeigler, Stopping Cross Sections for Energetic Ions in All Elements (Pergamon Press, New York, 1980).
11. P. K. Haff, Appl. Phys. Lett. 29, 473 (1976).
12. P. Sigmund and C. Claussen, J. Appl. Phys. 52, 990 (1981).
13. H. A. Bethe, Ann. Phys. (Paris) 5, 325 (1930).
14. C. C. Watson and T. A. Tombrello, submitted to Radiat. Eff. (1982).
15. W. T. Diamond, unpublished manuscript.
16. J. P. Biersack and E. Suntuev, Nucl. Instrum. Methods 132, 229 (1976); M. Szymoński, H. Overeijnder, and A. E. deVries, Radiat. Eff. 36, 189 (1978).

17. P. Børgesen and H. Sørensen, Phys. Rev. Lett., in press (1982).
18. J. Schou, contribution to this session.
19. F. Besenbacher, J. Bøttiger, O. Graversen, and J. L. Hansen, Nucl. Instrum. Methods 191, 221 (1981).
20. R. E. Johnson and M. Inokuti, manuscript in preprint form (1982).
21. T. A. Tombrello, contribution to the preceding session of this conference.
22. A. Albers, K. Wein, P. Dück, W. Treu, and H. Voit, Nucl. Instrum. Methods, in press (1982).
23. C. J. McNeal and R. D. Macfarlane, private communication.
24. G. J. Q. van der Peyl, K. Isa, J. Haverkamp. and P. G. Kistamaker, Nucl. Instrum. Methods, in press (1982).
25. R. D. Macfarlane and B. Sundqvist, private communication.
26. J. E. Griffith, Y. Qiu, M. H. Mendenhall, and T. A. Tombrello, submitted to Radiat. Eff. (1982), as quoted in [21].

7.3 Photon-Stimulated Ion Desorption from Condensed Molecules: N_2, CO, C_2H_2, CH_3OH, N_2O, D_2O, and NH_3

R.A. Rosenberg, V.Rehn, A.K. Green, and P.R. LaRoe
Michelson Laboratory, Physics Division, Naval Weapons Center
China Lake, CA 93555, USA

C.C. Parks
Lawrence Berkeley Laboratory and University of California
Berkeley, CA 94720, USA

Abstract

Results of photon-stimulated ion desorption measurements from the surfaces of solid molecular films are presented. In the photon-energy range < 35 eV, data on N_2, CO, C_2H_2, and CH_3OH illustrate the importance of including electron-correlation effects in describing the desorption process, while the low-energy threshold for NO^+ desorbed from N_2O clearly indicates a one-electron process. The N^+ yield from adsorbed N_2 resembles the O^+ yield from CO due to hole localization on the N atom nearest the surface. K-shell excitation spectra of NH_3 and D_2O show that excitations of Rydberg orbitals are severely perturbed by hydrogen bonding in the bulk and, to a lesser extent, on the surface. Similar spectra of N_2 show that reneutralization effects observed in N_2 adsorbed on Ni are due to interaction with the metal and are not intrinsic to the N_2 molecule.

1. Introduction and Summary

The recent discovery [1] of photon-stimulated ion desorption (PSID) from surfaces using synchrotron radiation has stimulated a great deal of experimental and theoretical interest. Many mechanisms[1] and models have been proposed to explain the experimental results, but many aspects of the phenomena remain to be understood before PSID can assume a role as a standard surface-science technique.

Photodissociative ionization studies of gas-phase materials have been ongoing for many years, and detailed information on molecular photoionization is available for many systems [2]. In order to take advantage of this wealth of knowledge, we have undertaken the study of the solid-state analog: condensed solid films of these gases. The results of these experiments are helping to bridge the gap between the gas-phase results and the complex substrate-adsorbate phenomena. Photodissociative ionization in the gas phase occurs via intramolecular decay channels; in molecular solids, these channels are modified by molecule-molecule, ion-molecule, or hydrogen-bond interactions, while in adsorbate systems, the complex chemisorption bond may be involved. By studying solids bonded by Van der Waals forces, we may come to understand the energetics and mechanisms of desorption in this situation intermediate between the gas phase and the adsorbed molecule, and to establish standards for energy thresholds and yields for various molecules.

---

[1]There are two commonly discussed mechanisms for PSID: the Menzel-Gomer-Redhead (MGR) or Franck-Condon-excitation mechanism and the Knotek-Feibelman (KF) or Auger-assisted Coulomb mechanism.

Thus far very little work has been reported on ion desorption from condensed molecular films by either photon or electron excitation. Using fixed-energy electrons on condensed hydrogen films, CLAMPETT and GOWLAND [3] observed hydrogen molecular-ion clusters. PRINCE and FLOYD [4] observed similar effects in condensed water and ammonia. MADEY and YATES [5] observed the angular distribution of protons from films of ice adsorbed on ruthenium. The first PSID experiment on a condensed molecular film was performed on amorphous ice films by ROSENBERG *et al.* [6], who reported the photon-energy dependence of hydrogen-ion desorption. Recently, STOCKBAUER *et al.* [7] have made electron-stimulated ion desorption (ESID) measurements on condensed methanol, which has also been studied by HANSON *et al.* [8] using PSID. ESID experiments have been reported recently by KELBER and KNOTEK [9] on condensed alkanes.

In this paper we report results of PSID excitation spectra and desorbed-ion mass analysis in the spectral range $14 \leqslant h\nu \leqslant 35$ eV on solid films of N_2O, N_2, CO, C_2H_2, and CH_3OH. In the spectral range 390 - 600 eV, we report nitrogen and oxygen K-shell excitation of PSID and total photoelectron yield (TPY) spectra for solid N_2, NH_3, and D_2O. Comparison of N_2 and CO results suggests that when N_2 is bonded to a surface and the outermost valence orbital is ionized, the charge localizes on the nitrogen atom nearest the surface. This localization results in the electron density distribution for an adsorbed N_2^+ ion to resemble that of the CO^+ ion [10]. Thus, the desorption spectrum of N^+ from adsorbed N_2 resembles that of O^+ from CO. The hydrogen ion yield from solid C_2H_2 (acetylene) shows good correlation with results of gas-phase photoemission of the $2\sigma_g$ and $2\sigma_u$ orbitals. PSID in C_2H_2 is shown to involve multielectron ("shake-up") excitations. In solid methanol (CH_3OH) there is good agreement between our data and that of HANSON *et al.* [8] which indicates that H^+ production proceeds via ionization of the 4a' orbital. For solid N_2O we have observed a desorption threshold of 14.0 eV, which is 0.8 eV *lower* than observed in the gas-phase photodissociative ionization process. Studies of ion desorption by K-shell excitation in D_2O, NH_3, and N_2 all show correlation between gas-phase spectra and both the TPY and PSID yield spectra from the condensed solid film. Solidification does not appreciably perturb the lowest excited molecular orbital, and its excitation closely resembles the gas-phase observations. The N^+ PSID yield of an adsorbed monolayer of N_2 on Ni(100), however, looks distinctly different from that of the solid film [11].

2. Experimental

The techniques used in these studies have been described previously [1,6], so only a brief description plus some recent improvements will be given here. The experimental apparatus is shown schematically in Fig. 1. Molecular films, typically hundreds of angstroms thick, are grown on an Al_2O_3 substrate cooled to 10-20 K. After completion of the initial film growth, the surface is continuously refreshed by slow condensation of the vapor from the doser tube at all times, while keeping the chamber pressure less than 1×10^{-9} torr. This technique was found to be essential to avoid surface contamination by the ambient at these low temperatures. Even with this precaution, all mass spectra exhibit a small H^+ signal whose origin is unknown. However, considering the relatively high yield cross section for H^+ ions, the concentration of this contaminant is estimated to be less than 1% on the refreshed surface.

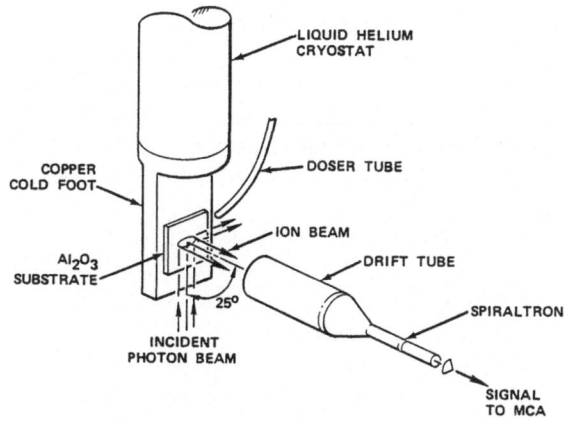

Fig. 1. Schematic diagram of the apparatus used for PSID on cryogenic surfaces. For these studies, a pair of microchannel plates replaced the spiraltron. Not shown is a channeltron used for total photoelectron yield measurements

The monochromatic synchrotron radiation was provided by the Stanford Synchrotron Radiation Laboratory. For photon energies less than 35 eV, the measurements were made on beam line I-2, which has a Seya-Namioka monochromator. We used 300-μm slits, giving a 2.5 Å band pass or an energy resolution of between 0.04 and 0.25 eV. For measurements taken at energies less than 17 eV (N_2O), an indium filter was used to eliminate second-order light. For photon energies $60 \leqslant h\nu \leqslant 800$ eV used in the K-shell excitation, experiments were performed on beam line III-1 which employs a 2-m "grasshopper" monochromator. The slit width used provided a resolution of 0.07 Å, corresponding to an energy resolution of 0.85 eV at 400 eV. The energy calibration for the "grasshopper" experiments was obtained from the results of gas-phase photoemission experiments [12], which resulted in a shift of 1.25 ± 0.3 eV at 400 eV for NH_3 and N_2. Unfortunately, calibration data were unavailable at higher energies, although we estimate a shift of about 2 eV at the O(K) edge, 530 eV.

Ions are detected by a set of microchannel plates (which replaced the spiraltron of Fig. 1) and mass analyzed by a time-of-flight method [1]. TPY measurements were performed using a positively biased channeltron approximately 3 cm from the sample. Either signal was normalized simultaneously. In the low-energy regime (< 35 eV), this was accomplished using the fluorescence from a sodium-salicylate-coated mesh, which has a relatively flat response in this energy regime [13]. In most cases it was possible to convert the normalized yield to an absolute ions-per-incident-photon scale by using photon-flux measurements taken subsequently with a National Bureau of Standards photodiode and assuming unity detector efficiency. The PSID detector efficiency, while unknown, is presumed to be high. At any rate, the relative yields can be compared from system to system. For K-shell studies, normalization was accomplished using the electron yield signal from a graphite-coated mesh.

3. Results and Discussion

Table 1 summarizes the systems investigated to date. A generalization can be made concerning the systems studied. Whereas many of the smaller molecules exhibit rich PSID mass spectra, all hydrogen-containing molecules investigated to date show only H^+ yield. In all the systems studied, the

249

Table 1 Photon-stimulated ion desorption from condensed molecular solids

A. Low-energy excitation studies (14 - 35 eV)

Solid	Products (maximum yield, ions per incident photon x 10^8)				
N_2	$N^+(1.6)$	$N_2^+(0.02)$	$N_3^+(0.06)$		
CO	$C^+(1.6)$	$O^+(0.06)$	$CO^+(0.06)$	$C_2O^+(0.02)$	$(CO)_2^+(0.02)$
O_2	$O^+(2)$	$O_2^+(0.02)$			
NO	$N^+(0.2)$	$O^+(0.4)$	$NO^+(0.4)$		
CO_2	$C^+(0.6)$	$O^+(0.6)$	$CO^+(2)$	$O_2^+(0.6)$	
N_2O	$N^+(2)$	$O^+(0.6)$	$N_2^+(1.0)$	$NO^+(2)$	$O_2^+(0.14)$
H_2O	H^+				
NH_3	$H^+(4)$				
C_2H_2	$H^+(36)$				
SiH_4	$H^+(18)$				
CH_3OH	$H^+(24)$				
$(CH_3)_2CO$	$H^+(6)$				

B. K-shell excitation studies (390 - 600 eV)

Solid	Edge	Product
N_2	N(K)	N^+
NH_3	N(K)	H^+
D_2O	O(K)	D^+

H^+ yield is significantly higher than any other species. These observations are probably a consequence of the higher cross section for H^+ PSID due to its lower reneutralization probability. The rest of this section is devoted to examining specific systems in detail.

3.1 Low-Energy Excitations (14 - 35 eV)

3.1.1 N2 and CO

Nitrogen is one of the most attractive candidates for these studies. It is a homonuclear, closed-shell, diatomic molecule which might lead one to believe that its behavior in the solid phase should closely mimic that of the gas phase. CO is isoelectronic with N_2 and is the prototype for many absorbate experiments. PSID experiments on chemisorbed molecular CO have been performed on Ru(001) at low energies [14] and on Ni(100) at the oxygen K-edge [15]. Photodissociative ionization studies on gaseous N_2 and CO

250

have been reported using the e, e + ion technique [16] and resonance lamps [17], and on CO using synchrotron radiation [18].

Numerous ultraviolet photoelectron spectroscopic (UPS) studies of gaseous CO and N_2 [19-21] have examined the satellite structure in the binding energy region above 20 eV. UPS results on solid N_2 and CO show gas-solid shifts of the molecular-orbital binding energies between 1 and 1.5 eV [22,23]. Recent calculations have assigned the satellite lines observed in the gas-phase UPS spectrum to various two-electron (shake-up) configurations of CO and N_2 [24,25]. There have been recent theoretical and experimental investigations relating the binding energies of chemisorbed N_2 and CO [10,26,27].

The photoyield of N^+ and N_3^+ ions from gaseous and solid N_2 is shown in Fig. 2. A prominent threshold and peak in the gaseous N^+ yield between 25 and 27 eV is suppressed in the case of solid N_2, whereas the thresholds and shapes of the N_3^+ curves agree and follow the N^+ yield curve from solid N_2. Figure 3 exhibits the C^+ yield from gaseous and solid CO as well as the O^+ yield from gaseous, solid, and chemisorbed CO. All the O^+ yield curves compare fairly well. The shapes of the C^+ curves correspond, but the threshold for solid CO occurs about 1.5 eV higher in energy.

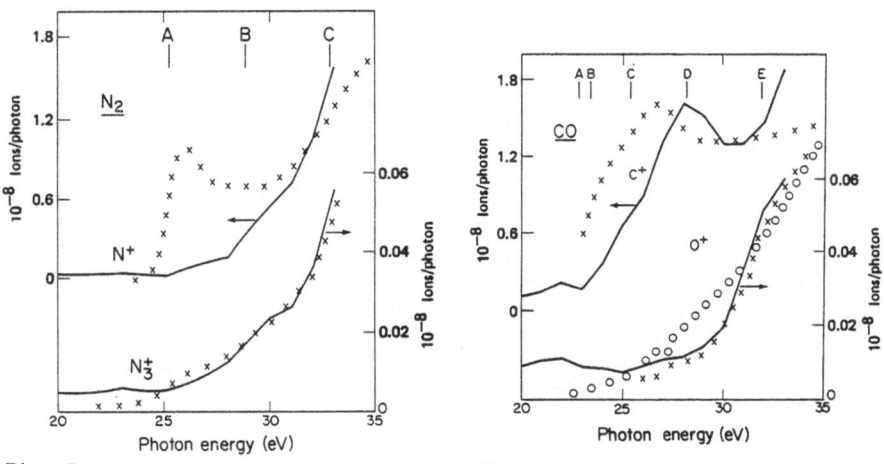

Fig. 2   Fig. 3

Fig. 2. Top: The N^+ photoyield from solid N_2 (solid line) compared with the N^+ photoyield of gaseous N_2 (x's) [16]. The scale for the N^+ yield of solid N_2 is to the left. The letters A-C mark positions of multielectron excitations observed in gas-phase N_2 [19]. Bottom: The N_3^+ photoyield of solid (solid line) and gaseous (x's) [33] N_2 dimers. The scale for the N_3^+ yield from solid N_2 is to the right. The N_3^+ data have been smoothed

Fig. 3. Top: The C^+ photoyield from solid CO (solid line) compared with the C^+ photoyield of gaseous CO (x's) [16]. The scale for the C^+ yield from solid CO is to the left. The letters A-E mark positions of multielectron excitations observed in gas-phase CO [19]. Bottom: The O^+ photoyield from solid CO (solid line) compared with the O^+ photoyield from gaseous CO (x's) [16] and from CO adsorbed on Ru(001) (o's) [14]. The scale for the O^+ yield from solid CO is to the right. The O^+ data have been smoothed

The cause for this difference in behavior of PSID from solid N_2 and CO may be understood by considering the ion-molecule interactions. The Van der Waals bond strength (0.008 eV for $(CO)_2$ and 0.007 eV for $(N_2)_2$ [28]) is far too weak to cause such effects. However, the ion-molecule interaction is much stronger: 0.97 ± 0.04 eV for $CO^+ \cdot CO$ and 0.90 ± 0.05 eV for $N_2^+ \cdot N_2$ [29]. Studies of $(N_2)_2$ indicate that its structure is probably "T" or "L" shaped [30,31]; thus, the solid N_2 surface probably has N_2 molecules protruding from the surface as is the case in chemisorbed N_2 on W(110) [26].

Solid N_2 and CO are isomorphic in crystal structure [32]. Both form cubic crystals below 35 K and 61.5 K, respectively, and hexagonal above. The four molecules in the cubic unit cell are centered on a face-centered cubic lattice and are oriented along each of four <111> directions. The intramolecular bond lengths are about 1.055 Å, compared with 1.094 and 1.130 Å for N_2 and CO, respectively, at room temperature. The nearest-neighbor intermolecular separations are 3.99 Å, clearly illustrating the molecular nature of the crystal. The relative orientations of near-neighbor molecules are similar to the dimer "L" arrangement, but with the molecular axes oriented at 109.5 deg to each other instead of 90 deg. Each molecule has twelve nearest neighbors, four oriented along each of the three other <111> directions. Nothing is known about the possible reconstructions of these molecular-crystal surfaces, but if we assume no reconstruction, it is clear that N_2 molecules would be expected to protrude from the surface.

The surface bond severely alters the symmetry of the N_2 molecule and perturbs the electronic structure. Bonding to the surface occurs primarily through the 5σ orbital [10]. (Here we have adopted the $C_{\infty v}$ notation for surface N_2.) Denoting N_a as the nitrogen atom closest to the surface and N_b as the nitrogen atom farthest from it (as in [10]), one finds that when the 5σ orbital is ionized, the hole is stabilized by localization on N_a. In this sense, electron density distribution of the $5\sigma^{-1}$ orbital resembles that of the $4\sigma^{-1}$ orbital of CO^+, and the surface N_2^+ electronically resembles CO^+, with N_a taking on carbon character and N_b taking on oxygen character [10]. PSID from the surface of solid N_2 produces N_b^+ ions which should resemble O^+ PSID from condensed (or gaseous) CO. Comparison of the O^+ yields of Fig. 3 with the N^+ yield from solid N_2 (Fig. 2) reveals this kind of similarity. Also, as pointed out in [16], a linear combination of the C^+ and O^+ yields resembles the gaseous N^+ yield.

Further evidence for this is found in the yields of N_3^+ ions from gaseous $(N_2)_2$ [33] and solid N_2 shown in Fig. 2. Both gas and solid compare well with each other and with the N^+ yield from solid N_2, which implies that a similar mechanism is responsible for the production of N_3^+ from $(N_2)_2$. The gas-phase experiments were performed so that production of N_3^+ via ion-molecule reactions ($N^+ + N_2 \rightarrow N_3^+$) is unlikely [33]; but, such a mechanism cannot be excluded from the solid N_2 results. Also, there is a possibility that some of the observed N_3^+ comes from heavier clusters than $(N_2)_2$ [33]. Theoretical work is necessary to understand both the mechanism of N_3^+ production and the similarity of the N_3^+ and N^+ yield spectra.

Another important question concerns the nature of the electronic states involved in the desorption process. The ground-state valence electronic configuration of CO and N_2 may be written as $3\sigma^2 4\sigma^2 1\pi^4 5\sigma^2$. In the vapor, the lower energy dissociative ionization processes (< 37 eV, $3\sigma^{-1}$ ionization) are thought to be due to predissociation involving multielectron transitions [16]. Both the lower-energy and K-shell excitations of the O^+ PSID from adsorbed CO are thought to involve multielectron transitions

[14,15,34]. In order to ascertain how such states may be involved in the condensed CO and N_2 PSID, we have marked the positions of the multielectron transitions observed in gas-phase UPS in the upper part of Figs. 2 and 3 [19].

As discussed earlier, condensation severely perturbs the PSID process in N_2, and a comparison with gas-phase energy levels is not particularly meaningful. For future discussion, it is important to note that the lowest threshold for N^+ production from N_2 vapor (marked A in Fig. 2) involves the $C^2\Sigma^+$ state, configuration $5\sigma^{-1}1\pi^{-1}2\pi^1$.

The shake-up energies observed in the UPS of CO (vapor) [19] are designated A-E in Fig. 3. In the C^+ yield curve, thresholds correlate approximately with energies A, B, C, and E, while in the O^+ yield, thresholds correspond to the C, D, and E energies. The final states of the shake-up excitations are assigned different designations and configurations by various authors [24,25,35,36]. However, there is agreement that either A or B results from excitation to the $C^2\Sigma^+$ state of main configuration $5\sigma^{-1}1\pi^{-1}2\pi^1$ as in N_2. In general, all these states involve ionization of one valence electron and excitation of one or two valence electrons to antibonding orbitals. Recently, it has been recognized that two-hole final states, particularly localized two-hole final states, may be very important in desorption of ions [37-39]. Since all these shake-up states have at least two holes in the valence shell, it is not surprising that they contribute heavily to desorption. More detailed calculations are needed in order to ascertain the manner in which these shake-up states interact with the dissociative continuum.

3.1.2 C2H2 (Acetylene)

Acetylene (H—C $\equiv$ C—H, a linear molecule) is isoelectronic with both CO and N_2; an important difference is the presence of hydrogen. Photodissociative ionization of gaseous C_2H_2 has been studied [2,40], but the yield of H^+ ions was not reported. Photoemission and e,2e studies have shown excitation energies in the 18 - 30 eV region [41-43]. Recent theoretical work has assigned these to one-electron and shake-up excitations [44].

The electronic configuration of acetylene is $(1\sigma_g)^2(1\sigma_u)^2(2\sigma_g)^2(2\sigma_u)^2$-$(3\sigma_g)^2(1\pi)^4$. Photoemission results [39] locate the $2\sigma_u$ orbital at 18.71 eV, the $2\sigma_g$ orbital at 23.65 eV, and a shake-up state at 27.53 eV. The positions of the $2\sigma_u$ and $2\sigma_g$ orbitals are marked in the H^+ yield from C_2H_2 of Fig. 4. There is a weak H^+ yield associated with ionization of the $2\sigma_u$

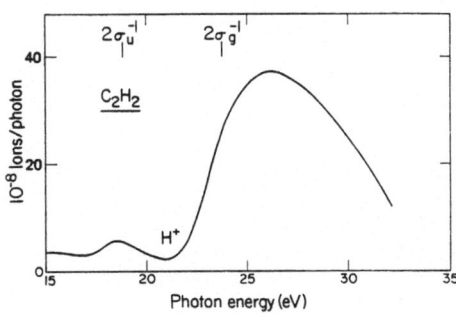

Fig. 4. H^+ photoyield as a function of photon energy from solid C_2H_2 (acetylene). The binding energies of the $2\sigma_u^1$ and $2\sigma_g^1$ orbitals of gaseous C_2H_2 [41] are marked

orbital and a strong yield threshold associated with $2\sigma_g$ ionization. No PSID yield threshold correlates with the shake-up excitation at 27.53 eV.

Calculations [44] indicate that the correlation interaction mixes the $2\sigma_g$ orbital primarily into two ionic final states: one at 23.5 eV and a second at 27.5 eV. The dominant configurations of the first are $(2\sigma_g)^{-1}$ and $(1\pi_u)^{-1}(2\sigma_u)^{-1}(2\pi_g)$, and the dominant configurations of the second are $(1\pi_u)^{-1}(2\sigma_u)^{-1}(2\pi_g)$, $(2\sigma_g)^{-1}$, and $(1\pi_u)^{-2}(4\sigma_g)$. Both these ionic states contain holes in *two* of the bonding orbitals, while populating one antibonding or nonbonding orbital. If the sum of the energies of hole-hole repulsion [37] and the antibonding repulsion is large enough, photodissociative ionization or PSID can result, provided that the excitation lifetime is sufficient and that the holes are localized in appropriate orbitals. Note that this mechanism includes Coulomb forces as in the case of the KF mechanism [45], and a Franck-Condon excitation as in the MGR mechanism [46, 47], but all made possible by a multielectron, shake-up or shake-off excitation. This mechanism [48,34] may be generally responsible for PSID in molecular solids at low photon energies where there is not enough energy for the Auger-assisted KF-type mechanism.

The correlation interaction appears to play a central role in the strong H^+ yield threshold near the $2\sigma_g$ ionization energy. At the same time, the $2\sigma_u$ is only weakly mixed by the correlation interaction [44], and only a weak H^+ yield threshold is observed near the $2\sigma_u$ ionization energy. The shake-up state at 27.5 eV is heavily mixed by the correlation interaction, containing about 25% of the $2\sigma_g$ [44]. However, no PSID threshold is observed near the 27.5 eV shake-up excitation energy.

Although extensive fracturing of C_2H_2 is observed in gas-phase photoionization [40], only H^+ is observed here. Fragmentation has been observed in the ESID of other hydrocarbon films [9,49].

3.1.3 CH₃OH (Methanol)

The PSID results are shown in Fig. 5, which depicts the H^+ yield from a CH_3OH thin solid film. Figure 5 also shows the PSID results of Ref. 8 observed at 90 K. The position of the 4a' orbital, marked in Fig. 5, is taken from UPS data [50]. Calculations [50] show that the 4a' is primarily derived from the C(2s) atomic orbital and contains strong C—H bonding character. Other orbitals in the energy range are the 3a' at 32.2 eV, which is derived mainly from the O(2s), and the 5a' at 17.6 eV which is primarily O(2p) and C(2p) [50]. No desorption threshold is found near

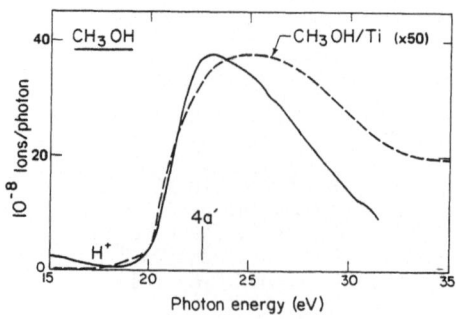

Fig. 5. Comparison of the absolute ion photoyield of H^+ ions from solid CH_2CH (methanol). The solid line represents this work; the dashed line represents the data of Ref. [8]

either of these energies. We conclude, therefore, that the threshold observed in Fig. 5 stems from 4a' ionization. Although correlation calculations have not been performed for this system, it is likely that these effects are important in describing 4a' ionization as in other hydrocarbon inner-valence-shell orbitals [44].

3.1.4 N2O

The NO^+ yield from solid N_2O is shown in Fig. 6. Although several other ions are observed in the mass spectrum of N_2O (Table 1), the low intensity of the indium-filtered light used below 18 eV allowed excitation spectra of only the high-yield NO^+ ion. To our knowledge, the threshold for NO^+ desorption at 14.1 eV is the lowest threshold yet observed in PSID and is lower than that observed in the gas-phase photodissociative ionization (14.77 eV). Note that the thermochemical threshold is 14.19 eV [51] for N_2O.

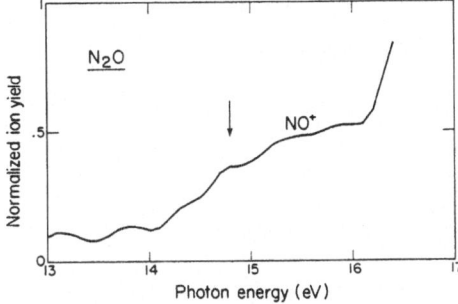

Fig. 6. NO^+ yield as a function of photon energy from solid N_2O. The data have been normalized to incident photon flux and smoothed. The arrow marks the position of the NO^+ threshold in gaseous N_2O [35]

The KF mechanism cannot operate at the low energy of the 14.1 eV threshold because of the absence of Auger transitions of such low energy. In the gas phase, dissociation is thought to take place by excitation to Rydberg levels which autoionize to repulsive states yielding NO^+ ions [51]. Due to the complexity of the gas-phase dissociation spectra, it is difficult to correlate with the NO^+ thresholds of 14.1 and 16.2 eV. The gas-phase spectrum does exhibit an abrupt increase in NO^+ yield at 16.55 eV, corresponding to $A^2\Sigma^+$ (100) ionization [51], which is the closest major feature to 16.2 eV.

3.2 K-Shell Excitations (390 - 600 eV)

By studying PSID and TPY excitations associated with the nitrogen or oxygen K-shells, we are able to distinguish surface and bulk effects of intermolecular interactions and to observe changes in the PSID mechanism as the available excitation energy is increased. A great deal of chemical and structural physics is represented in the x-ray absorption near-edge structure (XANES) and is reflected in the PSID and TPY near-edge spectra as well. The extended x-ray absorption fine structure (EXAFS) of molecular solids and their surfaces would also provide a wealth of structural information, but this is not reported here. In the following sections, we discuss the influences of molecular electronic structure and hydrogen bonding on the comparisons of the gas phase with solid molecular films of

NH$_3$, D$_2$O, and N$_2$. From these comparisons, information of importance in understanding molecular adsorption is generated, particularly in the case of N$_2$ (which may be extrapolated to CO) adsorbed on metals.

3.2.1 NH$_3$ and D$_2$O

In Figs. 7 and 8 are shown the H$^+$ yield from solid NH$_3$ and the D$^+$ yield from solid D$_2$O, respectively, which were the only ions observed. Also shown are TPY measurements from these solids and the results of gas-phase electron-loss spectroscopy (ELS) [52]. Comparison of the TPY and PSID shows that in each case the main structure in the PSID is broader than the TPY and has additional fine structure near the K-absorption edge. Similar broadening has been observed in the O(KVV) Auger spectrum of solid H$_2$O [53], while the Auger spectrum of solid NH$_3$ appears only slightly broadened in comparison to the gas phase [54]. The x-ray photoemission spectrum (XPS) of ice shows the O(1s) level to be twice as broad as the gas phase [55]. In the case of NH$_3$, the lower-energy feature lines up with the first excitation observed in the ELS spectrum. This would be true for D$_2$O also, if a monochromator correction of +1.8 eV were applied to the data on the solid, as discussed in Sec. 2.

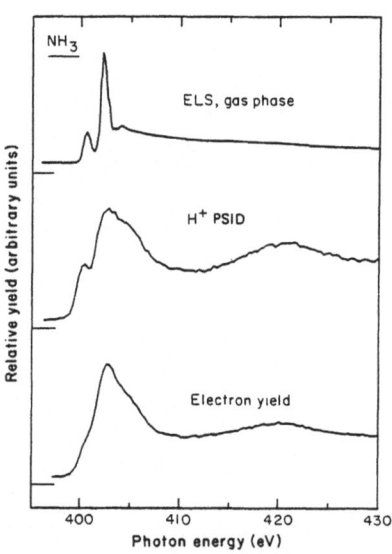

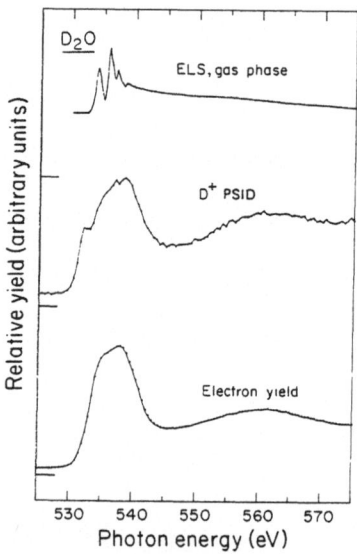

Fig. 7   Fig. 8

Fig. 7. Top: Electron energy-loss spectrum of gaseous NH$_3$ [52]. Middle: H$^+$ photoyield from solid NH$_3$. Bottom: Total photoelectron yield from solid NH$_3$. The solid-NH$_3$ spectra were shifted by +1.25 eV to compensate for monochromator-calibration error: the energies are now considered accurate to ±0.3 eV

Fig. 8. Top: Electron energy-loss spectrum of gaseous H$_2$O [52]. Middle: D$^+$ photoyield from solid D$_2$O. Bottom: Total photoelectron yield from solid D$_2$O. The spectra of solid D$_2$O should be shifted by +1.8±0.5 eV to compensate for a monochromator-calibration error

The sharp features observed in the gas-phase ELS spectra are attributed to Rydberg transitions [52]. In each spectrum, the first peak is due to a transition from the 1s- to a 3s-derived orbital, while the next two stem from transitions to 3p-like orbitals. The 1s electron-binding energies in gas-phase NH_3 and H_2O are 405.6 eV and 539.7 eV, respectively [55]. XPS on solid H_2O show a -2.3 eV shift of the 1s level upon condensation [56]; a similar shift may be expected for NH_3.

We may interpret the TPY data as representing the bulk photoabsorption spectrum [57]. Figures 7 and 8 show that the Rydberg excitations observed in the gas-phase ELS spectra are severely broadened in the solids and for D_2O shifted to higher energy. This shift upon condensation is in contrast with the XPS results but agrees with the shift observed in vacuum ultraviolet absorption [58].

Condensation is expected to perturb the larger, more diffuse Rydberg orbitals. Table 2 shows the calculated root-mean-square radii for several Rydberg orbitals of the H_2O molecule [59]. In solid D_2O, these orbitals should interact strongly with the two deuterons of the hydrogen bonds located 1.75 Å from the oxygen [32], accounting for the shifting and broadening of the Rydberg excitations. In solid ammonia, the hydrogen bonds are longer by 0.62 Å [32], suggesting a weaker interaction with the Rydberg states. This trend is confirmed by noting that the TPY data of Fig. 7 shows a considerably closer comparison with the gas phase than in the case of D_2O.

Table 2. Rydberg-orbital radii for H_2O (from Ref. [59])

Rydberg orbital	$\langle r^2 \rangle^{\frac{1}{2}}$ [Å]
$3sa_1$	2.21
$3pb_2$	2.56
$3pb_1$	3.65
$3pa_1$	3.91

We may interpret the PSID data as involving the electronic structure of surface molecules only. For both solids, the reduced coordination of the surface molecule reduces the perturbation of the gas-phase Rydberg excitations, as evident in Figs. 7 and 8. After accounting for the monochromator-calibration correction, the lowest-energy peak coincides with the first Rydberg excitation observed in the gas-phase ELS, the 1s → $3sa_1$ transition [52]. Both the lower coordination of the surface molecules and a possible surface reconstruction could contribute to the reduced perturbation of this Rydberg state in surface molecules. In solid H_2O, it has been suggested [6] that some PSID-active surface sites (e.g., site C, Fig. 4 of Ref. 6) are singly coordinated, while others are triply coordinated (e.g., sites B and E of the cited figure). The higher-energy 1s → 3p transitions are severely broadened in the PSID spectra, more closely resembling the TPY or bulk-like spectrum. This observation follows the trend in Rydberg-orbital size (Table 2), assuming the perturbation strength increases with increasing wave-function overlap.

Another possible explanation for the "bulk-like" portion of the PSID curve is based on the recently demonstrated phenomenon of x-ray-induced electron-stimulated ion desorption (XESID) [60]. In this phenomenon, secondary electrons produced by x-ray photoemission from the bulk cause desorption of ions from the surface. Such a process would tend to mimic the TPY spectrum and thereby contribute an apparent broadening of the higher Rydberg excitations in the PSID curve. At this time, we have no way of distinguishing this secondary mechanism from the primary PSID process.

3.2.2 N2

The N^+ PSID and TPY spectra from solid N_2 are shown in Fig. 9, along with the gas-phase ELS data [61]. Most of the structure in the gas-phase data is seen also in the TPY curve and, to a lesser extent, in the PSID spectrum. An important difference is that the first peak in the vapor data is at 400.96 eV, while the first peak in the TPY and PSID data is at 400.4 ± 0.3 eV. Although the width of the first peak is slightly greater in the PSID than in the TPY spectrum, we were unable to confirm this breadth difference due to lack of beam time.

The 0.5 eV energy shift upon solidification seems quite reasonable. A hole created by the N(1s) excitation is stabilized by polarization of the surrounding N_2 molecules, resulting in a kind of self-trapped hole or polaron. A 0.5 eV energy reduction is consistent with the ion-molecule interaction energy (0.9 eV) discussed in Sec. 3.1.1 [29]. Because PSID of N^+ may result from excitation of either of two inequivalent nitrogen atoms (see Sec. 3.1.1), one might expect a splitting of the peak in the N^+ yield. However, calculations for the Ni—N—N system indicate that this splitting would be less than 0.1 eV [27], beyond the resolving power of the monochromator.

As in the cases of NH_3 and D_2O, we must point out that the observed features of the N^+ PSID are consistent with the XESID mechanism. In con-

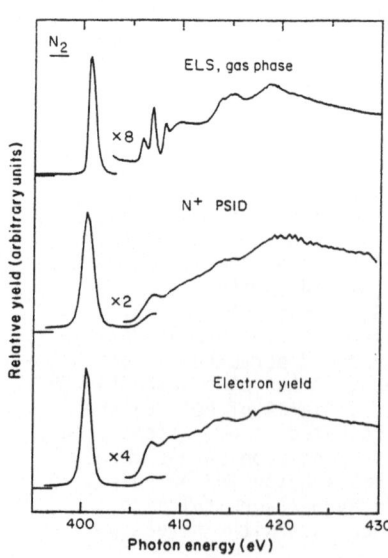

Fig. 9. Top: Electron energy-loss spectrum of gaseous N_2 61 . Middle: N^+ photo-yield from solid N_2. Bottom: Total photoelectron yield from solid N_2. The solid-N_2 spectra were shifted by 1.25 eV to account for monochromator-calibration error. The energies are now considered accurate to ±0.3 eV

trast with this are recent PSID results of N_2 adsorbed on Ni(100) [62] which show a slow, delayed onset of the N^+ yield above the N(1s) threshold, similar to the effect observed in the O^+ yield of CO on Ni(100) [15]. The N^+ PSID yield in the adsorbate case does not resemble the adsorbate partial photoelectron yield [15] or the results of Fig. 9.

This contrast in the two sets of condensed N_2 data has far-reaching implications. It indicates that the delayed onset seen in adsorbed N_2 and CO is not intrinsic to the molecule but is due to the chemisorption bond to the metal. This bond populates a normally empty 2π molecular orbital, which could lead to a higher reneutralization rate. Thus, an additional hole is required to break the N–N or C–O bond; the hole(s) may be acquired by the shake-up process [15,34]. Additional evidence for this is found in measurements of the total positive-ion yield produced by electron impact as a function of electron energy [62]. These data show a peak in the ion yield from N_2 at 400.8 ± 0.1 eV, in agreement with the main peak of Fig. 9.

The lowest-energy K-shell excitation of gaseous N_2 is a 1s → $2p\pi_g$ transition. This molecular-orbital final-state wave function would have a very small overlap with neighboring molecules nearly 4 Å away and would suffer little change upon condensation. The higher excitations involve larger wave functions of Rydberg final states, but would still have fairly small wave-function overlaps when compared to the hydrogen-bonded solids, for example. This accounts for the closer similarity of the TPY data of the solid to the gas-phase ELS data in the case of N_2 compared to either NH_3 or D_2O.

4. Acknowledgments

It is a pleasure to acknowledge the help of G. M. Loubriel of Sandia National Laboratories in the early stages of these experiments, D. Trevor of Lawrence Berkeley Laboratories and R. Jaeger of the Stanford Synchrotron Radiation Laboratory for helpful suggestions and for communicating results prior to publication. One of us (C.C.P.) acknowledges the guidance and encouragement of Prof. D. A. Shirley of the University of California, Berkeley. The financial support for this research was provided by the Naval Weapons Center Independent Research Fund, the Office of Naval Research, and the U.S. Department of Energy. Experiments were conducted at the Stanford Synchrotron Radiation Laboratory, which is supported by the National Science Foundation in cooperation with the Stanford Linear Accelerator Center and the U.S. Department of Energy.

5. References

[1] M.L. Knotek, V.O. Jones, and V. Rehn, Phys. Rev. Lett. 43, 300 (1979).
[2] For example, Joseph Berkowitz, *Photoabsorption, Photoionization, and Photoelectron Spectroscopy* (Academic, New York, 1979) and references therein.
[3] R. Clampett and L. Gowland, Nature 223, 815 (1969).
[4] R.H. Prince and G.R. Floyd, Chem. Phys. Lett. 43, 326 (1976).
[5] T.E. Madey and J.T. Yates, Jr., Chem. Phys. Lett. 51, 77 (1977).
[6] R.A. Rosenberg, V. Rehn, V.O. Jones, A.K. Green, C.C. Parks, and R.H. Stulen, Chem. Phys. Lett. 80, 488 (1981).
[7] R. Stockbauer, E. Bertel, and T.E. Madey, J. Chem. Phys. 76, 5639 (1982).
[8] D.M. Hanson, R. Stockbauer, and T.E. Madey, J. Chem. Phys. (to be published).
[9] J.A. Kelber and M.L. Knotek (submitted to Surf. Sci.).

[10]K. Hermann, P.S. Bagus, C.R. Brundle, and D. Menzel, Phys. Rev. B 24, 7025 (1981).

[11]R. Jaeger (private communication).

[12]P. Kobrin (private communication).

[13]J.A.R. Samson and G.N. Haddad, J. Opt. Soc. Am. 64, 1346 (1974).

[14]T.E. Madey, R. Stockbauer, S.A. Flodström, J.F. Van der Veen, F.-J. Himpsel, and D.E. Eastman, Phys. Rev. B 23, 6847 (1981).

[15]R. Jaeger, J. Stöhr, R. Treichler, and K. Baberschke, Phys. Rev. Lett. 47, 1300 (1981).

[16]G.R. Wight, M.J. Van der Wiel, and C.E. Brion, J. Phys. B 9, 675 (1976).

[17]P.L. Kronebusch and J. Berkowitz, Int. J. Mass Spectrosc. Ion Phys. 22, 283 (1976).

[18]T. Masuoka and J.A.R. Samson, J. Chem. Phys. 74, 1093 (1981).

[19]A.W. Potts and T.A. Williams, J. Electron Spectrosc. Rel. Phenom. 3, 3 (1974).

[20]A. Hamnett, W. Stoll, and C.E. Brion, J. Electron Spectrosc. Rel. Phenom. 8, 367 (1976).

[21]S. Krummacher, V. Schmidt, and F. Wuilleumier, J. Phys. B 13, 3993 (1980).

[22]D. Schmeisser and K. Jacobi, Chem. Phys. Lett. 62, 51 (1979).

[23]F.-J. Himpsel, P. Schwentner, and E.E. Koch, Phys. Status Solidi B 71, 615 (1975).

[24]P.S. Bagus and E.-K. Viinikka, Phys. Rev. A 15, 1486 (1977).

[25]P.W. Langhoff, S.R. Langhoff, T.N. Rescigno, J. Schirmer, L.S. Cederbaum, W. Domcke, and W. Von Niessen, Chem. Phys. 58, 71 (1981).

[26]E. Umbach, A. Schichl, and D. Menzel, Solid State Commun. 36, 93 (1980).

[27]C.R. Brundle, P.S. Bagus, D. Menzel, and K. Hermann, Phys. Rev. B 24, 7041 (1981).

[28]J.O. Hirschfelder, C.F. Curtiss, and R.B. Bird, *Molecular Theory of Gases and Liquids* (Wiley, New York, 1964), p. 1111.

[29]S.H. Linn, Y. Ono, and C.Y. Ng, J. Chem. Phys. 74, 3342 (1981).

[30]G.E. Ewing, Acc. Chem. Res. 8, 185 (1975).

[31]J.K. Burdett, J. Chem. Phys. 73, 2825 (1980).

[32]R.W.G. Wyckoff, *Crystal Structure* (Wiley, New York, 1963).

[33]D. Trevor, Ph.D. Thesis, University of California, LBL-11434 (1980) (unpublished) (private communication).

[34]D.E. Ramaker, Proc. this conference.

[35]M. Okuda and N. Jonathan, J. Electron. Spectrosc. Rel. Phenom. 3, 19 (1974).

[36]J. Schirmer, L.S. Cederbaum, W. Domcke, and W. Von Niessen, Chem. Phys. 26, 149 (1977).

[37]D.R. Jennison, J.A. Kelber, and R.R. Rye, Phys. Rev. B 25, 1384 (1982).

[38]P.J. Feibelman, Surf. Sci. 102, L51 (1981).

[39]R.E. Ramaker, C.T. White, and J.S. Murday, J. Vac. Sci. Technol. 18, 748 (1981).

[40]T. Hayaishi, S. Iwata, M. Sasanuma, E. Ishiguro, Y. Morioka, Y. Iida, and M. Nakamura, J. Phys. B. 15, 79 (1982).

[41]R.G. Cavell and D.A. Allison, J. Chem. Phys. 69, 159 (1978).

[42]D.G. Streets and A.W. Potts, J. Chem. Soc. Faraday II 70, 1505 (1974).

[43]A.J. Dixon, I.E. McCarthy, E. Weigold, and G.R.J. Williams, J. Electron. Spectrosc. Rel. Phenom. 12, 239 (1977).

[44]L.S. Cederbaum, W. Domcke, J. Schirmer, W. Von Niessen, G.H.F. Diereksen, and W.P. Kraemer, J. Chem. Phys. 69, 1591 (1978).

[45]M.L. Knotek and P.J. Feibelman, Phys. Rev. Lett. 40, 964 (1978).

[46]D. Menzel and R. Gomer, J. Chem. Phys. 41, 3311 (1964).

[47]P.A. Redhead, Can. J. Phys. 42, 886 (1964).

[48]R.A. Rosenberg, A. Green, V. Rehn, G. Loubriel, and C.C. Parks, Bull. Am. Phys. Soc. 26, 324 (1981).

[49]T.E. Madey and J.T. Yates, Jr., Surf. Sci. 76, 397 (1978).

[50]M.B. Robin and N.A. Kuebler, J. Electron Spectrosc. Rel. Phenom. 1, 13 (1972).

[51]J. Berkowitz and J.H.D. Eland, J. Chem. Phys. 67, 2740 (1977).

[52]G.R. Wight and C.E. Brion, J. Electron. Spectrosc. Rel. Phenom. 4, 25 (1974).

[53]R.R. Rye, T.E. Madey, J.E. Houston, and P.H. Holloway, J. Chem. Phys. 69, 1504 (1978).

[54]F.P. Larkins and A. Lubenfeld, J. Electron Spectrosc. Rel. Phenom. 15, 137 (1979).

[55]K. Siegbahn *et al.*, *ESCA Applied to Free Molecules* (North-Holland, Amsterdam, 1969).

[56]B. Baron and F. Williams, J. Chem. Phys. 64, 3896 (1976).

[57]A. Bianconi, Appl. Surf. Sci. 6, 392 (1980).

[58]M. Watanabe, H. Kitamura, and Y. Nakai, in *VUV Radiation Physics*, edited by E. E. Koch, R. Haensel, and C. Kunz (Pergamon/Vieweg, New York/ Braunschweig, 1974), p. 70.

[59]W.A. Goddard and W.J. Hunt, Chem. Phys. Lett. 24, 464 (1974).

[60]R. Jaeger, J. Stöhr, and T. Kendelewicz (in preparation).

[61]A.P. Hitchcock and C.E. Brion, J. Electron. Spectrosc. Rel. Phenom. 18, 1 (1980).

[62]G.C. King, J.W. McConkey, and F.H. Read, J. Phys. B. 10, L541 (1977).

7.4 Electron-Stimulated Desorption from Condense Branched Alkanes*

J.A. Kelber and M.L. Knotek

Sandia National Laboratories, Albuquerque, NM 87185, USA

1. Introduction

The electron-stimulated dissociation or desorption (ESD) of hydrocarbons is of relevance to the study of radiation-induced surface damage in organic solids and polymers, surface analysis, and radiation-induced surface chemistry. A number of such studies have been carried out on gas phase hydrocarbon systems [1]. Caution should be observed, however, in extrapolating gas phase results to the solid, since dissociation in the gas phase often occurs after a conversion of electronic excitation energy to vibrational energy which is inherently localized due to the isolation and finite extent of the molecule [2]. In an organic solid or polymer, however, the vibrational energy may be dissipated prior to dissociation, and the excitation energy must be localized in some other manner if dissociation is to occur. Condensed organic molecular solids offer the experimenter several other conveniences, including a well-characterized chemical structure, ease of desorption and the ability to systematically alter chemical structure as readily as in a gas phase study.

2. Experimental

The experimental methods employed in this study are detailed in a fuller account of this work which appears elsewhere [3]. Briefly, gases are adsorbed in multilayers on a liquid nitrogen cooled metal substrate, the targets are bombarded with electrons at excitation energies between 10 and 110 eV, and desorbed ions are detected by a quadrupole mass analyser using pulse counting techniques. One detail which should be noted, however, is that the electron excitation energy is a known quantity. The relationship of this quantity to work actually done on the system is uncertain because the exciting electron may end up either at the bottom of the conduction band of the sample or at the Fermi level of the metal substrate. This uncertainty is, for organic solids, generally between 3 and 5 eV [4]. While this point does not affect the conclusions of the present work, it does point out the need for photon-stimulated desorption studies in this area in order to establish detailed dissociation mechanisms from observed ion appearance potentials.

3. Results and Interpretation

H^+ and CH_3^+ yields from the materials studied are plotted as a function of electron excitation energy in Figure 1.

---

*This work was performed at Sandia National Laboratories supported by the U.S. Department of Energy under contract number DE-ACO4-76DPOO789.

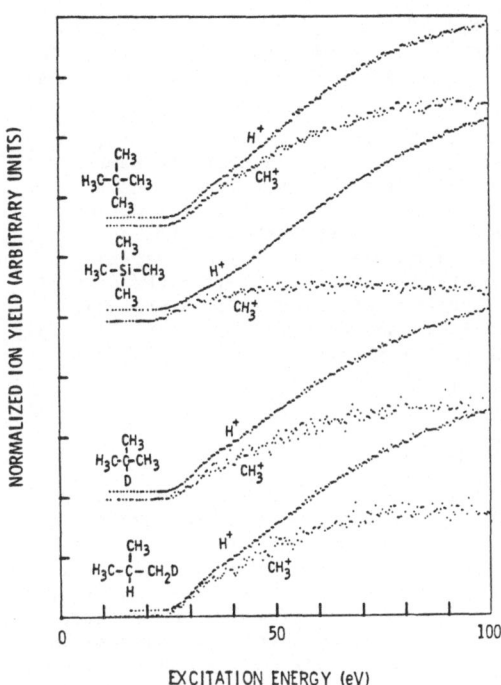

NORMALIZED ION YIELD (ARBITRARY UNITS)

EXCITATION ENERGY (eV)

In order to compare curve shapes, H$^+$ curves have been normalized to the same value at 100 eV, while the CH$_3^+$ curves have been normalized to the corresponding H$^+$ curves at 36 eV. There are four notable features of the data:

(A) In solid neopentane, the CH$_3^+$ emission-threshold energy (appearance potential) is 25 ($\pm$ 1 eV, compared to 16 eV in the gas phase [5]), indicating that dissociation to CH$_3^+$ occurs by a different pathway in the solid than in the gas. The statistical pathway [2] is shut down in the solid, as vibrational energy is no longer localized.

(B) In each material, H$^+$ and CH$_3^+$ have identical appearance potentials, demonstrating that the same initial excitation ultimately leads to both C-C (Si-C) and C-H bond breaking.

(C) In the neopentane and deuterated isobutane systems, the H$^+$ and CH$_3^+$ appearance potential is 25 ($\pm$ 1 eV). Assuming the exciting electron ends up in the bottom of the sample conduction band, this threshold corresponds to an excitation near the edge of the C2s band [6]. If the electron tunnels to the substrate Fermi level, the threshold corresponds to an excitation deeper within the C2s shell. The same holds for tetramethylsilane, except the corresponding shell has both C2s and Si3s character [7].

(D) In changing the environment around the central carbon atom (neopentane vs. isobutane), no visible change is observed in the shape of the H$^+$ and CH$_3^+$ ion yield curves (Fig.1), indicating that the dissociation/desorption process is unchanged. Substitution of Si for C at the central atom site

263

produces a noticeable change in the shape of the CH_3^+ curve (Fig.1) which now reaches a maximum in intensity near 70 eV and declines in intensity at higher energies. The shape of the H^+ curve, however, is still quite similar to that observed for the alkanes.

Observation (D) suggests that the central atom is not significantly involved in the desorption process, while (B) demonstrates that the same initial excitation breaks both C-H and C-C (Si-C) bonds. Taken together, these two observations imply that the excitations responsible for dissociation are localized on methyl groups. In order to further examine this possibility, the (mass 2^+) production as a function of electron excitation energy was observed for $(CH_3)_3CD$ - a deuterium atom attached to the central carbon atom - and $(CH_3)_2(CH_2D)CH$ - a deuterium atom bonded to a primary carbon atom. The results are shown in Fig.2 and compared to (mass 2^+), i.e., H_2^+, production in neopentane and tetramethylsilane.

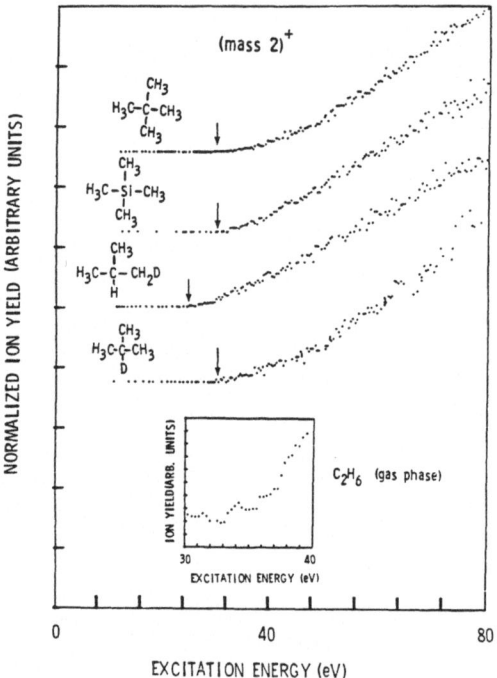

Fig.2. H_2^+ and D^+ yields as a function of electron excitation energy. The curves have been normalized to equal intensity at 80 eV. Gas phase ethane data are from Ref. 8. Arrows mark emission threshold energies and are guides to the eye. The similarity near threshold, of the neopentane and tetramethylsilane data, to gas phase ethane indicates that the H_2^+ emission process is similar in the gas and condensed phase. The curve shape and emission threshold energy of $(CH_3)_3CD$ is similar to neopentane and tetramethylsilane, indicating no D^+ desorption from this material. The data for $(CH_3)_2(CH_2D)CH$ have the same threshold as H^+ from this material ($\sim$ 25 eV) and a less concave upwards shape near threshold, indicating that both D^+ and H_2^+ are desorbed from this material

The (mass 2+) curves have been normalized to the same value at 80 eV. In neopentane and tetramethylsilane, the H_2 appearance potential is $\sim$ 30 eV and the curve is concave upward immediately after threshold, in qualitative agreement with the ethane gas phase results [8]. The (mass 2^+)/(mass 1^+) intensity ratio at 108 eV is 1/33 ($\pm \sim$ 3%). For $(CH_3)_2(CH_2D)CH$, however, different behavior is observed. The threshold is 25 eV, corresponding to H^+ dissociation from this compound. The curve shape near threshold is linear, not concave upward and the (mass 2^+)/(mass 1^+) ratio at 108 eV is 1/24, indicating that the (mass 2^+) yield curve for this material is composed of both D^+ and H_2^+. D^+, however, is not desorbed as readily as H^+

from the central position, or the (mass 2^+)/(mass 1) ratio would be ~ 1/33 + 1/9 >> 1/24. For $(CH_3)_3CD$, on the other hand, the threshold is again near 30 eV, the curve is concave upward immediately after threshold and the the (mass 2^+)/(mass 1^+) ratio is again 1/33 eV. This demonstrates that no D^+ is observed to be desorbed from the central carbon site. These result are what one would expect if the excitations which cause dissociation are localized on methyl groups. One cannot completely rule out alternative explanations such as, for example, a large isotope effect at the central carbon site. The observation of D^+ desorption from the methyl carbon site and not the central carbon site is, however, consistent with the evidence discussed above which indicates that the central atom is not significantly involved in the desorption process.

The evidence shows that H+ and CH_3^+ are desorbed by excitations which are localized on methyl groups and that these excitations are initiated by creating a hole in the C2s level (with the electron promoted perhaps to a Rydberg or vacuum level). In these systems, the molecular orbitals composed of C2s (or in tetramethylsilane, C2s and Si3s) atomic orbitals have no significant hydrogen character. It is then difficult to understand how a single hole in the C2s level could result in both H+ and CH_3^+ desorption. The only alternative is that this C2s hole decays to either a single-valence-hole or multi-valence-hole final state. A single hole final state can be ruled out, because such states may be directly excited at ~ 10-18 eV excitation energy, where no ion desorption is observed. Additional support for the multi-hole model comes from recent studies of the core-valence-valence Auger spectra of neopentane [9] and linear alkanes [10]. These studies show that highly correlated two-hole states exist on $-CH_2$ groups in the alkanes (2) and $-CH_3$ groups in neopentane. Such states are localized in the sense that, because of hole-hole correlation, the two holes remain on one site or functional group for times long compared to single-hole hopping times. These studies show that such localized states do not exist on the central carbon atom of neopentane [9]. Thus, in isobutane, we may expect localized multi-hole states, and therefore desorption from methyl groups but not necessarily from the central carbon atom.

4. Summary and Conclusions

H^+, CH_3^+, H_2^+ and D^+ desorption have been measured as a function of electron excitation energy for solid neopentane, tetramethylsilane and two deuterated isomers of isobutane. the evidence shows that C-C (or Si-C) and C-H bonds are broken by electronic excitations localized on methyl groups, in contrast to CH_3^+ production in gas-phase neopentane [5], and that these excitations are the final states of decay processes initiated by creation of a hole in the C2s level, or, in tetrametlylsilane, the C2s/Si3s level. This is in accord with other evidence which shows that localized multi-valence hole states result in C-H, C-C, Si-C and Si-H [11] dissociation, and that such states may be excited either directly or by shakeup, by decay from a C2s hole, or by decay for a C1s core hole [11]. It is apparent then, that dissociation and desorption of ions from covalent materials is a multi-(electron) hole mechanism, and that the means of localizing the excitation energy in such systems involves multi-hole correlation.

5. Future Work

The model discussed above suggests several avenues for further work. One experiment which would confirm the occurrence or absence of induced bond breaking at the central atom site is the examination of ESD of H^+ from

$(CD_3)_3CH$. If the above model is correct, little or no H^+ should be observed at the excitation energies examined in this work. The theoretical description of hole-hole correlation and localization [9,10] indicates that the polarization response of the material to the excitation is an important factor in determining the lifetime of the localized resonance (the greater the polarization response, the shorter the lifetime). An interesting test of this description and its relevance to ESD of covalent materials would be to compare the ESD of H^+ from hydrocarbons and their partially fluorinated analogs (e.g., $(CF_3)_3CH$, C_6H_6 vs $C_6F_3H_3$, etc.). The inductive effect of the fluorines should decrease the polarization response of the system and influence H^+ desorption. Another series of experiments which would further elucidate the role of the central atom in the ESD of H^+ from these materials is the observation of ESD from $(CH_3)_4Si$, $(CH_3)_4$ Sn, and $(CH_3)_4Ge$ at energies corresponding to core level evaluations in the central atom.

References

[1] See, for example, R. Locht, J. L. Oliver and J. Momigny, Chem. Phys. 43, 425 (1979)

[2] C. Lifshitz, Adv. Mass Spec. 7, 3 (1977)

[3] J. Kelber and M. Knotek, Surface Science (in press)

[4] J. J. Pireaux, R. Caudano, S. Svensson, E. Basilier, P. A. Malmqvist, U. Gelius and K. Siegbahn, J. dePhysique, 38, 1213 (1977)

[5] O. Osberghaus and R. Taubert, Z. fuer Phys. Chem. 4, 264 (1955)

[6] B. E. Mills and D. A. Shirley, J. Amer. Chem. Soc. 99, 3585 (1977)

[7] W. B. Perry and W. L. Jolly, J. Elec. Spec. Rel. Phen. 4, 219 (1977)

[8] I .H. Suzuki and K. Maeda, Mass. Spec. 26, 53 (1978)

[9] D. R. Jennison, J. A. Kelber and R. R. Rye, Phys. Rev. B25, 1384 (1982)

[10] J. A. Kelber and D. R. Jennison, J. Vac. Sci. and Tech. 20, 848 (1982)

[11] H. H. Madden, D. R. Jennison, M. M. Traum, G. Margaritondo and W. Stoffel, Phys. Rev. B (to be published). See also B. B. Pate, M. H. Hecht, C. Binns, I. Lindau and W. E. Spicer, J. Vac. Sci. and Tech., PSCI-9 Proceedings Issue (in press)

7.5 PSD and ESD of Condensed Films: Relevance to the Mechanism of Ion Formation and Desorption

R. Stockbauer, E. Bertel[1] and T.E. Madey

Surface Science Division, National Bureau of Standards
Washington, DC 20234, USA

We report here results of photon- and electron-stimulated desorption experiments performed to study the electronic processes leading to ion formation in condensed films of the hydrogen-containing molecules water [1], methanol (CH_3OH) [2], and cyclohexane (C_6H_{12}) [3].

Both water and methanol contain OH groups and form hydrogen-bonded films. Using ESD and time-of-flight (TOF) mass spectrometry, the dominant ions from condensed layers of both molecules are H^+, with yields of higher mass ions < 1% of H^+. Cluster ions from the water film [$(H_2O)_nH^+$] were at the 0.1% level or less. This is in contrast to the gas-phase mass spectra where high mass ions are dominant. These results were independent of the film thickness unlike the results from C_6H_{12} discussed below.

PSD of ions from the condensed methanol was studied over the photon energy range 15 to 75 eV. The threshold for H^+ desorption was at 18 eV, with the maximum yield at 25 eV. The 4a' level of methanol has its onset near the 18 eV threshold. In order to determine if the H^+ desorption was due to C-H or O-H bond cleavage (or both), we studied low-energy ESD of isotopically labelled CH_3OH films [4]. It was found that low-energy (<70 eV) electron excitation and fragmentation of CH_3OH results overwhelmingly in H^+ ejection by C-H bond cleavage. This is additional evidence for involvement of the 4a' level since it is dominantly a C-H bonding orbital. In contrast, the gas-phase fragmentation shows H^+ arising from both O-H and C-H bond cleavage [5].

PSD of H^+ from H_2O films [6] has higher threshold and peak yield energies (25 eV and 30 eV, respectively) than CH_3OH, and may arise from excitation of a 2-hole shake-up state of H_2O [7].

Preliminary data for ESD of C_6H_{12}, a non-hydrogen-bonded film, demonstrates that higher mass fragments are more abundant than for H_2O and CH_3OH. In a study of the ESD ion yield vs. film thickness for C_6H_{12} on Cu(100) we observed only H^+ ions for thin layers. These ions show a broad kinetic energy distribution with a peak near 4 eV. The H^+ signal saturates near a dose of 10L, where the heavy fragments are first observed. These heavy ions ($C_2H_n^+$ -$C_5H_n^+$) have a sharp energy distribution peaked near 2 eV. The heavy fragment signal continues to increase to saturation at 50L. Coverage dependence of the signal was independent of the incident electron current density showing that these fragments are not due to electron beam damage nor to surface charging. These results are qualitatively consistent with a model by which more massive

---

[1] Permanent address: Institute of Physical Chemistry, University of Innsbruck,
   A-   Innsbruck, Austria

ions are preferentially neutralized close to the metal surface. Their lower kinetic energy and high mass gives them a much lower velocity than the H^+ and hence are more susceptible to reneutralization. This reneutralization rate decreases with increasing film thickness causing an increased yield of higher mass fragments.

The absence of heavy mass fragments from the water and methanol films may be due to hydrogen-bonding. This could provide an effective de-excitation mechanism for higher mass fragments at all film thicknesses. The absence could also be due to protron transfer reactions occurring at the surface before or during desorption which would effectively neutralize the ions [8]. Finally we note that hydrogen bonding is not a significant process in C_6H_{12} nor do protron transfer processes occur with high probability.

EB wishes to acknowledge fellowship support provided by the Max Kade Foundation, NY, during his stay at NBS. This work was supported in part by the Office of Naval Research.

References

1. R. Stockbauer, D. M. Hanson, S. A. Flodström and T. E. Madey, Phys. Rev. B (in press).

2. D. M. Hanson, R. Stockbauer and T. E. Madey, J. Chem. Phys. (in press).

3. R. Stockbauer, E. Bertel and T. E. Madey (to be submitted).

4. R. Stockbauer, E. Bertel and T. E. Madey, J. Chem. Phys. 76, 5639 (1982).

5. M. D. Burrows, S. R. Ryan, W. E. Lamb, Jr., and L. C. McIntyre, Jr., J. Chem. Phys. 71, 4931 (1979).

6. R. A. Rosenberg, V. Rehn, V. O. Jones, A. K. Green, C. C. Parks, G. Loubriel and R. H. Stulen, Chem. Phys. Lett. 80, 488 (1981).

7. D. E. Ramaker, (submitted).

8. K. R. Ryan, L. W. Sieck, and J. H. Futrell, J. Chem. Phys. 41, 111 (1964).

Index of Contributors

Chemistry and Physics of Solid Surfaces IV

Editors: **R. Vanselow, R. Howe**
1982. 247 figures. XIII, 496 pages
(Springer Series in Chemical Physics,
Volume 20)
ISBN 3-540-11397-5

Chemistry and Physics of Solid Surfaces IV presents
selected review articles predominantly from the
area of solid/gas interfaces. These articles are written by internationally recognized experts, the invited speakers of the International Summer Institute
in Surface Science (ISISS). Volume IV covers the
following topics: photoemission, Auger spectroscopy, SIMS ellipsometry, Doppler shift laser spectroscopy, analytical electron microscopy, He ion
diffraction as a surface probe, low energy electron
diffraction, Monte Carlo simulations of chemisorbed overlayers, critical phenomena of overlayers,
surface defects, theoretical aspects of surfaces,
EELS of surface vebrations, electronic aspects of
adsorption rates, thermal desorption, field desorption, surface and internal segregation.
This series helps to bring researchers in academia
and industry up-to-date and bridges the gab between conventional surface science textbooks and
specialized conference proceedings. Extensive literature references are provided together with a detailed subject index.

Dynamics of Gas-Surface Interaction

Proceedings of the International School on
Material Science and Technology, Erice, Italy,
July 1–15, 1981

Editors: **G. Benedek, U. Valbusa**
1982. 132 figures. XI, 282 pages
(Springer Series in Chemical Physics,
Volume 21)
ISBN 3-540-11693-1

The recent great advance in molecular beam techniques, triggered off by aerospace research, has
open new perspectives in surface physics. This
book, originated by International School held at
the Majorana Centre in Erice, offers an updated
review of studies on the structure and dynamics of
solid surfaces and on gas-surface interaction by
means of molecular beams comparing them with
other techniques such as LEED, neutron and optical spectroscopies.
Novel topics treated in the book are the spectroscopy of surface phonons by atom scattering, the
investigation of inelastic bound-state resonances,
the characterization of adsorbates by atom scattering and surface-enhanced Raman scattering, and
the detection of charge-density waves at the surface
of layered crystals. All physicists and chemists, gra-
duate students included, involved in the study of
gas-surface interactions will find this book useful.

Inelastic Particle-Surface Collisions

Proceedings of the Third International Workshop
on Inelastic Ion-Surface Collisions
Feldkirchen-Westerham, Federal Republic
of Germany, September 17–19, 1980

Editors: **E. Taglauer, W. Heiland**
1981. 194 figures. VIII, 329 pages
(Springer Series in Chemical Physics,
Volume 17)
ISBN 3-540-10898-X

The proceedings of the 3rd International Workshop on Inelastic Ion-Surface Collisions provides a
state-of-the-art assessment of the field of inelastic
ion surface collisions and some related subjects of
atomic collision physics. The topic is reviewed by
14 leading scientists representing different areas of
research. Original papers give first-hand information on current research.
The book provides new information (theoretical
and experimental) on secondary electron emission,
interaction of metastables with surfaces, electron
and photon stimulated desorption, formation of
ions, and excited particles by scattering or sputtering.

L. Kannoo, J. Bourgoin

Point Defects in Semiconductors I

Theoretical Aspects

With a Foreword by J. Friedel
1981. 87 figures. XVII, 265 pages
(Springer Series in Solid-State Sciences,
Volume 22)
ISBN 3-540-10518-2

Contents: Atomic Configuration of Point
Defects. – Effective Mass Theory. – Simple Theory
of Deep Levels in Semiconductors. – Many-Electron Effects and Sophisticated Theories of Deep
Levels. – Vibrational Properties and Entropy. –
Thermodynamics of Defects. – Defect Migration
and Diffusion. – References. – Subject Index.

Springer-Verlag
Berlin
Heidelberg
NewYork

Electron Spectroscopy for Surface Analysis

Editor: **H. Ibach**
1977. 123 figures, 5 tables. XI, 255 pages
(Topics in Current Physics, Volume 4)
ISBN 3-540-08078-3

Contents: *H. Ibach:* Introduction. –
D. Roy, J. D. Carette: Design of Electron
Spectrometers for Surface Analysis. –
J. Kirschner: Electron-Excited Core
Level Spectroscopies. – *M. Henzler:*
Electron Diffraction and Surface Defect
Structure. – *B. Feuerbacher, B. Fitton:*
Photoemission Spectroscopy. –
H. Froitzheim: Electron Energy Loss
Spectroscopy.

M. A. Van Hove, S. Y. Tong
Surface Crystallography by LEED

Theory, Computation and Structural Results
1979. 19 figures, 2 tables. IX, 286 pages
(Springer Series in Chemical Physics,
Volume 2)
ISBN 3-540-09194-7

"...This is an excellent book for anyone
seriously interested in LEED who
would like to be able to perform his
own sophisticated calculations. It is
clearly and carefully written, and all the
nuances of the techniques are tho-
roughly covered. If any book will make
the practical calculation of LEED intensity profiles widely available, this is
the one."

American Scientist

"...The extensive experience of the
authors, the welltested nature of the
programs, and the many practical suggestions in the text make this book an
essential reference for workers studying
surface structure by LEED and a
valuable reference for quantitative surface theory generally."

Contemporary Physics

Structural Studies of Surfaces

With contributions by **K. Heinz,
K. Müller, T. Engel, K.-H. Rieder**
1982. 120 figures. VII, 180 pages
(Springer Tracts in Modern Physics,
Volume 91)
ISBN 3-540-10964-1

The book presents a comprehensive
description of classical and recently
developed methods for measuring low-
energy electron intensities diffracted by
ordered surfaces (LEED). It shows how
the shortcomings of the classical
methods stimulated the development of
new measuring procedures which are
more precise, more rapid, and more
practicable.

Theory of Chemisorption

Editor: **J. R. Smith**
1980. 116 figures, 8 tables. XI, 240 pages
(Topics in Current Physics, Volume 19)
ISBN 3-540-09891-7

Contents: *J. R. Smith:* Introduction. –
S. C. Ying: Density Functional Theory of
Chemisorption of Simple Metals. –
J. A. Appelbaum, D. R. Hamann: Chemi-
sorption on Semiconductor Surfaces. –
F. J. Arlinghaus, J. G. Gay, J. R. Smith:
Chemisorption on d-Band Metals. –
B. Kunz: Cluster Chemisorption. –
T. Wolfram, S. Ellialtioğly: Concepts of
Surface States and Chemisorption on
d-Band Perovskites.– *T. L. Einstein,
J. A. Hertz, J. R. Schrieffer:* Theoretical
Issues in Chemisorption.

Springer-Verlag
Berlin
Heidelberg
New York